Fundamentals of Magnetism

T0217863

Mathias Getzlaff

Fundamentals of Magnetism

With 292 Figures and 18 Tables

 Springer

Prof. Dr. Mathias Getzlaff
Univ. Düsseldorf
Inst. Angewandte Physik
Universitätsstr. 1
40225 Düsseldorf
Germany
getzlaff@uni-duesseldorf.de

ISBN-13 978-3-642-06827-0   e-ISBN-13 978-3-540-31152-2

Springer is a part of Springer Science+Business Media
springer.com
© Springer-Verlag Berlin Heidelberg 2010

Cover design: eStudio Calamar

To my wife Anette

Preface

Magnetism is a fascinating subject which is known since a few thousand years. This means that this phenomenon was already observed before recorded history began.

In the Near East region some ores were found to be "attractive" or "magnetic". In Europe, the use of this iron ore called Lodestone for navigation in a compass is unquestionably dated to about 1300. But, the first incontrovertible mention of a magnetic device used for establishing direction is to be found in a Chinese manuscript dated about 1040. The Lodestone was located into a spoon that was placed on a plate being of bronze. Rather than navigation this pointer was used for geomancy being a technique for aligning buildings in order to be in harmony with the forces of nature.

Today our understanding of magnetism is closely related to the concept of spin which arises from the relativistic description of an electron in an external electromagnetic field and becomes manifest in the Dirac equation. This concept results in the spin magnetic moment and the orbital magnetic moment which is due to the motion of electronic charges.

As throughout history magnetism is closely related to applications. A lot of common today's devices would be unthinkable without the forefront research areas in magnetism. One example is given by read heads in hard disks which allowed a tremendous enhancement of storage density. They are based on the discovery of the Giant Magnetoresistance (GMR) in 1988 by P. Grünberg and A. Fert. And also today technology is driven by the aim to develop devices which are smaller, faster, and cheaper than every one before.

Already these few annotations prove that the field of magnetism was exciting in the past. Nevertheless, in the future we will discover many new effects which will be beyond our imagination and will have a strong influence on several things "around us". Thus, they will shape our daily lives.

To my opinion most of the new discoveries will be related to low-dimensional systems.

Therefore, the first part of this book deals with the fundamentals of magnetism whereas the second one is devoted to magnetic phenomena of systems which are reduced in at least one-dimension.

In *Chap. 1* we will start the introduction by summarizing some basic terms of magnetism, give a classification of magnetic materials, and schematically show why magnetism cannot be explained using classical mechanics and electrodynamics.

Chapter 2 deals with magnetic moments of free, i.e. isolated, atoms in a magnetic field. After the discussion of atomic dia- and paramagnetism we will develop some simple rules which allow to determine the magnitude of magnetic moments concerning an atom of a given chemical element.

Using the model of free electrons we will see in *Chap. 3* that collective magnetism can be stabilized above 0 K without applying an external magnetic field in a solid state due to interaction between magnetic moments.

Chapter 4 is devoted to the question how magnetic moments in different atoms can interact with each other.

The description of different types of magnetic order in the solid state, i.e. collective magnetism (ferromagnetism, antiferromagnetism, ferrimagnetism, helical order), is given in *Chap. 5*.

The vanishing of long range ferromagnetic order above a critical temperature is related to a breaking of a symmetry and can be characterized by a phase transition. This distinct temperature dependence of specific parameters like the magnetization is discussed in *Chap. 6*.

In *Chap. 7* we will learn about the influence of anisotropy on the magnetic behavior which can be induced by, e.g., the shape and the crystalline structure of a sample.

Chapter 8 deals with magnetic domains being uniformly magnetized regions which exhibit a parallel orientation of all magnetic moments and the boundaries between them, the domains walls.

In *Chap. 9* we will discuss the "answer" of the magnetization concerning the magnitude and the direction if a varying external magnetic field is applied.

The next six chapters are devoted to magnetism in reduced dimensions. In *Chap. 10* we will start our discussion with the behavior of single magnetic atoms on a surface and proceed with ensembles consisting of only a few atoms, so-called clusters, in *Chap. 11*. The properties of larger agglomerations, nanoparticles, will be presented in *Chap. 12*. The behavior of wires with dimensions in the nanometer regime will be explained in *Chap. 13*. The next two chapters contain the description of magnetic thin film systems. In *Chap. 14* we will restrict our discussion to single thin metallic films and explain the properties of multilayers in *Chap. 15*.

The influence of the spin arrangement in a magnetic layer on the electrical resistance will be discussed in *Chap. 16*.

This effect called magnetoresistance can be rather huge and thus acts as the basis to realize technologically important applications like read heads in hard disks and various sensors which will be presented in *Chap. 17*.

Several chapters exhibit problems in order to improve the respective knowledge. The solutions are given at the end of the book.

In the appendices abbreviations, symbols, and important constants are listed.

In this textbook we make use of the units and definitions of the système international SI. In some figures the properties are given in the cgs-system (centimeter, gram, second). This observation and problem should additionally sensitize the reader that different "languages" are often spoken in the field of magnetism. A comparison of important quantities in both systems is also given in order to round off this textbook.

Last but not least it is a pleasure for me to thank all the people which significantly imparted my knowledge on this wide field of magnetism: Joachim Bansmann, Matthias Bode, Ulrich Heinzmann, Gerd Schönhense, and Roland Wiesendanger.

Düsseldorf, June 2007   *Mathias Getzlaff*

Contents

1

Introduction

In this introductory chapter we will summarize some basic terms of magnetism and explain the difference of magnetic fields inside and outside of materials. In the following a short characterization of different classes of magnetism (diamagnetism, paramagnetism, ferromagnetism, antiferromagnetism, ferrimagnetism) will be given. At the end we will see that magnetization above $0\,\mathrm{K}$ *cannot* be understood using classical mechanics. Thus, we will have to deal with quantum physics in order to explain magnetic properties.

1.1 Fundamental Terms

Magnetic Moment and Magnetic Dipole

In classical electromagnetism the magnetic moment μ can be explained using the picture of a current loop. Assuming a current I around an infinitely small loop exhibiting an area of $\mathrm{d}A$ the corresponding magnetic moment $\mathrm{d}\mu$ amounts to:

$$\mathrm{d}\boldsymbol{\mu} = I\,\mathrm{d}\boldsymbol{A} \tag{1.1}$$

The direction of the vector area is given by the right-hand rule. Summing up the magnetic moments of this "small" loops allows to calculate the magnetic moment μ for a loop of finite size:

$$\boldsymbol{\mu} = \int \mathrm{d}\boldsymbol{\mu} = I \int \mathrm{d}\boldsymbol{A} \tag{1.2}$$

because the currents of neighboring loops cancel each other only leaving the current running around the perimeter of the finite-sized loop.

The magnetic dipole is equivalent to a magnetic moment of a current loop in the limit of a small area but finite moment. The energy of a magnetic moment is given by:

$$E = -\mu_0 \boldsymbol{\mu} \cdot \boldsymbol{H} = -\mu_0 \mu\, H \cos\theta \tag{1.3}$$

with θ being the angle between the magnetic moment $\boldsymbol{\mu}$ and an external magnetic field $\boldsymbol{H}$ and μ_0 the magnetic permeability of free space.

Magnetization

The magnitude of the magnetization $\boldsymbol{M}$ is defined as the total magnetic moment per volume unit:

$$M = \mu \frac{N}{V} \tag{1.4}$$

Usually, it is given on a length scale which is large enough that an averaging over at least several atomic magnetic moments is carried out. Under this condition the magnetization can be considered as a smoothly varying vector field. In vacuum no magnetization $\boldsymbol{M}$ occurs.

Magnetic Induction

The response of a material when applying an external magnetic field $\boldsymbol{H}$ is called magnetic induction or magnetic flux density $\boldsymbol{B}$. The relationship between $\boldsymbol{H}$ and $\boldsymbol{B}$ is a characteristic property of the material itself. In vacuum we have a linear correlation between $\boldsymbol{B}$ and $\boldsymbol{H}$:

$$\boldsymbol{B} = \mu_0 \boldsymbol{H} \tag{1.5}$$

But inside a magnetic material $\boldsymbol{B}$ and $\boldsymbol{H}$ may differ in magnitude and direction due to the magnetization $\boldsymbol{M}$:

$$\boldsymbol{B} = \mu_0(\boldsymbol{H} + \boldsymbol{M}) \tag{1.6}$$

In the following we simply refer to both as the "magnetic field" due to the common usage in literature. The letter itself directly explains which term of both is meant.

Magnetic Susceptibility and Permeability

If the magnetization $\boldsymbol{M}$ is parallel to an external magnetic field $\boldsymbol{H}$:

$$\boldsymbol{M} = \chi \boldsymbol{H} \tag{1.7}$$

with χ being the magnetic susceptibility the material is called *linear material*. In this situation, a linear relationship between $\boldsymbol{B}$ and $\boldsymbol{H}$ remains present:

$$\boldsymbol{B} = \mu_0(1 + \chi)\boldsymbol{H} \tag{1.8}$$
$$= \mu_0 \mu_r \boldsymbol{H} \tag{1.9}$$

with $\mu_r = 1 + \chi$ being the relative permeability. Typical values for the relative permeability are:

$$\text{in vacuum: } \mu_r = 1$$
$$\text{in matter generally: } \mu_r \geq 1$$
$$\text{possible in matter: } \mu_r \approx 100.000$$

1.2 Classification of Magnetic Material

A rough classification into three classes is carried out by means of the susceptibility χ.

- Diamagnetism
  Diamagnetism is purely an induction effect. An external magnetic field H induces magnetic dipoles which are oriented antiparallel with respect to the exciting field due to Lenz's rule. Therefore, the diamagnetic susceptibility is negative:

$$\chi^{\mathrm{dia}} = \mathrm{const.} < 0 \tag{1.10}$$

  Diamagnetism is a property of *all* materials. It is only relevant in the absence of para- and collective magnetism.
  A few examples of diamagnetic materials are:
  - nearly all organic substances,
  - metals like Hg,
  - superconductors below the critical temperature. These materials are ideal diamagnets, i.e. $\chi^{\mathrm{dia}} = -1$ (Meißner–Ochsenfeld effect)
- Paramagnetism
  The susceptibility of paramagnetic materials is characterized by:

$$\chi^{\mathrm{para}} > 0 \tag{1.11}$$
$$\chi^{\mathrm{para}} = \chi^{\mathrm{para}}(T) \tag{1.12}$$

A crucial precondition for the appearance of paramagnetism is the existence of permanent magnetic dipoles. These are oriented in the external magnetic field H. The orientation may be hindered by thermal fluctuations. The magnetic moments can be localized or of itinerant nature:

- Localized moments
  They are caused by electrons of an inner shell which is only partially filled. Typical examples are:
  - $4f$ electrons in rare earth metals
  - $5f$ electrons in actinides

  This class of material exhibits the so-called Langevin paramagnetism. The susceptibility χ^{Langevin} depends on the temperature. At high temperatures the Curie law is valid:

$$\chi^{\mathrm{Langevin}}(T) = \frac{c}{T} \tag{1.13}$$

- Itinerant moments
  Nearly free electrons in the valence band carry a permanent magnetic moment of $1\mu_B$ with μ_B being the Bohr magneton. This type is called Pauli paramagnetism. The corresponding susceptibility is nearly independent on temperature:

$$\frac{\partial \chi^{\mathrm{Pauli}}}{\partial T} \approx 0 \tag{1.14}$$

The magnitudes of these susceptibilities are very different:

$$\chi^{\text{Pauli}} \ll \chi^{\text{Langevin}} \qquad (1.15)$$

- Collective magnetism
  The susceptibility exhibits a significantly more complicated functionality of different parameters compared to dia- and paramagnetism:

$$\chi^{\text{coll}} = \chi^{\text{coll}}(T, \boldsymbol{H}, \text{"history"}) \qquad (1.16)$$

The collective magnetism is a result of an exchange interaction between permanent magnetic dipoles which can solely be explained by quantum mechanics.

For materials showing collective magnetism a critical temperature T^* occurs which is characterized by the observation that a *spontaneous magnetization* is present below T^*, i.e. the magnetic dipoles exhibit an orientation which is *not* enforced by external magnetic fields.

The magnetic moments can again be localized (e.g. Gd, EuO, ...) or itinerant(e.g. Fe, Co, Ni).

Collective magnetism is divided into three subclasses:

- Ferromagnetism
   The critical temperature T^* is called Curie temperature T_C
   · $0 < T < T_C$
     The magnetic moments exhibit a preferential orientation ($\nwarrow\uparrow\nearrow\nwarrow\uparrow$).
   · $T = 0$
     All magnetic moments are aligned parallel ($\uparrow\uparrow\uparrow\uparrow\uparrow$).
   Ferromagnetic materials exhibiting itinerant magnetic moments are called band or itinerant ferromagnets.

- Ferrimagnetism
   The lattice decays into two ferromagnetic sublattices A and B exhibiting different magnetization:

$$\boldsymbol{M}_A \neq \boldsymbol{M}_B \qquad (1.17)$$

whereas

$$\boldsymbol{M} = \boldsymbol{M}_A + \boldsymbol{M}_B \neq 0 \qquad \text{for} \quad T < T_C \qquad (1.18)$$

with $\boldsymbol{M}$ being the total magnetization.

- Antiferromagnetism
   Antiferromagnetism is a special situation of the ferrimagnetism with the critical temperature T^* being called Néel temperature T_N and is characterized by:

$$|\boldsymbol{M}_A| = |\boldsymbol{M}_B| \neq 0 \qquad \text{for} \quad T < T_N \qquad (1.19)$$

and

$$\boldsymbol{M}_A = -\boldsymbol{M}_B \qquad (\text{e.g.} \uparrow\downarrow\uparrow\downarrow\uparrow) \qquad (1.20)$$

Therefore, the total magnetization vanishes:

$$M = M_A + M_B \equiv 0 \tag{1.21}$$

The collective magnetism merges into the paramagnetism above the critical temperature T^* with the corresponding characteristic behavior of the inverse susceptibility.

The temperature dependence of the magnetization and magnetic susceptibility for various types of magnetic materials is summarized in Fig. 1.1. The values

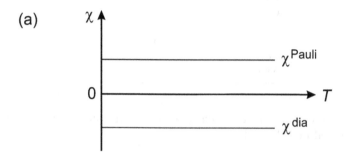

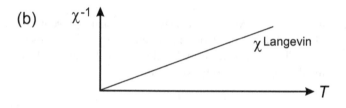

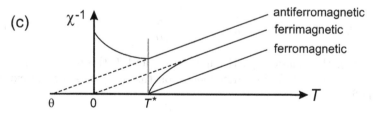

Fig. 1.1. Temperature dependence of the magnetic susceptibility χ and inverse magnetic susceptibility χ^{-1} in the case of (**a**) diamagnetism and Pauli paramagnetism, (**b**) Langevin paramagnetism, (**c**) ferromagnetism, antiferromagnetism, and ferrimagnetism with T^* being the critical temperature and θ the paramagnetic Curie temperature

Table 1.1. Curie temperature T_C of ferro- and ferrimagnetic and Néel temperature T_N of antiferromagnetic materials

Material	T_C [K]	T_N [K]
Fe	1043	
Co	1388	
Ni	627	
Gd	293	
Dy	88	
EuO	69	
Fe_3O_4	853	
CrO_2	387	
Cr		311
CoO		293
NiO		525

of the critical temperature for some selected ferro-, ferri- and antiferromagnet materials are listed in Table 1.1.

1.3 Bohr–van Leeuwen Theorem

Magnetism, i.e. dia-, para-, as well as collective magnetism, presents a quantum mechanical effect which *cannot* be explained using classical mechanics and electrodynamics. The proof shall only be given as a sketch. With μ being the magnetic moment of a single atom, we obtain:

$$\mu = -\frac{\partial H}{\partial B} \tag{1.22}$$

with H being the classical Hamilton function and B an external magnetic field. The average value $\langle \mu \rangle$ is given by

$$\langle \mu \rangle = \frac{k \cdot T}{Z} \cdot \frac{\partial Z}{\partial B} \tag{1.23}$$

with Z being the classical partition function and T the temperature. It is possible to show that Z does not depend on the magnitude of the external magnetic field:

$$Z \neq Z(B) \tag{1.24}$$

Therefore, we can conclude:

$$\langle \mu \rangle \equiv 0 \tag{1.25}$$

2

Magnetism of Atoms

This chapter deals with the properties of magnetic moments of free, i.e. isolated, atoms in a magnetic field. Any interaction of the atoms will therefore be neglected.

This situation seems to be rather simple but we will see that the description of magnetic behavior is rather complicated due to the interaction already of all the electrons in one atom. At the end we will develop some rules which allow to determine the magnitude of the magnetic moment concerning an atom of a given chemical element.

2.1 Atoms in a Magnetic Field

The Hamiltonian $\mathcal{H}_0$ of a single atom which contains Z electrons is given by

$$\mathcal{H}_0 = \sum_{i=1}^{Z} \left(\frac{p_i^2}{2m} + V_i \right) \tag{2.1}$$

with $p_i^2/2m$ being the kinetic energy and V_i the potential energy of electron i. The situation becomes more complex if an external magnetic field $\boldsymbol{B}$ is present which is given by:

$$\boldsymbol{B} = \boldsymbol{\nabla} \times \boldsymbol{A} \tag{2.2}$$

with $\boldsymbol{A}$ being the magnetic vector potential. This vector potential is chosen in such a way that the magnetic field is homogeneous within the atom and the Coulomb gauge

$$\boldsymbol{\nabla} \cdot \boldsymbol{A} = 0 \tag{2.3}$$

is valid. In this situation the magnetic vector potential can be written as:

$$\boldsymbol{A}(\boldsymbol{r}) = \frac{1}{2} \left(\boldsymbol{B} \times \boldsymbol{r} \right) \tag{2.4}$$

The corresponding kinetic energy amounts to:

$$E_{\text{kin}} = \frac{1}{2m} \left(p + eA\left(r \right) \right)^2 \tag{2.5}$$

$$= \frac{1}{2m} \left(p^2 + e\left(p \cdot A + A \cdot p \right) + e^2 A \cdot A \right) \tag{2.6}$$

Due to the Coulomb gauge we obtain:

$$p \cdot A = A \cdot p \tag{2.7}$$

As a result the Hamiltonian $\mathcal{H}^i$ of electron i is given by:

$$\mathcal{H}^i = \frac{p_i^2}{2m} + V_i + \frac{e}{m} A \cdot p + \frac{e^2}{2m} A \cdot A \tag{2.8}$$

The last term can be written as a function of the external magnetic field:

$$\frac{e^2}{2m} A \cdot A = \frac{e^2}{2m} \left(\frac{1}{2} \left(B \times r \right) \right)^2 \tag{2.9}$$

$$= \frac{e^2}{8m} \left(B \times r \right)^2 \tag{2.10}$$

For the third term in (2.8) one gets accordingly:

$$A \cdot p = \frac{1}{2} \left(B \times r \right) \cdot p \tag{2.11}$$

$$= \frac{1}{2} \left(r \times p \right) \cdot B \tag{2.12}$$

$$= \frac{1}{2} \hbar L \cdot B \tag{2.13}$$

with $\hbar L$ being the orbital angular momentum. Thus, we can express the Hamiltonian $\mathcal{H}^i$ as:

$$\mathcal{H}^i = \frac{p_i^2}{2m} + V_i + \mu_B L \cdot B + \frac{e^2}{8m} \left(B \times r_i \right)^2 \tag{2.14}$$

with $\mu_B = e\hbar/2m$ being the Bohr magneton. The consideration of the electron spin (angular momentum) S results in an additional term $\mu_B g\, S \cdot B$ with $g \approx 2$ being the g-factor of an electron. The complete Hamiltonian is therefore given by

$$\mathcal{H} = \sum_{i=1}^{Z} \left(\frac{p_i^2}{2m} + V_i \right) + \mu_B \left(L + gS \right) \cdot B + \frac{e^2}{8m} \sum_{i=1}^{Z} \left(B \times r_i \right)^2 \tag{2.15}$$

$$= \mathcal{H}_0 + \mathcal{H}_1 \tag{2.16}$$

The part $\mathcal{H}_1$ represents the modification due to the external magnetic field B and amounts to:

$$\mathcal{H}_1 = \mu_B \left(\boldsymbol{L} + g\boldsymbol{S} \right) \cdot \boldsymbol{B} + \frac{e^2}{8m} \sum_{i=1}^{Z} \left(\boldsymbol{B} \times \boldsymbol{r}_i \right)^2 \tag{2.17}$$

$$= \mathcal{H}_1^{\text{para}} + \mathcal{H}_1^{\text{dia}} \tag{2.18}$$

The first term $\mathcal{H}_1^{\text{para}}$ is known as the paramagnetic term, the second one $\mathcal{H}_1^{\text{dia}}$ as the diamagnetic term.

2.2 Atomic Diamagnetism

Each material exhibits diamagnetic behavior, i.e. a weak negative magnetic susceptibility. Consequently, an external magnetic field induces a magnetic moment which is oriented antiparallel to the external field. This behavior is explained in classical physics using Lenz's rule. But, magnetism cannot be understood within the framework of classical physics as shown in the Bohr–van Leeuwen theorem (see Chap. 1.3).

Therefore, we use a quantum mechanical approach. For simplification it is assumed that all electronic shells are filled. Then the orbital as well as spin angular momentum vanish:

$$\boldsymbol{L} = \boldsymbol{S} = 0 \tag{2.19}$$

Consequently, we obtain:

$$\mu_B \left(\boldsymbol{L} + g\boldsymbol{S} \right) \cdot \boldsymbol{B} \equiv 0 \tag{2.20}$$

i.e. the paramagnetic term of (2.18) is zero. Additionally, we assume that the external magnetic field $\boldsymbol{B}$ is parallel to the z-axis:

$$\boldsymbol{B} = (0, 0, B) \tag{2.21}$$

Due to

$$\boldsymbol{B} \times \boldsymbol{r}_i = B \cdot \begin{pmatrix} -y_i \\ x_i \\ 0 \end{pmatrix} \tag{2.22}$$

we get

$$\left(\boldsymbol{B} \times \boldsymbol{r}_i \right)^2 = B^2 \left(x_i^2 + y_i^2 \right) \tag{2.23}$$

Consequently, an energy shift of the ground state energy occurs due to the diamagnetic term which amounts to:

$$\Delta E_0 = \frac{e^2 B^2}{8m} \sum_i \langle 0 \left| x_i^2 + y_i^2 \right| 0 \rangle \tag{2.24}$$

with $| 0 \rangle$ being the wave function of the ground state. Atoms in the ground state with filled electron shells exhibit spherically symmetric electronic wave functions:

$$\langle x_i^2 \rangle = \langle y_i^2 \rangle = \langle z_i^2 \rangle = \frac{1}{3}\langle r_i^2 \rangle \tag{2.25}$$

Thus, we obtain:

$$\Delta E_0 = \frac{e^2 B^2}{12m} \sum_i \langle 0 \left| r_i^2 \right| 0 \rangle \tag{2.26}$$

Using the Helmholtz free energy F which is given by:

$$F = E - TS \tag{2.27}$$

with S being the entropy we obtain at $T = 0$

$$M = -\frac{\partial F}{\partial B} \tag{2.28}$$

$$= -\frac{N}{V}\frac{\partial \Delta E_0}{\partial B} \tag{2.29}$$

$$= -\frac{N}{V}\frac{e^2 B}{6m} \sum_i \langle r_i^2 \rangle \tag{2.30}$$

with N being the number of atoms in volume V. Assuming a linear material and a relative permeability $\mu_r \approx 1$ we can write:

$$\chi = \frac{\mu_0 M}{B} \tag{2.31}$$

due to $M = \chi H$ and $B = \mu_0 H$. Thus:

$$\chi = -\frac{N}{V}\frac{\mu_0 e^2}{6m} \sum_i \langle r_i^2 \rangle \tag{2.32}$$

From this expression we obtain the following consequences:

- The susceptibility is negative: $\chi^{\text{dia}} < 0$.
- Only the outermost shells significantly contribute due to $\chi \propto \langle r_i^2 \rangle$.
- The temperature dependence is negligible.

2.3 Atomic Paramagnetism

Paramagnetism is related to a positive magnetic susceptibility, i.e. the magnetization M is orientated parallel to an external magnetic field B.

The situation considered above was characterized by no unpaired electrons which implied a vanishing magnetic moment without an external magnetic field.

Now, we assume a non-vanishing magnetic moment due to unpaired electrons. Without an external magnetic field no favored orientation of the magnetic moments occurs and the resulting magnetization tends to zero. But, applying an external magnetic field leads to the existence of a preferential orientation, i.e. $M \neq 0$. The total magnetization depends on the magnitude of the external magnetic field and on temperature:

$$M \propto \frac{B}{T} \tag{2.33}$$

Semiclassical Consideration

Without loss of generality we can assume that the external magnetic field is oriented along the z-direction, i.e. $B = (0, 0, B)$. The energy of the magnetic moments exhibiting an angle between θ and $\theta + d\theta$ with respect to the z-axis and therefore to the external magnetic field direction is given by:

$$E = -\mu B \cos \theta \tag{2.34}$$

with μ being the magnetic moment. The net magnetic moment along B amounts to:

$$\mu_z = \mu \cos \theta \tag{2.35}$$

Under the assumption that all angles θ are found with the same probability the fraction with values between θ and $\theta+d\theta$ is proportional to $2\pi \sin \theta d\theta$ for a unit sphere with a surface of 4π and is determined by:

$$\frac{2\pi \sin \theta d\theta}{4\pi} = \frac{1}{2} \sin \theta d\theta \tag{2.36}$$

The probability dw of an angle between θ and $\theta + d\theta$ at a temperature T is the product of the geometrical factor $\frac{1}{2} \sin \theta d\theta$ and the Boltzmann factor:

$$e^{-E/kT} = e^{\mu B \cos \theta / kT} \tag{2.37}$$

This results in:

$$dw = \frac{1}{2} \sin \theta \, e^{\mu B \cos \theta / kT} d\theta \tag{2.38}$$

Thus, the averaged magnetic moment along B is given by

$$\langle \mu_z \rangle = \frac{\int \mu_z dw}{\int dw} \tag{2.39}$$

$$= \frac{\int_0^\pi \mu \cos \theta \, e^{\mu B \cos \theta / kT} \frac{1}{2} \sin \theta d\theta}{\int_0^\pi e^{\mu B \cos \theta / kT} \frac{1}{2} \sin \theta d\theta} \tag{2.40}$$

The replacements of:

$$y = \frac{\mu B}{kT} \tag{2.41}$$

$$x = \cos \theta \tag{2.42}$$

lead to:

$$dx = -\sin\theta d\theta \tag{2.43}$$

$$\int\limits_0^\pi \rightarrow \int\limits_{-1}^1 \tag{2.44}$$

This substitution results in:

$$\langle\mu_z\rangle = \mu \cdot \frac{\int\limits_{-1}^1 x\,e^{xy}dx}{\int\limits_{-1}^1 e^{xy}dx} \tag{2.45}$$

Due to:

$$\int e^{xy}dx = \frac{1}{y}e^{xy} \tag{2.46}$$

$$\int x\,e^{xy}dx = \frac{1}{y^2}e^{xy}(xy-1) \tag{2.47}$$

we obtain:

$$\frac{\langle\mu_z\rangle}{\mu} = \frac{\int\limits_{-1}^1 x\,e^{xy}dx}{\int\limits_{-1}^1 e^{xy}dx} \tag{2.48}$$

$$= \frac{y\,e^y + y\,e^{-y} - e^y + e^{-y}}{y\,e^y - y\,e^{-y}} \tag{2.49}$$

$$= \frac{e^y + e^{-y}}{e^y - e^{-y}} - \frac{1}{y} \tag{2.50}$$

$$= \coth y - \frac{1}{y} \tag{2.51}$$

$$=: L(y) \tag{2.52}$$

This function $L(y)$ is the so-called Langevin function which is shown in Fig. 2.1. Assuming a small external magnetic field or a high temperature, i.e. $y \ll 1$ (see (2.41)), we can approximate:

$$\coth y = \frac{1}{y} + \frac{y}{3} + \mathcal{O}(y^3) \tag{2.53}$$

Therefore:

$$L(y) = \frac{y}{3} + \mathcal{O}(y^3) \tag{2.54}$$

$$\approx \frac{y}{3} \tag{2.55}$$

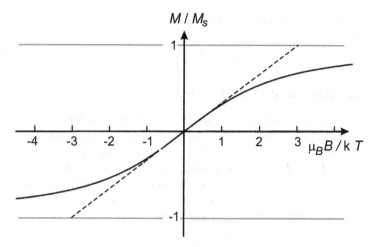

Fig. 2.1. The relative magnetization of a classical paramagnet can be characterized using the Langevin function $L(y) = \coth y - 1/y$. For small values of y the Langevin function can be approximated by $y/3$ being indicated by the *dashed line*

Thus, a linear behavior of the Langevin function on y occurs for small values of y. Let n the number of magnetic moments per unit cell. The saturation magnetization M_S is reached if all magnetic moments are parallel:

$$M_S = n\mu \qquad (2.56)$$

The magnetization M along $\boldsymbol{B}$ amounts to:

$$M = n\langle\mu_z\rangle \qquad (2.57)$$

For the relative magnetization, i.e. the magnetization with respect to the saturation magnetization, we obtain:

$$\frac{M}{M_S} = \frac{n\langle\mu_z\rangle}{n\mu} = L(y) \overset{y \ll 1}{\approx} \frac{y}{3} = \frac{\mu B}{3kT} \qquad (2.58)$$

For small external magnetic fields the susceptibility can be written as:

$$\chi = \frac{M}{H} \approx \frac{\mu_0 M}{B} \qquad (2.59)$$

Thus:

$$\chi = \frac{\mu_0 \mu B M_S}{3kTB} \qquad (2.60)$$

$$= \frac{n\mu_0\mu^2}{3k} \cdot \frac{1}{T} \qquad (2.61)$$

$$= \frac{c}{T} \qquad (2.62)$$

The last equation represents the Curie law:

$$\chi = \frac{c}{T} \quad \text{with} \quad c = \frac{n\mu_0\mu^2}{3k} \tag{2.63}$$

Quantum Mechanical Consideration

In the following we carry out the analogous calculation for a quantum mechanical system and substitute the classical magnetic moments by quantum mechanical spins possessing $J = 1/2$. J is defined using the eigenvalue of $\boldsymbol{J}$ which is given by $J(J+1)$ with $\boldsymbol{J}$ being the total angular momentum. Only two values are allowed for the z-component of the magnetic moment (magnetic quantum number): $m_J = \pm 1/2$, i.e. they are aligned parallel or antiparallel with regard to B, respectively. The energy is given by:

$$E = g m_J \mu_B B \tag{2.64}$$

Using the values for electrons ($g = 2$ and $m_J = \pm 1/2$) we obtain:

$$\langle \mu_z \rangle = \frac{-\mu_B \, e^{\mu_B B/kT} + \mu_B \, e^{-\mu_B B/kT}}{e^{\mu_B B/kT} + e^{-\mu_B B/kT}} \tag{2.65}$$

$$= \mu_B \tanh\left(\frac{\mu_B B}{kT}\right) \tag{2.66}$$

The substitution of:

$$y = \frac{\mu_B B}{kT} \tag{2.67}$$

leads to the relative magnetization:

$$\frac{M}{M_S} = \frac{\langle \mu_z \rangle}{\mu_B} = \tanh y \overset{y \ll 1}{\approx} y \tag{2.68}$$

This function (see Fig. 2.2) is different compared to the Langevin function (cf. Fig. 2.1) but the shape looks similar.

Now, we consider the general situation that J is an integer or has an half-integer value. The discussion is carried out using the partition function Z:

$$Z = \sum_{m_J=-J}^{J} \exp(m_J g_J \mu_B B/kT) = \sum_{m_J=-J}^{J} e^{x m_J} \tag{2.69}$$

using:

$$x = \frac{g_J \mu_B B}{kT} \tag{2.70}$$

Thus, we obtain:

$$\langle m_J \rangle = \frac{\sum m_J \, e^{x m_J}}{\sum e^{x m_J}} \tag{2.71}$$

$$= \frac{1}{Z} \cdot \frac{\partial Z}{\partial x} \tag{2.72}$$

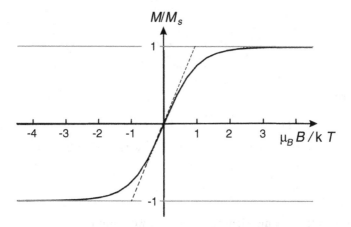

Fig. 2.2. The relative magnetization of a paramagnet with a spin of $1/2$ is given by a $\tanh y$ function with $y = \mu_B B/kT$. For small values of y the $\tanh y$ function can be approximated by y being indicated by the *dashed line*

For the magnetization we can write:

$$M = n g_J \mu_B \langle m_J \rangle \tag{2.73}$$

$$= \frac{n g_J \mu_B}{Z} \cdot \frac{\partial Z}{\partial x} \tag{2.74}$$

$$= \frac{n g_J \mu_B}{Z} \cdot \frac{\partial Z}{\partial B} \cdot \frac{\partial B}{\partial x} \tag{2.75}$$

Using the relationship:

$$\frac{\partial \ln Z}{\partial B} = \frac{1}{Z} \cdot \frac{\partial Z}{\partial B} \tag{2.76}$$

and (2.70) we obtain:

$$\frac{\partial B}{\partial x} = \frac{kT}{g_J \mu_B} \tag{2.77}$$

Thus, the magnetization is given by:

$$M = nkT \cdot \frac{\partial \ln Z}{\partial B} \tag{2.78}$$

The calculation of the partition function leads to:

$$Z = \sum_{m_J = -J}^{J} e^{x m_J} \tag{2.79}$$

$$= e^{-Jx} + e^{-(J-1)x} + \cdots + e^{(J-1)x} + e^{Jx} \tag{2.80}$$

$$= e^{-Jx} + e^{-Jx} e^{x} + e^{-Jx} e^{2x} + \cdots$$
$$+ e^{-Jx} e^{(2J-1)x} + e^{-Jx} e^{2Jx} \tag{2.81}$$

$$= e^{-Jx} \left(1 + e^{x} + e^{2x} + \cdots + e^{(2J-1)x} + e^{2Jx} \right) \tag{2.82}$$

With:

$$b = e^{-Jx} \tag{2.83}$$

$$t = e^{x} \tag{2.84}$$

we can write:

$$Z = b \left(1 + t + t^2 + \cdots + t^{2J} \right) \tag{2.85}$$

$$= b + bt + bt^2 + \cdots + bt^{M-1} \quad \text{with} \quad M = 2J + 1 \tag{2.86}$$

$$= \sum_{j=1}^{M} bt^{j-1} \tag{2.87}$$

which represents a geometrical series. Thus, we obtain:

$$Z = b \cdot \frac{1 - t^M}{1 - t} \tag{2.88}$$

$$= b \cdot \frac{1 - t^{2J+1}}{1 - t} \tag{2.89}$$

$$= e^{-Jx} \cdot \frac{1 - e^{(2J+1)x}}{1 - e^{x}} \tag{2.90}$$

$$= \frac{e^{-Jx} - e^{Jx} e^{x}}{1 - e^{x}} \tag{2.91}$$

$$= \frac{e^{Jx} e^{x/2} - e^{-Jx} e^{-x/2}}{e^{x/2} - e^{-x/2}} \tag{2.92}$$

$$= \frac{\frac{1}{2} \left(e^{(2J+1)x/2} - e^{-(2J+1)x/2} \right)}{\frac{1}{2} \left(e^{x/2} - e^{-x/2} \right)} \tag{2.93}$$

Due to:

$$\sinh x = \frac{1}{2} \left(e^{x} - e^{-x} \right) \tag{2.94}$$

the partition function amounts to:

$$Z = \frac{\sinh \left((2J + 1) x/2 \right)}{\sinh \left(x/2 \right)} \tag{2.95}$$

With

$$M_S = n g_J \mu_B J \tag{2.96}$$

and using (2.74) as well as setting

$$y = xJ = g_J \mu_B J B / kT \tag{2.97}$$

we obtain for the relative magnetization:

$$\frac{M}{M_S} = \frac{1}{J} \cdot \frac{1}{Z} \cdot \frac{\partial Z}{\partial x} \tag{2.98}$$

$$= \frac{1}{Z} \cdot \frac{\partial Z}{\partial y} \tag{2.99}$$

Setting

$$a = \frac{y}{2J} = \frac{xJ}{2J} = \frac{x}{2} \tag{2.100}$$

we get:

$$\partial y = 2J \partial a \tag{2.101}$$

and thus

$$\frac{M}{M_S} = \frac{1}{Z} \cdot \frac{1}{2J} \cdot \frac{\partial Z}{\partial a} \tag{2.102}$$

with

$$Z = \frac{\sinh\left[(2J+1)a\right]}{\sinh a} \tag{2.103}$$

Using

$$\frac{\mathrm{d}}{\mathrm{d}x} \sinh x = \cosh x \tag{2.104}$$

we can write

$$\frac{\partial Z}{\partial a} = \frac{(2J+1)\cosh[(2J+1)a]\sinh a - \sinh[(2J+1)a]\cosh a}{\sinh^2 a} \tag{2.105}$$

resulting in

$$\frac{1}{Z} \cdot \frac{\partial Z}{\partial a} = (2J+1)\frac{\cosh[(2J+1)a]}{\sinh[(2J+1)a]} - \frac{\cosh a}{\sinh a} \tag{2.106}$$

$$= (2J+1)\coth[(2J+1)a] - \coth a \tag{2.107}$$

Inserting (2.107) into (2.102) yields

$$\frac{M}{M_S} = \frac{1}{2J} \cdot \frac{1}{Z}\frac{\partial Z}{\partial a} \tag{2.108}$$

$$= \frac{2J+1}{2J}\coth\left[\frac{2J+1}{2J}y\right] - \frac{1}{2J}\coth\left[\frac{y}{2J}\right] \tag{2.109}$$

$$= B_J(y) \tag{2.110}$$

This expression describes the Brillouin function $B_J(y)$ which is shown in Fig. 2.3. In the following some special situations will be discussed:

- $J = \infty$

  Expanding (2.109) into a Taylor series using

$$\coth x = \frac{1}{x} + \frac{1}{3}x + \cdots \quad \text{for} \quad x \ll 1 \tag{2.111}$$

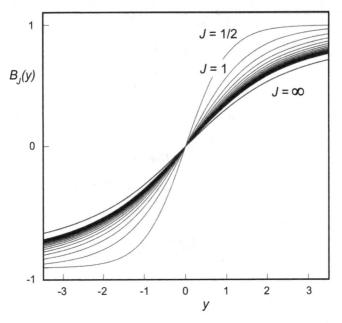

Fig. 2.3. Brillouin function $B_J(y)$ for different half-integer values of the magnetic moment quantum number J describing the magnetization of a paramagnet

results in

$$B_J(y) = \frac{2J+1}{2J} \coth\left[\frac{2J+1}{2J}y\right] - \frac{1}{2J}\left(\frac{2J}{y} + \frac{1}{3}\cdot\frac{y}{2J} + \cdots\right) \quad (2.112)$$

$$= \frac{2+\frac{1}{J}}{2}\coth\left[\frac{2+\frac{1}{J}}{2}y\right] - \frac{1}{y} - \frac{y}{12J^2} - \cdots \quad (2.113)$$

Therefore we obtain:

$$B_\infty(y) = \coth y - \frac{1}{y} = L(y) \quad (2.114)$$

This means that $J = \infty$ describes the situation in the semiclassical picture.
- $J = 1/2$

The corresponding Brillouin function is given by:

$$B_{1/2}(y) = \frac{2\cdot\frac{1}{2}+1}{2\cdot\frac{1}{2}}\coth[2y] - \coth y \quad (2.115)$$

$$= 2\coth[2y] - \coth y \quad (2.116)$$

Using the addition theorem:

$$\coth(2x) = \frac{\coth^2 x + 1}{2\coth x} \quad (2.117)$$

we get:

$$B_{1/2}(y) = \frac{\coth^2 y + 1}{\coth y} - \coth y \tag{2.118}$$

$$= \frac{\coth^2 y + 1 - \coth^2 y}{\coth y} \tag{2.119}$$

$$= \frac{1}{\coth y} \tag{2.120}$$

$$= \tanh y \tag{2.121}$$

This result coincides with our former consideration for magnetic moments with $m_J = 1/2$ (see (2.68)).

- $y \ll 1$

Typical experimental values are:

$$J = \frac{1}{2}, \quad g_J = 2, \quad B = 1\,\text{T}, \quad T = 300\,\text{K} \tag{2.122}$$

which leads to:

$$y = \frac{g_J \mu_B J B}{kT} \approx 2 \cdot 10^{-3} \ll 1 \tag{2.123}$$

This means that in most cases $y \ll 1$ is valid except the situation of very low temperatures or extremely high external magnetic fields. The Brillouin function can then be expressed as:

$$B_J(y) = \frac{2J+1}{2J} \coth\left[\frac{2J+1}{2J}y\right] - \frac{1}{2J}\coth\left[\frac{y}{2J}\right] \tag{2.124}$$

$$= \frac{2J+1}{2J}\left(\frac{2J}{2J+1} \cdot \frac{1}{y} + \frac{1}{3} \cdot \frac{2J+1}{2J}y + \mathcal{O}(y^3)\right)$$
$$\quad - \frac{1}{2J}\left(\frac{2J}{y} + \frac{y}{6J} + \mathcal{O}(y^3)\right) \tag{2.125}$$

$$= \frac{1}{y} + \frac{(2J+1)^2}{12J^2}y - \frac{1}{y} - \frac{1}{12J^2}y + \mathcal{O}(y^3) \tag{2.126}$$

$$\overset{y\ll 1}{\approx} \frac{4J^2+4J}{12J^2}y \tag{2.127}$$

$$= \frac{J+1}{3J}y \tag{2.128}$$

For low magnetic fields the magnetic susceptibility can thus be expressed as using (2.96) and (2.128):

$$\chi = \frac{M}{H} \tag{2.129}$$

$$= \frac{\mu_0 M_S B_J}{B} \tag{2.130}$$

$$= \frac{\mu_0 n g_J \mu_B J (J+1) g_J \mu_B J B}{3BJkT} \tag{2.131}$$

$$= \frac{n\mu_0}{3kT} \cdot g_J^2 \mu_B^2 J(J+1) \tag{2.132}$$

$$= \frac{n\mu_0 \mu_{\text{eff}}^2}{3kT} \tag{2.133}$$

with

$$\mu_{\text{eff}} = g_J \mu_B \sqrt{J(J+1)} \tag{2.134}$$

being the effective magnetic moment. This expression can be written as:

$$\chi = \frac{c_{\text{Curie}}}{T} \tag{2.135}$$

with

$$c_{\text{Curie}} = n\mu_0 g_J^2 \mu_B^2 J(J+1)/3k \tag{2.136}$$

being the classical Curie's law. For low magnetic fields the effective magnetic moment amounts to

$$\mu_{\text{eff}} = g_J \mu_B \sqrt{J(J+1)} \tag{2.137}$$

whereas for high fields the saturation magnetization is reached with

$$\mu_{\text{eff}} = g_J \mu_B J \tag{2.138}$$

Generally, both values are not identical except the case of $J \to \infty$, i.e. within the semiclassical picture.

The question arises how g_J, the Landé – g-factor, can be estimated. The magnetic moment is given by:

$$\boldsymbol{\mu} = \mu_B \left(g_L \boldsymbol{L} + g_S \boldsymbol{S} \right) \tag{2.139}$$

with g_L the g-factor for orbital angular momentum and g_S that for spin angular momentum. Using the total angular momentum $\boldsymbol{J} = \boldsymbol{L} + \boldsymbol{S}$ we can write

$$\boldsymbol{\mu} = \mu_B g_J \boldsymbol{J} \tag{2.140}$$

i.e. g_J is the projection of $g_L \boldsymbol{L} + g_S \boldsymbol{S}$ on $\boldsymbol{J}$. Therefore:

$$\boldsymbol{\mu} \boldsymbol{J} = \mu_B \left(g_L \boldsymbol{L} \boldsymbol{J} + g_S \boldsymbol{S} \boldsymbol{J} \right) \tag{2.141}$$

$$= \mu_B g_J \boldsymbol{J}^2 \tag{2.142}$$

Due to:

$$\boldsymbol{J} = \boldsymbol{L} + \boldsymbol{S} \tag{2.143}$$

we obtain:

$$\boldsymbol{L}^2 = \left(\boldsymbol{J} - \boldsymbol{S} \right)^2 = \boldsymbol{J}^2 + \boldsymbol{S}^2 - 2\boldsymbol{S}\boldsymbol{J} \tag{2.144}$$

and thus:

$$\boldsymbol{S}\boldsymbol{J} = \frac{1}{2} \left(\boldsymbol{J}^2 + \boldsymbol{S}^2 - \boldsymbol{L}^2 \right) \tag{2.145}$$

Analogously, we can calculate:

$$LJ = \frac{1}{2}\left(J^2 + L^2 - S^2\right) \tag{2.146}$$

Inserting into (see (2.142)):

$$g_L LJ + g_S SJ = g_J J^2 \tag{2.147}$$

and using that the eigenvalues of S^2 are $S(S+1)$, of L^2 are $L(L+1)$, and J^2 are $J(J+1)$ we obtain

$$g_J = g_L \frac{J(J+1) + L(L+1) - S(S+1)}{2J(J+1)} \tag{2.148}$$

$$+ g_S \frac{J(J+1) + S(S+1) - L(L+1)}{2J(J+1)} \tag{2.149}$$

with L being the orbital angular momentum quantum number, S the spin angular momentum quantum number, and J the total angular momentum quantum number. Using $g_L = 1$ and $g_S = 2$ yields:

$$g_J = \frac{3}{2} + \frac{S(S+1) - L(L+1)}{2J(J+1)} \tag{2.150}$$

which allows to determine g_J.

2.4 Hund's Rules for the Ground State of Atoms

Filled electronic shells have a vanishing total angular momentum. But, if an atom possesses electrons in unfilled shells spin and orbital angular momentum can be different to zero. The z-component of the orbital angular momentum exhibits values of $\hbar L$ with $-L \leq m_L \leq L$, that of the spin angular momentum of $\hbar S$ with $-S \leq m_S \leq S$. Totally, there are $(2L+1)\cdot(2S+1)$ combinations for the coupling of $J = L+S$ following $|L-S| \leq J \leq L+S$. Different configurations are energetically not identical. The question arises which configuration represents the energetic ground state. The answer is given using the following empirical rules which are known as Hund's rules. They are listed with decreasing importance, i.e. before applying the second rule the first one must be satisfied. The same holds for the third rule accordingly.

1. *S has the maximum value but must be compatible with the Pauli exclusion principle*
   With ℓ being the orbital angular momentum of the not-completed shell and p the number of electrons within this shell (i.e. $p < 2(2\ell + 1)$) we have:

$$S = \frac{1}{2}\left[(2\ell + 1) - |2\ell + 1 - p|\right] \tag{2.151}$$

2. *L has the maximum value but must be compatible with the Pauli exclusion principle and rule 1*

   L is given by:

$$L = S \cdot |2\ell + 1 - p| \tag{2.152}$$

Each multiplet being constructed in accordance with rules 1 and 2 possesses

$$\sum_{J=|L-S|}^{L+S} (2J + 1) = (2L + 1)(2S + 1) \tag{2.153}$$

different states. Mostly, only that states are important which fulfill the third rule:

3. *J is given by:*

$$J = |L - S| \tag{2.154}$$

if the shell is filled with less than one half (i.e. $p \leq 2\ell + 1$) or

$$J = L + S \tag{2.155}$$

if the shell is filled with more than one half (i.e. $p \geq 2\ell + 1$). This can be written as:

$$J = S \cdot |2\ell - p| \tag{2.156}$$

The configuration which is found using Hund's rules is described by the following expression being called term symbol:

$$^{2S+1}L_J \tag{2.157}$$

with $2S + 1$ named multiplicity. $L = 0, 1, 2, 3, \ldots$ is labelled by S, P, D, F, G, ... For the special case of a filled shell with $p = 2 \cdot (2\ell + 1)$ we therefore obtain:

$$S = 0 \quad L = 0 \quad J = 0 \tag{2.158}$$

Consequently, the total momentum of the entire atom is given by the momentum of the not-completed shell.

On the one hand Hund's rules are able to predict the ground state of an ion. But we must keep in mind that they are on the other hand not suited to describe any excited state and do also not allow the calculation of the energetic difference between the ground state and the excited states. If the ground state and the first excited state are energetically located closed to each other the reliability of the predicted values may be reduced.

The predictions of the total angular momentum quantum number J, of the angular momentum quantum number L, and of the spin angular momentum quantum number S for the magnetically relevant $3d$ ions using Hund's rules are given in Fig. 2.4 whereas the corresponding predictions concerning the $4f$ ions are shown in Fig. 2.5.

From theoretical point of view, μ_{eff}^2 can be predicted using Hund's rules due to:

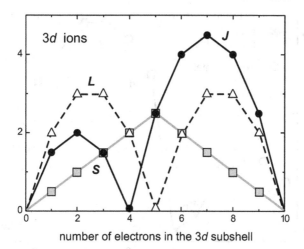

Fig. 2.4. Values of S, L, and J for $3d$ ions according to Hund's rules

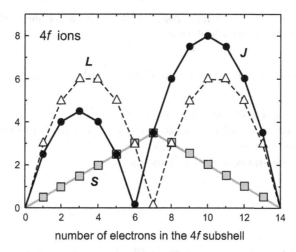

Fig. 2.5. Values of S, L, and J for $4f$ ions according to Hund's rules

$$\mu_{\text{eff}} = \mu_B g_J \sqrt{J(J+1)} \tag{2.159}$$

in combination with

$$g_J = \frac{3}{2} + \frac{S(S+1) - L(L+1)}{2J(J+1)} \tag{2.160}$$

The measurement of the magnetic susceptibility allows the determination of the effective magnetic moment:

$$\chi = \frac{n\mu_0 \mu_{\text{eff}}^2}{3kT} \tag{2.161}$$

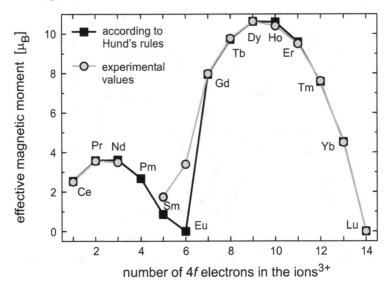

Fig. 2.6. Effective magnetic moment per μ_B of $4f^{3+}$ ions according to Hund's rules in comparison to the experimental value

The comparison between experiment and theoretical calculation (see Fig. 2.6) yields:

- generally a fair agreement for $4f$ elements.
- Exceptions are Sm and Eu due to excited states which are energetically located closed to the ground state.
- Discrepancies also occur for $3d$ elements.

Problems

2.1. Determine the paramagnetic susceptibility per unit volume of an ideal gas at room temperature assuming that $J = 1$ and $g = 2$ (being the values for molecular oxygen).

2.2. Calculate the Landé – g-factor g_J for Ce which is a $4f$ ion with one electron in the $4f$ subshell.

2.3. Term symbols
(a) Determine the term symbol for the ground state of an Fe ion $3d^6$.
(b) Determine the term symbol for the ground state of a Gd ion $4f^7$.

3

Solid State Magnetism

In the previous chapter we have discussed the properties of isolated and localized magnetic moments. Now we allow their interaction and will concentrate on the solid state. We will not deal with magnetic behavior of molecules in the gas phase and fluids in this chapter.

Metals exhibit conduction electrons which are delocalized. These electrons (so-called itinerant electrons) can move nearly free inside the metal. Thus, we will start our discussion concerning solid state magnetism with the model of free electrons. The magnetic moments in the solid state can be

- localized or
- carried by the delocalized conduction electrons.

Both situations result in dia- and paramagnetism. But now, collective magnetism like ferromagnetism is additionally allowed which becomes manifest in spontaneously spin split electronic states and will be discussed at the end of this chapter.

3.1 Model of Free Electrons

The model of free electrons only represents an estimation but allows a plausible description of the most important properties.

The assumptions being made are that volume electrons of the constituent atoms become conduction electrons and that these electrons move about freely through the volume of the metal. The periodic potential of the crystal lattice is not taken into account.

The description of the electronic states is carried out by planar waves. Each state can be occupied with 2 electrons due to the Pauli exclusion principle. In the ground state all states exhibiting a wave vector k within the Fermi sphere $|k| \leq k_F$ are occupied whereas all states being outside are unoccupied (see Fig. 3.1). The distance in k-space between different states is given by $2\pi/L$

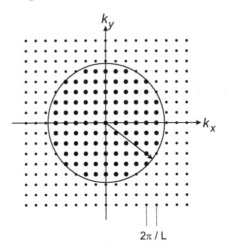

Fig. 3.1. Electron states are separated by $2\pi/L$ being doubly occupied and containing a volume of $(2\pi/L)^3$. States are only occupied within the Fermi sphere exhibiting a radius of k_F (see *circle*) whereas they are unoccupied for $|k| > k_F$

and the volume of the sample by $V = L^3$. The number of states between k and $k + \mathrm{d}k$ amounts to $4\pi k^2 \mathrm{d}k$. The density of states (DOS) is given by:

$$g(k)\mathrm{d}k = 2 \cdot \frac{1}{(2\pi/L)^3} \cdot 4\pi k^2 \mathrm{d}k \tag{3.1}$$

$$= \frac{Vk^2}{\pi^2}\mathrm{d}k \tag{3.2}$$

The factor of two is due to the two spin states of electrons. For $T = 0\,\mathrm{K}$ the states are occupied up to k_F with N electrons:

$$N = \int\limits_0^{k_F} g(k)\mathrm{d}k = \frac{V}{\pi^2}\int\limits_0^{k_F} k^2\mathrm{d}k = \frac{Vk_F^3}{3\pi^2} \tag{3.3}$$

Thus, we obtain:

$$k_F^3 = 3\pi^2 n \tag{3.4}$$

with $n = N/V$ being the number of electrons per volume. The highest occupied energy levels possess the Fermi energy E_F:

$$E_F = \frac{\hbar^2 k_F^2}{2m_e} \tag{3.5}$$

as kinetic energy. This results in:

$$k_F^3 = \left(\frac{2m_e}{\hbar^2}\right)^{3/2} \cdot E_F^{3/2} \tag{3.6}$$

The density of states as a function of energy can be derived from the relation:

$$n = \int_0^{E_F} g(E)dE = \frac{k_F^3}{3\pi^2} \tag{3.7}$$

$$= \frac{1}{3\pi^2} \left(\frac{2m_e}{\hbar^2}\right)^{3/2} \cdot E^{3/2} \tag{3.8}$$

Therefore:

$$g(E_F) = \frac{dn}{dE}\bigg|_{E=E_F} \tag{3.9}$$

$$= \frac{3n}{2E_F} \tag{3.10}$$

$$= \frac{m_e k_F}{\pi^2 \hbar^2} \tag{3.11}$$

i.e.

$$g(E_F) \propto m_e \tag{3.12}$$

This quantity m_e represents the effective electron mass which can be larger than the mass of a free electron.

The previous considerations were carried out for $T = 0\,\mathrm{K}$. For a non-vanishing temperature $T > 0\,\mathrm{K}$ the density of states remains unchanged whereas the occupation becomes influenced which is characterized by the Fermi function $f(E)$:

$$f(E) = \frac{1}{1 + e^{E-\mu/kT}} \tag{3.13}$$

with μ being the chemical potential. At $T = 0\,\mathrm{K}$ the Fermi function is given by:

$$f(E) = \begin{cases} 0 & \text{for} \quad E > \mu \\ 1 & \text{for} \quad E < \mu \end{cases} \tag{3.14}$$

and exhibits a step-like behavior. At higher temperatures a smoothing out occurs (see Fig. 3.2). The step function is an adequate approximation for most of the metals (situation of the degenerate limit). For $E - \mu \gg kT$ the Fermi function is determined by the Boltzmann distribution:

$$f(E) \propto e^{E-\mu/kT} \tag{3.15}$$

due to:

$$\frac{1}{1 + e^x} \longrightarrow \frac{1}{e^x} \quad \text{for} \quad x \to 0 \tag{3.16}$$

This situation is called the non-degenerate limit. For $T > 0\,\mathrm{K}$ we get:

$$n = \int_0^{E_F} g(E)f(E)dE \tag{3.17}$$

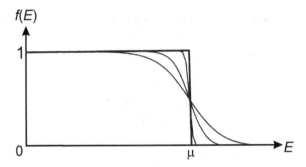

Fig. 3.2. Fermi function $f(E)$ as being defined by (3.13). At $T = 0\,\mathrm{K}$ a step-like behavior is present whereas for higher temperatures a smoothing out occurs

The energy dependence of $g(E)$ is shown in Fig. 3.3, that of $f(E)g(E)$ for different temperatures in Fig. 3.4. At $T = 0\,\mathrm{K}$ we have $\mu = E_F$; for higher temperatures $T > 0\,\mathrm{K}$ the chemical potential can be developed into a Taylor series:

$$\mu = E_F \left(1 - \frac{\pi^2}{12} \left(\frac{kT}{E_F} \right)^2 + \mathcal{O}\left(\left(\frac{kT}{E_F} \right)^4 \right) \right) \tag{3.18}$$

At room temperature ($T \approx 300\,\mathrm{K}$) most of the metals exhibit a deviation of about 0.01% between E_F and μ. This means that the Fermi energy and the chemical potential are nearly identical.

The Fermi surface is characterized by states in the k-space with $E(k) = \mu$. For semiconductors or insulators, the chemical potential is located in band gaps or gaps of the density of states, respectively. Thus, no electrons are present at the Fermi surface whereas metals exhibit electrons at the Fermi surface.

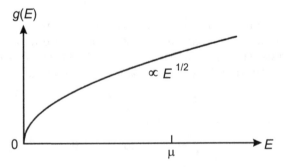

Fig. 3.3. Density of states $g(E)$ of a free electron gas being $\propto E^{1/2}$

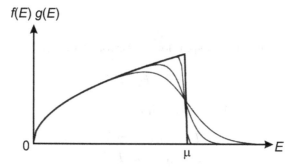

Fig. 3.4. Energy dependence of the function $f(E) \, g(E)$ shown in Figs. 3.2 and 3.3

3.2 Pauli Paramagnetism

A doubled occupation of each state in k-space due to two possible spin states occurs in metals, i.e. each electron is "spin up" or "spin down". Applying an external magnetic field results in an increase or decrease of the electron energy by $\pm g \mu_B B m_S$ being dependent on the corresponding spin. This leads to the paramagnetic susceptibility of the electron gas ("Pauli paramagnetism").

Neglecting the orbital momentum at $T = 0 \, \mathrm{K}$ results in $g = 2$ and a step-like behavior of the Fermi function. The external magnetic field induces a splitting of both spin subbands of $2 g \mu_B B m_S \approx 2 \mu_B B$ (see Fig. 3.5). Assuming that $g \mu_B B$ represents a fairly small energy results in a small splitting of both bands. The increasing number of spin up electrons per unit volume amounts to:

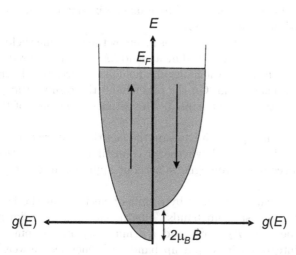

Fig. 3.5. Density of states for spin up and spin down electrons when applying an external magnetic field B. The splitting of both bands amounts to $2 \mu_B B$

$$n_\uparrow = \frac{1}{2}g(E_F)\mu_B B \tag{3.19}$$

and the decreasing number of spin down electrons to:

$$-n_\downarrow = \frac{1}{2}g(E_F)\mu_B B \tag{3.20}$$

The magnetization M is therefore given by:

$$M = \mu_B(n_\uparrow - n_\downarrow) \tag{3.21}$$
$$= g(E_F)\mu_B^2 B \tag{3.22}$$

and the magnetic (Pauli) susceptibility results in:

$$\chi^{\text{Pauli}} = \frac{M}{H} \approx \frac{\mu_0 M}{B} \tag{3.23}$$
$$= \mu_0\mu_B^2 g(E_F) \tag{3.24}$$
$$= \frac{3n\mu_0\mu_B^2}{2E_F} \tag{3.25}$$

due to $g(E_F) = 3/2 \cdot n/E_F$ (see (3.10)).

3.3 Spontaneously Spin Split States

The magnetic moment per iron atom in the solid state amounts to $2.2\mu_B$. The non-integral value demonstrates that a description with localized magnetic moments fails. This situation can be described by band or itinerant ferromagnetism being characterized by a magnetization due to a spontaneous spin splitting of the valence bands.

One approach for the explanation is given by the molecular field theory. All spins are influenced by an identical mean field λM which is caused by all other electrons. On the one hand, the molecular field magnetizes the electron gas due to the Pauli paramagnetism. On the other hand the resulting magnetization is responsible for the molecular field. This situation reminds of the chicken-egg-scenario.

A more promising approach bases on the fact that nature tries to minimize energy. Thus, we have to look whether it is possible to decrease the energy of a system if it becomes ferromagnetic without applying an external magnetic field.

This situation can be realized by a shift of electrons at the Fermi surface from spin down into spin up bands. This means that spin down electrons with energies between $E_F - \delta E$ and E_F must perform a spin-flip and are subsequently integrated in a spin up band with energies between $E_F + \delta E$ and E_F (see Fig. 3.6). The energy gain per electron amounts to δE and the number of electrons being moved to $1/2g(E_F)\delta E$. Thus, the increase of kinetic energy is given by:

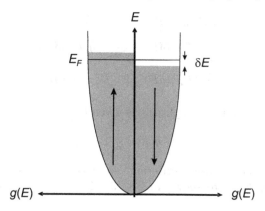

Fig. 3.6. Density of states for spin up and spin down electrons exhibiting a spontaneous spin splitting without applying an external magnetic field

$$\Delta E_{\mathrm{kin}} = \frac{1}{2} g(E_F)(\delta E)^2 \tag{3.26}$$

This situation does not look favorable but it is possible that the increase of kinetic energy is overcompensated due to the exchange of the magnetization with the molecular field as shown in the following. After the spin flip the number of spin up and spin down electrons are given by:

$$n_\uparrow = \frac{1}{2} n + \frac{1}{2} g(E_F) \delta E \tag{3.27}$$

$$n_\downarrow = \frac{1}{2} n - \frac{1}{2} g(E_F) \delta E \tag{3.28}$$

with n being the number of electrons at the Fermi energy in the paramagnetic case. Because each electron carries a magnetic moment of $1\mu_B$ the magnetization can be written as:

$$M = \mu_B(n_\uparrow - n_\downarrow) \tag{3.29}$$

The potential or molecular field energy amounts to:

$$\Delta E_{\mathrm{pot}} = -\frac{1}{2} \mu_0 M \cdot \lambda M \tag{3.30}$$

$$= -\frac{1}{2} \mu_0 \lambda M^2 \tag{3.31}$$

$$= -\frac{1}{2} \mu_0 \mu_B^2 \lambda (n_\uparrow - n_\downarrow)^2 \tag{3.32}$$

Introducing $U = \mu_0 \mu_B^2 \lambda$ which is a measure of the Coulomb energy we obtain:

$$\Delta E_{\mathrm{pot}} = -\frac{1}{2} U \cdot (g(E_F)\delta E)^2 \tag{3.33}$$

The total change in energy amounts to:

$$\Delta E = \Delta E_{\text{kin}} + \Delta E_{\text{pot}} \tag{3.34}$$

$$= \frac{1}{2}g(E_F)(\delta E)^2 - \frac{1}{2}U\left(g(E_F)\delta E\right)^2 \tag{3.35}$$

$$= \frac{1}{2}g(E_F)(\delta E)^2 \left(1 - U \cdot g(E_F)\right) \tag{3.36}$$

Therefore, a spontaneous spin splitting is given for $\Delta E < 0$, i.e.

$$U \cdot g(E_F) \geq 1 \tag{3.37}$$

which is the so-called "Stoner criterion" for ferromagnetism. Values of U, $g(E_F)$, and $U \cdot g(E_F)$ are shown in Fig. 3.7 for the first 50 elements. We

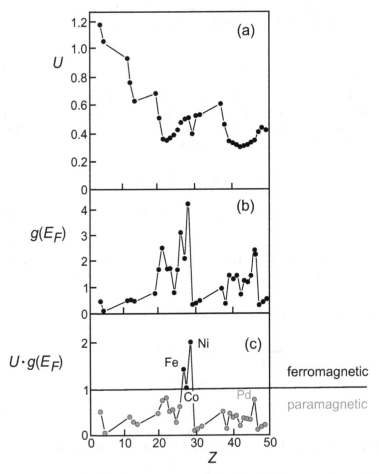

Fig. 3.7. Values of the Stoner parameter U, density of states per atom $g(E_F)$ at the Fermi energy, and $U \cdot g(E_F)$ as a function of the atomic number Z. Only the elements Fe, Co, and Ni fulfill the Stoner criterion and are ferromagnetic

directly see that only Fe, Co, and Ni exhibit a value of $U \cdot g(E_F) > 1$ which is mainly caused by the large density of states directly at the Fermi energy.

If the Stoner criterion is fulfilled a splitting of the spin up and spin down bands by Δ occurs without applying an external magnetic field. This value Δ represents the exchange splitting. For $U \cdot g(E_F) < 1$ no spontaneous magnetization is present but nevertheless the magnetic susceptibility may be different from the paramagnetic value.

In the following this change of the magnetic susceptibility is discussed taking into consideration an applied magnetic field and the electronic interactions. The resulting energy shift induces a magnetization:

$$M = \mu_B(n_\uparrow - n_\downarrow) \tag{3.38}$$

$$= \mu_B g(E_F)\delta E \tag{3.39}$$

The corresponding total energy shift amounts to:

$$\Delta E = \frac{1}{2}g(E_F)(\delta E)^2 (1 - U \cdot g(E_F)) - MB \tag{3.40}$$

$$= \frac{M^2}{2\mu_B^2 g(E_F)} (1 - U \cdot g(E_F)) - MB \tag{3.41}$$

The minimum is reached for $\partial \Delta E / \partial M = 0$:

$$\frac{M}{\mu_B^2 g(E_F)} \cdot (1 - U \cdot g(E_F)) - B = 0 \tag{3.42}$$

which leads to:

$$M = B \cdot \frac{\mu_B^2 g(E_F)}{1 - U \cdot g(E_F)} \tag{3.43}$$

Thus, we can calculate the susceptibility to be:

$$\chi = \frac{\mu_0 M}{B} \tag{3.44}$$

$$= \frac{\mu_0 \mu_B^2 g(E_F)}{1 - U \cdot g(E_F)} \tag{3.45}$$

$$= \frac{\chi^{\text{Pauli}}}{1 - U \cdot g(E_F)} \tag{3.46}$$

i.e. $\chi > \chi^{\text{Pauli}}$ due to the Coulomb interaction. This situation is called Stoner enhancement.

An example is given for Palladium. For this element the Stoner criterion is not fulfilled but $U \cdot g(E_F)$ is significantly larger than zero (see Fig. 3.7c). Thus, it exhibits an enhanced Pauli susceptibility. In other words Pd is "nearly" ferromagnetic.

The properties of an itinerant or band ferromagnet are exemplarily shown for Fe. The calculated density of states (DOS) without exchange splitting

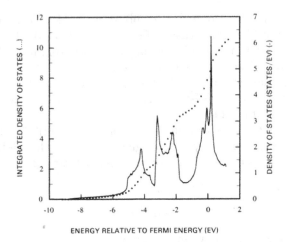

Fig. 3.8. Calculation of the density of states as a function of energy with respect to the Fermi energy E_F for Fe neglecting the exchange splitting, i.e. for the paramagnetic situation (Reprinted from [1] p. 85. Copyright 1978, with permission from Elsevier)

(i.e. for the paramagnetic case) is shown in Fig. 3.8. A high DOS occurs directly at the Fermi energy E_F, i.e. ferromagnetism seems to be probable due to the Stoner criterion. In comparison, the DOS of Pd (see Fig. 3.9) also exhibits a high magnitude which points to a "nearly" magnetic behavior. A very different situation is present for noble metals like Cu (see Fig. 3.10). The low DOS at E_F indicates a non-magnetic element.

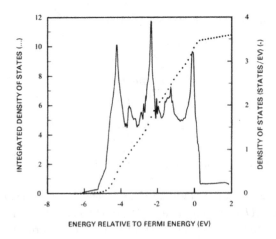

Fig. 3.9. Calculation of the density of states as a function of energy with respect to the Fermi energy E_F for Pd which also exhibits a large value at E_F compared to Fe (Reprinted from [1] p. 145. Copyright 1978, with permission from Elsevier)

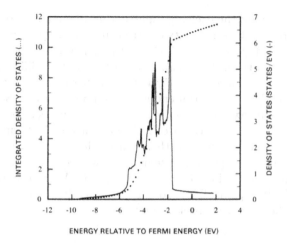

Fig. 3.10. Calculation of the density of states as a function of energy with respect to the Fermi energy E_F for the noble metal Cu. No d-states are present at E_F which results in a small magnitude of the total DOS. (Reprinted from [1] p. 97. Copyright 1978, with permission from Elsevier)

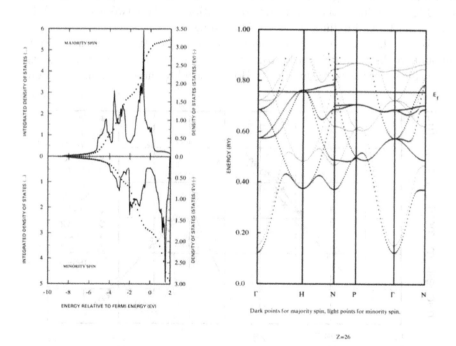

Fig. 3.11. Left: Calculation of the spin resolved density of states as a function of energy with respect to the Fermi energy E_F for Fe taking into account the exchange splitting. **Right**: Calculated spin resolved band structure of Fe. Majority bands are characterized by dark points, minority bands by light points. (Reprinted from [1] pp. 169 and 170. Copyright 1978, with permission from Elsevier)

Taking into account the exchange splitting results in a spin resolved DOS which behaves different for majority electrons (exemplarily shown for Fe: upper curve in the left part of Fig. 3.11) and minority electrons (lower curve in the left part of Fig. 3.11 for Fe). Due to the shift in different direction two properties become important. First, the number of majority and minority electrons directly at the Fermi energy is no more identical. Second, more majority electrons below E_F are present than minority electrons.

The spin resolved band structure (right part of Fig. 3.11) directly shows the exchange splitting of bands with majority character (dark points) and minority character (light points). Exemplarily, we see that a majority band crosses the Fermi energy along Γ - H also known as Δ direction which corresponds to the Fe(100) plane. This crossing can directly be probed using, e.g., spin resolving photoelectron spectroscopy.

3.4 Magnetism of 3d Transition Metals and Alloys

The variation of magnetic moments for different compositions in $3d$ metals and alloys is quite regular. This representation is called Slater–Pauling curve (see Fig. 3.12) due to the significant contributions of these two scientists, John Slater and Linus Pauling, to its understanding.

We see that the average magnetic moment μ per transition metal atom amounts to 2.2 μ_B for Fe, to 1.7 μ_B for Co, and to 0.6 μ_B for Ni. For Fe$_{50}$Ni$_{50}$ this value is very close to that of Co. Both, this alloy and Co, exhibit the same

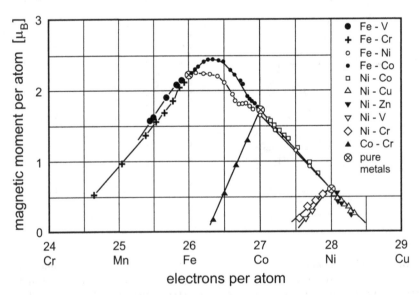

Fig. 3.12. Saturation magnetization of ferromagnetic alloys as a function of the electron concentration (Adapted from [2])

average atomic number or electrons per atom which is 27 or average number of valence electrons being $n_{\mathrm{val}} = 9$. Looking at the right side of the curve we observe that the magnetic moment per atom increases nearly linearly from zero at about 60% Cu in Ni for the data on the right to the maximum. The slope amounts to -1. The same behavior occurs for alloys consisting of FeCo, FeNi, and NiCo with $n_{\mathrm{val}} > 8.6$. The magnetic moment per atom reaches a maximum with about 2.5 μ_B near the average electron concentration of about 26.5 electrons per atom for metals and alloys. Assuming one $4s$ electron per atom the number of d electrons amounts to $n_d = 7.5$.

Using a simple band model which is known as the rigid band model we are able to understand the most important differences between the magnetic behavior of metallic Fe and metallic Ni. It assumes that the s and d bands are rigid in shape with varying atomic number. This allows to model the magnetic behavior of different alloys by only moving the Fermi energy E_F up or down through the majority and minority band according to the number of electrons being present. This model is of course not correct but we can understand some trends in physical properties.

Fe possesses 8 valence electrons being in $3d$ and $4s$ states whereas Ni has 10. Different measurements prove that Fe has slightly less than one electron that can be called free or itinerant ($4s^{0.95}$). The remaining 7.05 electrons occupy the more localized $3d$ band. The number of d electrons with respect to each spin subband is therefore:

$$n_d^\uparrow + n_d^\downarrow = 7.05 \qquad (3.47)$$

The observed magnetic moment of 2.2 μ_B per atom yields:

$$n_d^\uparrow - n_d^\downarrow = 2.2 \qquad (3.48)$$

Thus, we obtain that 4.62 of these 7.05 $3d$ electrons are spin up and 2.43 are spin down electrons. On the other hand Ni exhibits 0.6 free electrons ($4s^{0.6}$) and $3d^{9.4}$. The magnetic moment of 0.6 μ_B results in 5 spin up electrons and 4.4 spin down electrons or 0.6 spin down holes; therefore, the spin up band is fully occupied.

Metals which exhibit an exchange splitting that is less than the energy difference between the Fermi energy E_F and the top of the d band are called *weak ferromagnets*. They have, by definition, holes in both the majority and minority spin band. These metals are found on the left side of the maximum in the Slater–Pauling curve. One example is Fe. Metals that have an exchange splitting larger than this difference are called *strong ferromagnets*. They only have holes in the minority band and are located on the right side of the peak in the Slater–Pauling curve. One example is given by Ni.

As one example using the rigid-band model to understand magnetic properties of alloys let us consider Ni substituted in Fe with the composition given by $Fe_{1-x}Ni_x$. The average amount of the valence electrons is $n_{\mathrm{val}} = 8(1-x) + 10x$. Thus, the Fermi energy moves up through the rigid bands with increasing Ni content.

The magnetic moment per atom of an alloy is given by:

$$\mu = (n^\uparrow - n^\downarrow)\mu_B \tag{3.49}$$

In transition metal alloys the spins are mostly due to d electrons which results in:

$$\mu = (n_d^\uparrow - n_d^\downarrow)\mu_B \tag{3.50}$$

Generally, both $n_d^\uparrow$ and $n_d^\downarrow$ can vary on alloying. However, the magnetic moment per atom may be simply calculated if the Fermi energy E_F lies above the top of the spin up band, i.e. for strong ferromagnets. For such systems we have $n_d^\uparrow = 5$. Thus, the magnetic moment per atom amounts to $\mu = (5 - n_d^\downarrow)\mu_B$. Due to $n_d^\downarrow = n_d - 5$ we have $\mu = (10 - n_d)\mu_B$.

For strong ferromagnets this relation represents a straight line exhibiting a slope of -1 and thus adequately describes the data on the right side of the Slater–Pauling curve. This equation also explains why the average moment of Co should be very close to that of $Ni_{50}Fe_{50}$. Both exhibit the same valence electron concentration and thus the same value of n_d.

Further, this consideration explains the observation of non-integral average magnetic moments in $3d$ alloys. Additionally, we understand why the maximum in the Slater–Pauling curve is reached with a value of 2.5 μ_B. The Fermi level is stable when it coincides with the density of states minimum near the center of the minority spin band and the majority $3d$ band is full ($n_d = 7.5$). For a lower d electron concentration than 7.5 majority as well as minority states are (partially) empty and the magnetic moment decreases with decreasing n_d. This situation is realized in weak ferromagnets.

As already mentioned above the rigid-band model is a rather naive picture due to the assumption that the band structure and the shape of the curve describing the density of states do change with alloy composition in reality.

Another problem is that this model gives the averaged value of the magnetic moment of an alloy. It does not allow to determine the magnetic moment of each constituent individually. It is known that these element specific values may vary in a different way in an alloy which is shown in Fig. 3.13 for the example of FeCo, CoNi, and FeNi with different stoichiometry. In the FeNi alloy (closed and open squares) the magnetic moment of each Fe atom as well as that of each Ni atom nearly remain constant. This results in a linear increase of the averaged magnetic moment with increasing amount of Fe in the alloy and in a value of the slope being -1 (gray squares). In the FeCo alloy (black and white circles) the magnetic moment of each Co atom does not change with varying composition whereas that of each Fe atom exhibits different values if the composition changes and leads to a maximum (gray circles). This latter observation directly gives evidence that the slope in the Slater–Pauling curve cannot be constant over the whole range.

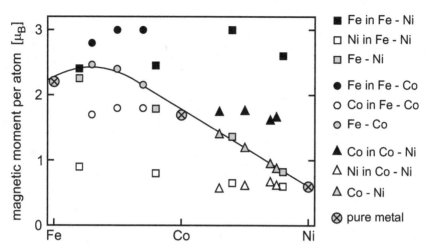

Fig. 3.13. Magnetic moments per atom in 3d transition metal alloys as a function of the electron concentration. *Full* and *open* symbols denote element specific magnetic moments in the alloy, *gray* symbols refer to the average value weighted with the respective concentration. The corresponding line is comparable to the Slater–Pauling curve (see Fig. 3.12). (Data taken from [3] and [4])

Problems

3.1. Ferromagnetic alloys
Assume a magnetic moment per atom of 1 μ_B. Determine the relative composition of the following alloys using the Slater–Pauling curve:
(a) Ni–Co
(b) Co–Cr
(c) Fe–Cr

4

Magnetic Interactions

In this chapter we will discuss different types of magnetic interaction which are responsible for properties which base on the fact that magnetic moments "feel" each other. As a result magnetic long range order can occur.

We will start our considerations with the interaction of two single magnetic dipoles. Subsequently, we will discuss the exchange interaction which is a pure quantum mechanical effect and is due to the Coulomb interaction and Pauli exclusion principle. Depending on the distance between the magnetic moments we distinguish between the direct and indirect exchange.

The situation that the electrons of neighboring magnetic atoms directly interact is called "direct exchange" because the interaction is mediated without needing intermediate atoms. If the overlap of the involved wave functions is only small (e.g. for rare earth metal atoms with their localized $4f$ electrons) then the direct exchange does not represent the dominating mechanism for magnetic properties. For this class of systems the indirect exchange interaction is responsible for magnetism.

4.1 Magnetic Dipole Interaction

The energy of two magnetic dipoles μ_1 and μ_2 separated by the vector r is given by:

$$E = \frac{\mu_0}{4\pi r^3} \left(\mu_1 \cdot \mu_2 - \frac{3}{r^2} (\mu_1 \cdot r)(\mu_2 \cdot r) \right) \tag{4.1}$$

and thus depends on their distance and relative orientations.

For an estimation of this energy we choose typical values with $\mu_1 = \mu_2 = 1 \ \mu_B$ and $r = 2$ Å. Additionally, we assume $\mu_1 \uparrow\uparrow \mu_2$ and $\mu \uparrow\uparrow r$. This situation results in an energy of:

$$E = \frac{\mu_0 \mu_B^2}{2\pi r^3} = 2.1 \cdot 10^{-24} \mathrm{J} \tag{4.2}$$

The corresponding temperature ($E = kT$) is far below $1\,\mathrm{K}$. But, the order temperature typically reaches values of several $100\,\mathrm{K}$ for a lot of ferromagnetic materials. Therefore, the magnetic dipole interaction is too small to cause ferromagnetism.

4.2 Direct Exchange

Let us assume a rather simple model with only two electrons which exhibit position vectors r_1 and r_2. Furthermore, we consider that the total wave function is composed of the product of single electron states $\psi_a(r_1)$ and $\psi_b(r_2)$. The electrons belonging to are undistinguishable. Therefore, the wave function squared must be invariant for the exchange of both electrons. Because the electrons are fermions the Pauli exclusion principle must be fulfilled which leads to an antisymmetric wave function. Taking into consideration the spin of the electrons two possibilities are given: a symmetric spatial part ψ in combination with an antisymmetric spin part χ or an antisymmetric spatial part in combination with a symmetric spin part. The first situation represents a singlet state with $S_{\mathrm{total}} = 0$, the second one a triplet state with $S_{\mathrm{total}} = 1$. The corresponding total wave functions are given by:

$$\psi_S = \frac{1}{\sqrt{2}}\left(\psi_a(r_1)\psi_b(r_2) + \psi_a(r_2)\psi_b(r_1)\right) \cdot \chi_S \tag{4.3}$$

$$\psi_T = \frac{1}{\sqrt{2}}\left(\psi_a(r_1)\psi_b(r_2) - \psi_a(r_2)\psi_b(r_1)\right) \cdot \chi_T \tag{4.4}$$

The energies of the singlet and triplet states amount to:

$$E_S = \int \psi_S^* \mathcal{H} \psi_S \mathrm{d}V_1 \mathrm{d}V_2 \tag{4.5}$$

$$E_T = \int \psi_T^* \mathcal{H} \psi_T \mathrm{d}V_1 \mathrm{d}V_2 \tag{4.6}$$

taking into account normalized spin parts of the singlet and triplet wave functions, i.e.

$$S^2 = (S_1 + S_2)^2 = S_1^2 + S_2^2 + 2S_1 \cdot S_2 \tag{4.7}$$

Thus, we obtain:

$$S_1 \cdot S_2 = \frac{1}{2}S_{\mathrm{total}}(S_{\mathrm{total}} + 1) - \frac{1}{2}S_1(S_1 + 1) - \frac{1}{2}S_2(S_2 + 1) \tag{4.8}$$

$$= \frac{1}{2}S_{\mathrm{total}}(S_{\mathrm{total}} + 1) - \frac{3}{4} \quad \text{due to} \quad S_1 = S_2 = \frac{1}{2} \tag{4.9}$$

$$= \begin{cases} -\frac{3}{4} & \text{for} \quad S_{\mathrm{total}} = 0 \quad \text{(singlet)} \\[2mm] +\frac{1}{4} & \text{for} \quad S_{\mathrm{total}} = 1 \quad \text{(triplet)} \end{cases} \tag{4.10}$$

The effective Hamiltonian can be expressed as:

$$\mathcal{H} = \frac{1}{4}(E_S + 3E_T) - (E_S - E_T)\boldsymbol{S}_1 \cdot \boldsymbol{S}_2 \tag{4.11}$$

The first term is constant and often included in other energy contributions. The second term is spin dependent and the important one concerning ferromagnetic properties.

Let us define the exchange constant or exchange integral J by:

$$J = \frac{E_S - E_T}{2} = \int \psi_a^*(\boldsymbol{r}_1)\psi_b^*(\boldsymbol{r}_2)\mathcal{H}\psi_a(\boldsymbol{r}_2)\psi_b(\boldsymbol{r}_1)\mathrm{d}V_1\mathrm{d}V_2 \tag{4.12}$$

Then, the spin dependent term in the effective Hamiltonian can be written as:

$$\mathcal{H}_{\mathrm{spin}} = -2J\boldsymbol{S}_1 \cdot \boldsymbol{S}_2 \tag{4.13}$$

If the exchange integral J is positive then $E_S > E_T$, i.e. the triplet state with $S_{\mathrm{total}} = 1$ is energetically favored. If the exchange integral J is negative then $E_S < E_T$, i.e. the singlet state with $S_{\mathrm{total}} = 0$ is energetically favored.

We see that this situation considering only two electrons is relatively simple. But, atoms in magnetic systems exhibit a lot of electrons. The Schrödinger equation of these many-body systems cannot be solved without assumptions. The most important part of such an interaction like the exchange interaction mostly apply between neighboring atoms. This consideration leads within the Heisenberg model to a term in the Hamiltonian of:

$$\mathcal{H} = -\sum_{ij} J_{ij}\boldsymbol{S}_i \cdot \boldsymbol{S}_j \tag{4.14}$$

with J_{ij} being the exchange constant between spin i and spin j. The factor 2 is included in the double counting within the sum. Often a good approximation is given by:

$$J_{ij} = \begin{cases} J & \text{for nearest neighbor spins} \\ 0 & \text{otherwise} \end{cases} \tag{4.15}$$

Generally, J is positive for electrons at the same atom whereas it is often negative if both electrons belong to different atoms.

4.3 Indirect Exchange

The different classes of indirect exchange significantly depend on the kind of magnetic material.

- Superexchange interaction
  This type of indirect exchange interaction occurs in ionic solids. The exchange interaction between non-neighboring magnetic ions is mediated by

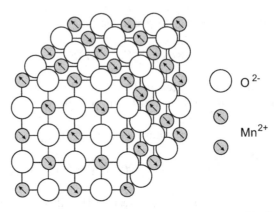

Fig. 4.1. Crystal and magnetic structures of MnO

means of a non-magnetic ion which is located in-between. The distance between the magnetic ions is too large that a direct exchange can take place.

An example of an antiferromagnetic ionic solid is MnO (see Fig. 4.1). Each Mn^{2+} ion exhibits 5 electrons in its d shell with all spins being parallel due to Hund's rule. The O^{2-} ions possess electrons in p orbitals which are fully occupied with their spins aligned antiparallel. There are two possibilities for the relative alignment of the spins in neighboring Mn atoms. A parallel alignment leads to a ferromagnetic arrangement whereas an antiparallel alignment causes an antiferromagnetic arrangement. That configuration is energetically favored which allows a delocalization of the involved electrons due to a lowering of the kinetic energy (see Fig. 4.2). In the antiferromagnetic case the electrons with their ground state given in (a) can be exchanged via excited states shown in (b) and (c) leading to a delocalization. For ferromagnetic alignment with the corresponding ground state presented in (d) the Pauli exclusion principle forbids the arrangements shown in (e) and (f). Thus, no delocalization occurs. Therefore, the antiferromagnetic coupling between two Mn atoms is energetically favored as depicted in Fig. 4.1. It is important that the electrons of the oxygen atom are located within the same orbital, i.e. the atom must connect the two Mn atoms.

- RKKY exchange interaction

The RKKY exchange interaction (RKKY: Ruderman, Kittel, Kasuya, Yosida) occurs in metals with localized magnetic moments. The exchange is mediated via the valence electrons; it is therefore not a direct interaction. The coupling is characterized by the distance dependent exchange integral $J_{\mathrm{RKKY}}(r)$:

$$J_{\mathrm{RKKY}}(r) \propto F(2k_F r) \tag{4.16}$$

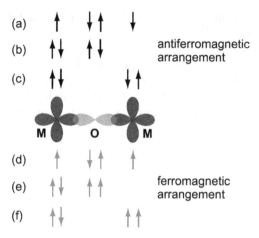

Fig. 4.2. Occurrence of a super exchange interaction in a magnetic oxide. The arrows represent the spins of the electrons being involved into the interaction between the metal (M) and oxygen (O) atom

with

$$F(x) = \frac{\sin x - x \cos x}{x^4} \qquad (4.17)$$

(see Fig. 4.3). This type of exchange coupling is long range and anisotropic which often results in complicated spin arrangements. Additionally, it possesses an oscillating behavior. Thus, the type of coupling, ferro- or antiferromagnetic nature, is a function of the distance between the magnetic

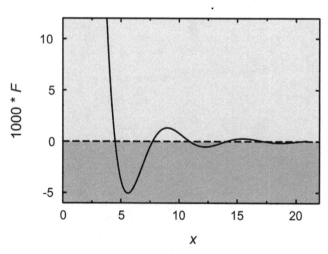

Fig. 4.3. Graphical representation of the function $F(x)$ defined in (4.17) using $x = 2k_F r$. Positive values (*light gray area*) lead to a ferromagnetic coupling whereas negative ones (*dark gray area*) result in an antiferromagnetic arrangement

moments. One example is represented by rare earth metals with their localized $4f$ electrons.

- Double exchange

  In some oxides the magnetic ions exhibit mixed valencies, i.e. different oxidation states occur which results in a ferromagnetic arrangement. One example is given by magnetite (Fe_3O_4) which includes Fe^{2+} as well as Fe^{3+} ions. A detailed discussion is given in Chap. 16.4 concerning the colossal magnetoresistance.

5

Collective Magnetism

In the chapter before we learned about the various magnetic interactions between magnetic moments in the solid state. In the following different magnetic ground states will be discussed which are caused by these interactions and which result in collective magnetism. A schematic overview is given in Fig. 5.1. Ferromagnets exhibit magnetic moments which are aligned parallel to each other. In antiferromagnets adjacent magnetic moments are oriented in

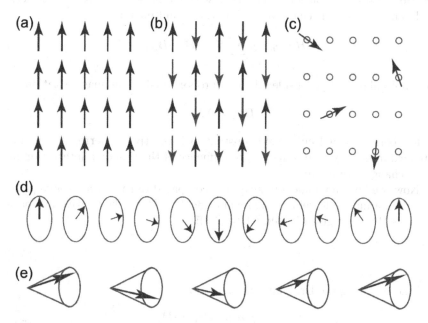

Fig. 5.1. Different arrangements of magnetic moments for ordered magnetic systems: (**a**) ferromagnets, (**b**) antiferromagnets, (**c**) spin glasses, (**d**) helical arrangement, (**e**) spiral arrangement

the opposite direction. Magnetic moments in a spin glass are frozen out with a random orientation. A helical or spiral arrangement is given if the magnetic moments are aligned parallel in a plane but the direction varies from plane to plane in such a way that the vector of the magnetic moment moves on a circle or a cone, respectively.

5.1 Ferromagnetism

Ferromagnetism is characterized by a spontaneous magnetization even without applying an external magnetic field. At $T = 0\,\mathrm{K}$ all magnetic moments are aligned parallel which is caused by the exchange interaction. The magnetically relevant part of the Hamiltonian for a ferromagnetic system with an additional external magnetic field $\boldsymbol{B}$ is given by:

$$\mathcal{H} = -\sum_{ij} J_{ij} \boldsymbol{S}_i \cdot \boldsymbol{S}_j + g_J \mu_B \sum_i \boldsymbol{S}_i \cdot \boldsymbol{B} \tag{5.1}$$

with $J_{ij} > 0$ for nearest neighbors.

We start our discussion with the approximation, which is known as the model of Weiss, that the interaction of a magnetic ion with its neighbors is described using a molecular field $\boldsymbol{B}_{\mathrm{mf}}$ which represents an inner magnetic field. This situation is characterized by the Hamiltonian:

$$\mathcal{H} = g_J \mu_B \sum_i \boldsymbol{S}_i \cdot (\boldsymbol{B} + \boldsymbol{B}_{\mathrm{mf}}) \tag{5.2}$$

The molecular field is considered to be proportional to the magnetization:

$$\boldsymbol{B}_{\mathrm{mf}} = \lambda \boldsymbol{M} \tag{5.3}$$

with λ being the molecular field constant. λ is positive for ferromagnets and often exhibits large values due to the influence of the Coulomb interaction on the exchange interaction.

Now, we have an analogous situation compared to that of a paramagnetic system in a magnetic field of the magnitude $\boldsymbol{B} + \boldsymbol{B}_{\mathrm{mf}}$. The relative magnetization is therefore given by (cf. Chap. 2.3):

$$\frac{M}{M_S} = B_J(y) \tag{5.4}$$

with $B_J(y)$ being the Brillouin function and

$$y = \frac{g_J \mu_B J (B + \lambda M)}{kT} \tag{5.5}$$

Without an external magnetic field (i.e. $B = 0$) the temperature dependence of the magnetization is given by:

$$M(T) = M_S \cdot B_J \left(\frac{g_J \mu_B J \lambda M(T)}{kT} \right) \tag{5.6}$$

The trivial solution is $M(T) = 0$ due to $B_J(0) = 0$. The general solution can be obtained using the expression:

$$\frac{M}{M_S} = \frac{M}{M_S} \cdot \frac{g_J \mu_B J \lambda M}{kT} \cdot \frac{kT}{g_J \mu_B J \lambda M} \tag{5.7}$$

$$= y \cdot \frac{kT}{g_J \mu_B J \lambda M_S} \tag{5.8}$$

and corresponds to the intersection points of a line through origin with $B_J(y)$ in a diagram representing M/M_S as a function of y due to $M/M_S = B_J(y)$ (see Fig. 5.2). Only one solution occurs at high temperatures which is given at $y = 0$ with $M(T) = 0$. If the temperature is low enough three solutions exist with $M = 0$ and $M = \pm a$ for a specific $a > 0$.

The magnetization $M \neq 0$ vanishes at a temperature $T = T_C$ which is characterized by a slope of the straight line being equal to the slope of B_J in the origin. This procedure can be used to obtain the critical temperature T_C graphically.

The "mathematical" determination of the Curie temperature T_C is carried out making use of the condition that the derivations with respect to y must be equal for

$$\frac{M}{M_S} = y \cdot \frac{kT}{g_J \mu_B J \lambda M_S} \tag{5.9}$$

and

$$\frac{M}{M_S} = B_J(y) \tag{5.10}$$

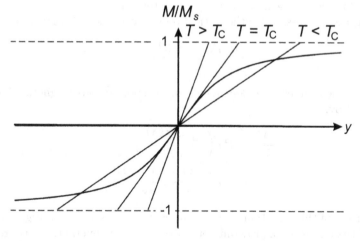

Fig. 5.2. Graphical method to determine the relative magnetization without external magnetic field

for small values of y. The derivation of (5.9) is given by:

$$\frac{dM/M_S}{dy} = \frac{kT}{g_J\mu_B J\lambda M_S} \qquad (5.11)$$

Due to:

$$B_J(y) = y \cdot \frac{J+1}{3J} + \mathcal{O}(y^3) \qquad (5.12)$$

the derivation of (5.10) can be approximated by:

$$\frac{dM/M_S}{dy} = \frac{dB_J}{dy} = \frac{J+1}{3J} \qquad (5.13)$$

for small values of y. Thus, the critical temperature amounts to:

$$T_C = \frac{(J+1)g_J\mu_B\lambda M_S}{3k} \qquad (5.14)$$

Using $M_S = ng_J\mu_B J$ (see (2.96)) and $\mu_{\text{eff}} = g_J\mu_B\sqrt{J(J+1)}$ (see (2.134)) we obtain:

$$T_C = \frac{n\lambda\mu_{\text{eff}}^2}{3k} \qquad (5.15)$$

as the Curie temperature in molecular field approximation. The molecular field can be estimated by:

$$B_{\text{mf}} = \lambda M_S = \frac{3kT_C}{g_J\mu_B(J+1)} \qquad (5.16)$$

Using typical values $J = 1/2$ and $T_C = 1000\,\text{K}$ the magnitude of the molecular field amounts to $B_{\text{mf}} \approx 1500\,\text{T}$. This extremely high magnetic field gives evidence for the strength of the magnetic interaction.

Using the expression for the critical temperature (see (5.14)) results in

$$\frac{g_J\mu_B\lambda}{k} = \frac{3T_C}{(J+1)M_S} \qquad (5.17)$$

Therefore, the temperature dependence of the spontaneous magnetization can be derived by:

$$\frac{M}{M_S} = B_J\left(\frac{g_J\mu_B J\lambda M}{kT}\right) \qquad (5.18)$$

$$= B_J\left(\frac{3J}{J+1} \cdot \frac{M}{M_S} \cdot \frac{T_C}{T}\right) \qquad (5.19)$$

The solution of this equation for different values of J is shown in Fig. 5.3. The shape of the curves is slightly different nevertheless a general trend is present:

- $T > T_C$: The magnetization vanishes $M = 0$
- $T < T_C$: $M \neq 0$

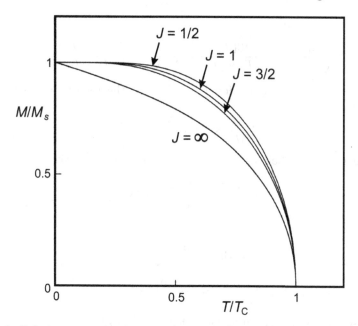

Fig. 5.3. Relative magnetization as a function of the reduced temperature T/T_C for different values of J

- $T = T_C$: The magnetization exhibits a continuous behavior but it is not continuously differentiable. Thus, we find a phase transition of second order.

The value $t = T/T_C$ is called reduced temperature.

Now, we want to look for the behavior of the spontaneous magnetization near the critical temperature T_C which was found to be (see (5.19)):

$$\frac{M}{M_S} = B_J\left(\frac{3J}{J+1} \cdot \frac{M}{M_S} \cdot \frac{T_C}{T}\right) \tag{5.20}$$

The argument of the Brillouin function is small for $T \to T_C$. Thus, $B_J(y)$ can be developed into a Taylor series:

$$B_J(y) = \frac{J+1}{3J}y - \frac{(J+1)(2J^2+2J+1)}{90J^3}y^3 + \mathcal{O}(y^5) \tag{5.21}$$

Neglecting higher orders we obtain:

$$\frac{M}{M_S} = \frac{J+1}{3J} \cdot \frac{3J}{J+1} \cdot \frac{M}{M_S} \cdot \frac{T_C}{T}$$

$$- \frac{(J+1)(2J^2+2J+1)}{90J^3} \cdot \frac{27J^3}{(J+1)^3} \cdot \left(\frac{M}{M_S}\right)^3 \cdot \left(\frac{T_C}{T}\right)^3 \tag{5.22}$$

$$= \frac{M}{M_S} \cdot \frac{T_C}{T}\left(1 - \left(\frac{M}{M_S}\right)^2 \cdot \left(\frac{T_C}{T}\right)^2 \cdot \frac{3(2J^2+2J+1)}{10(J+1)^2}\right) \tag{5.23}$$

which results in:

$$\left(\frac{M}{M_S}\right)^2 = \frac{10(J+1)^2}{3(J^2+(J+1)^2)} \cdot \left(1 - \frac{T}{T_C}\right) \cdot \left(\frac{T}{T_C}\right)^2 \quad (5.24)$$

The last term becomes unity for $T \to T_C$ and the relative magnetization amounts to:

$$\frac{M}{M_S} \propto \left(1 - \frac{T}{T_C}\right)^{1/2} \quad (5.25)$$

In the next step the behavior near $T = 0\,\mathrm{K}$ is examined. For $T \to 0\,\mathrm{K}$ we obtain:

$$y = \frac{3J}{J+1}\frac{M}{M_S}\frac{T_C}{T} \to \infty \quad (5.26)$$

With:

$$\coth x = \frac{e^x + e^{-x}}{e^x - e^{-x}} \quad (5.27)$$

$$= \frac{1 + e^{-2x}}{1 - e^{-2x}} \quad (5.28)$$

we can use an approximation of $\coth x$ for $x \to \infty$ to be:

$$\coth x = (1 + e^{-2x})(1 + e^{-2x}) \quad (5.29)$$

$$= 1 + 2e^{-2x} + e^{-4x} \quad (5.30)$$

Thus, we obtain for $x \to \infty$:

$$\coth x = 1 + 2e^{-2x} \quad (5.31)$$

Both results can be used to estimate $B_J(y)$ for $T \to 0$ K:

$$2JB_J(y) = (2J+1)\coth\left(\frac{2J+1}{2J}y\right) - \coth\frac{y}{2J} \quad (5.32)$$

which simplifies for $y \to \infty$ to:

$$2JB_J(y) = (2J+1)(1 + 2e^{-\frac{2J+1}{J}y}) - (1 + 2e^{-\frac{y}{J}}) \quad (5.33)$$

$$= 2J + 1 + 2(2J+1)e^{-\frac{y}{J}}e^{-2y} - 2e^{-\frac{y}{J}} - 1 \quad (5.34)$$

and can be approximated by:

$$2JB_J(y) = 2J - 2e^{-\frac{y}{J}} \quad (5.35)$$

Thus, we obtain:

$$\frac{M}{M_S} = 1 - (1/J)e^{-\frac{3}{J+1}\frac{T_C}{T}\frac{M}{M_S}} \quad (5.36)$$

$$\approx 1 - (1/J)e^{-c/T} \quad (5.37)$$

with c being a constant. Thus, the magnetization M approaches M_S exponentially for $T \to 0\,\mathrm{K}$. This result does not correctly describe the situation for all ferromagnets. For low temperatures the physics behind is more complicated compared to this simple picture and will be discussed in Chap. 6.5 in more detail.

For the discussion of the magnetic susceptibility we consider a small external magnetic field B applied at $T > T_C$ which results in a small magnetization. Thus, the argument of $B_J(y)$ can be approximated by:

$$B_J(y) = \frac{J+1}{3J}y \tag{5.38}$$

The relative magnetization approximately amounts to:

$$\frac{M}{M_S} = \frac{J+1}{3J} \cdot \frac{g_J\mu_B J(B + \lambda M)}{kT} \tag{5.39}$$

$$= \frac{g_J\mu_B(J+1)}{3k} \cdot \frac{B + \lambda M}{T} \tag{5.40}$$

Using (5.14) we obtain:

$$\frac{M}{M_S} = \frac{T_C}{\lambda M_S} \cdot \frac{B + \lambda M}{T} \tag{5.41}$$

Therefore, we can write:

$$\frac{M}{M_S}\left(1 - \frac{T_C}{T}\right) = \frac{T_C}{T} \cdot \frac{B}{\lambda M_S} \tag{5.42}$$

Using this equation the magnetic susceptibility can be determined to be:

$$\chi = \lim_{B \to 0} \frac{\mu_0 M}{B} = \frac{\mu_0 T_C}{\lambda} \cdot \frac{1}{T - T_C} = \frac{c}{T - T_C} \tag{5.43}$$

with c being a constant which represents the Curie–Weiss law .

Now, we drop the restriction of an only small magnitude of the external magnetic field and allow any values. Due to (see (5.5)):

$$y = \frac{g_J\mu_B J(B + \lambda M)}{kT} \tag{5.44}$$

a shift of the linear functions (cf. Fig. 5.2) to larger y-values occurs in the M/M_S-y-diagram (see Fig. 5.4). It is obvious that a solution with $M \neq 0$ exists for all temperatures. The phase transition has been vanished. Therefore, ferromagnetic materials exhibit a non-vanishing magnetization even above the critical temperature if an external magnetic field is applied (see Fig. 5.5).

The molecular field being characterized by λ with $B_{\mathrm{mf}} = \lambda M$ is related to the exchange interaction characterized by J_{ij}. Now, let us assume that the

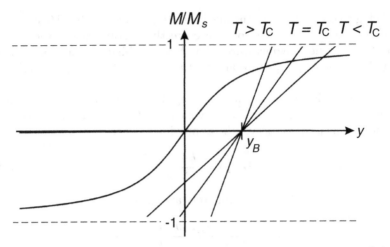

Fig. 5.4. Graphical method to determine the relative magnetization applying an external magnetic field with $y_B = g_J \mu_B J B / kT$

exchange interaction only occurs between the z nearest neighbors with the constant strength $\mathcal{J}$. The molecular field can then be expressed as:

$$B_{\mathrm{mf}} = \frac{2}{g_J \mu_B} \sum_j J_{ij} S_j \qquad (5.45)$$

$$= \frac{2}{g_J \mu_B} \cdot z \mathcal{J} \cdot S \qquad (5.46)$$

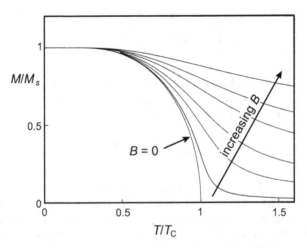

Fig. 5.5. Relative magnetization as a function of the reduced temperature for $J = 1/2$ being calculated for different magnitudes of an external magnetic field B. A vanishing magnetization and thus a phase transition only occurs for $B = 0$

Due to $\lambda M = \lambda n g_J \mu_B J$ for $M = M_S$ and $J = S$ because of the assumption $L = 0$ we obtain:

$$\frac{2}{g_J \mu_B} z \mathcal{J} S = \lambda n g_J \mu_B S \tag{5.47}$$

and thus:

$$\lambda = \frac{2 z \mathcal{J}}{n g_J^2 \mu_B^2} \tag{5.48}$$

The critical temperature was expressed as (see (5.15)):

$$T_C = \frac{n \lambda \mu_{\text{eff}}^2}{3k} \tag{5.49}$$

with $\mu_{\text{eff}} = g_J \mu_B \sqrt{J(J+1)}$. Thus, we get:

$$T_C = \frac{n \lambda g_J^2 \mu_B^2 J(J+1)}{3k} \tag{5.50}$$

Using (5.48) results in:

$$T_C = \frac{2 n g_J^2 \mu_B^2 J(J+1) z \mathcal{J}}{3 k n g_J^2 \mu_B^2} \tag{5.51}$$

$$= \mathcal{J} \cdot \frac{2 z J(J+1)}{3k} \tag{5.52}$$

i.e. T_C scales with the strength of the exchange interaction.

5.2 Antiferromagnetism

The simplest situation concerning antiferromagnetic behavior is given for magnetic moments of nearest neighbors which are aligned antiparallel. We have two different possibilities to describe this situation within the model of Weiss:

- A negative exchange interaction is considered between nearest neighbors.
- The lattice is divided into two sublattices:
  - Each exhibits a ferromagnetic arrangement and
  - an antiparallel orientation of the magnetization between both sublattices is present.

The latter situation is schematically shown in Fig. 5.6.

We start our discussion of antiferromagnetism using the model of Weiss with the assumption that no external magnetic field is applied and that the molecular field of one sublattice (labelled with "1") is proportional to the magnetization of the other one (labelled with "2") and vice versa. Thus, both molecular fields can be expressed as:

$$\boldsymbol{B}_{\text{mf}}^{(1)} = -|\lambda| \boldsymbol{M}_2 \tag{5.53}$$

$$\boldsymbol{B}_{\text{mf}}^{(2)} = -|\lambda| \boldsymbol{M}_1 \tag{5.54}$$

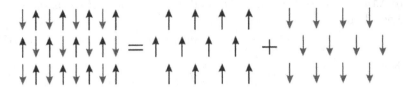

Fig. 5.6. The lattice of an antiferromagnet can be composed of two sublattices each being ferromagnetically ordered

with a negative molecular field constant λ. The magnetization of the sublattices can then be written as:

$$M_1 = M_S B_J \left(-\frac{g_J \mu_B J |\lambda| M_2}{kT} \right) \tag{5.55}$$

$$M_2 = M_S B_J \left(-\frac{g_J \mu_B J |\lambda| M_1}{kT} \right) \tag{5.56}$$

Both sublattices exhibit an antiparallel orientation but the same magnitude of their magnetization:

$$|\boldsymbol{M}_1| = |\boldsymbol{M}_2| \equiv M \tag{5.57}$$

Each relative magnetization amounts to:

$$\frac{M}{M_S} = B_J \left(\frac{g_J \mu_B J |\lambda| M}{kT} \right) \tag{5.58}$$

with the properties discussed in Chap. 5.1 and the temperature dependence shown in Fig. 5.3. The magnetization of each sublattice consequently vanishes at a transition temperature T_N, the Néel temperature, which is given by:

$$T_N = \frac{(J+1) g_J \mu_B |\lambda| M_S}{3k} \tag{5.59}$$

$$= \frac{n |\lambda| \mu_{\text{eff}}^2}{3k} \tag{5.60}$$

Analogously to Chap. 5.1 the magnetic susceptibility above T_N for small external fields can be calculated to be:

$$\chi = \lim_{B \to 0} \frac{\mu_0 M}{B} \propto \frac{1}{T + T_N} \tag{5.61}$$

The positive sign in the denominator is due to $B = -|\lambda| M$ instead of $B = \lambda M$ for the ferromagnetic situation. Equation (5.61) reflects the Curie–Weiss law with a substitution of $-T_C$ by $+T_N$.

Above the transition temperature the magnetic susceptibility can be expressed as:

$$\chi \propto \frac{1}{T - \theta} \tag{5.62}$$

with θ being the Weiss temperature or paramagnetic Curie temperature, respectively, and amounts to:

- $\theta = 0$ for a paramagnet (see (2.135))
- $\theta > 0$ for a ferromagnet with $\theta = T_C$ (see (5.43))
- $\theta < 0$ for an antiferromagnet with $\theta = -T_N$ (see (5.61))

In the next step, the magnetization of the own lattice is additionally considered for the magnitude of the molecular field in each sublattice:

$$B_{\mathrm{mf}}^{(1)} = v_1 \boldsymbol{M}_1 + w_1 \boldsymbol{M}_2 \tag{5.63}$$

$$B_{\mathrm{mf}}^{(2)} = v_2 \boldsymbol{M}_1 + w_2 \boldsymbol{M}_2 \tag{5.64}$$

The magnitude of the magnetization is equal but the magnetization exhibits the opposite direction within each sublattice. This allows the simplification:

$$v_1 = w_2 \quad \text{and} \quad v_2 = w_1 \tag{5.65}$$

and we obtain:

$$B_{\mathrm{mf}}^{(1)} = v_1 \boldsymbol{M}_1 + v_2 \boldsymbol{M}_2 \tag{5.66}$$

$$B_{\mathrm{mf}}^{(2)} = v_2 \boldsymbol{M}_1 + v_1 \boldsymbol{M}_2 \tag{5.67}$$

Without an external magnetic field we have $\boldsymbol{M}_1 = -\boldsymbol{M}_2$ which allows us to express the relative magnetization of, e.g., sublattice 1 as:

$$\frac{M_1}{M_S} = B_J \left(\frac{g_J \mu_B J (v_1 - v_2) M_1}{kT} \right) \tag{5.68}$$

Thus, the Néel temperature is given by (cf. (5.59)):

$$T_N = \frac{g_J \mu_B (J + 1)(v_1 - v_2) M_S}{3k} \tag{5.69}$$

$$= \frac{n(v_1 - v_2)\mu_{\mathrm{eff}}^2}{3k} \tag{5.70}$$

$$= c(v_1 - v_2) \tag{5.71}$$

Now, we assume temperatures *above* T_N and a small external magnetic field $\boldsymbol{B}$. Thus, the argument of the Brillouin function can be approximated as expressed in (5.38). In this situation the magnetization of the sublattices amounts to:

$$\boldsymbol{M}_1 = \frac{c}{T}(\boldsymbol{B} + v_1 \boldsymbol{M}_1 + v_2 \boldsymbol{M}_2) \tag{5.72}$$

$$\boldsymbol{M}_2 = \frac{c}{T}(\boldsymbol{B} + v_2 \boldsymbol{M}_1 + v_1 \boldsymbol{M}_2) \tag{5.73}$$

Due to the external magnetic field the total magnetization $\boldsymbol{M} = \boldsymbol{M}_1 + \boldsymbol{M}_2$ does not vanish and is given by:

$$M = \frac{c}{T}(2B + (v_1 + v_2)M) \tag{5.74}$$

which results in:

$$M = \frac{2Bc}{T - c(v_1 + v_2)} \tag{5.75}$$

For the magnetic susceptibility we therefore obtain:

$$\chi \propto \frac{2c}{T - c(v_1 + v_2)} \propto \frac{1}{T - \theta} \tag{5.76}$$

with $\theta = c(v_1 + v_2)$. This means that two characteristic temperatures T_N and θ exist which are correlated by (see (5.71)):

$$\frac{T_N}{\theta} = \frac{v_1 - v_2}{v_1 + v_2} \tag{5.77}$$

An identical behavior of the susceptibilities

$$\chi \propto \frac{1}{T + T_N} \quad \text{and} \quad \chi \propto \frac{1}{T - \theta} \tag{5.78}$$

is only given if $v_1 = 0$.

Now, we consider the behavior of an antiferromagnet *below* the Néel temperature in an external magnetic field B and start our discussion with a vanishing temperature $T = 0$ K, i.e. we neglect thermal fluctuations.

If a *weak* external magnetic field B is aligned parallel to the magnetization of one sublattice and hence antiparallel to that of the other one the inner field of one sublattice is enhanced whereas the inner field of the other sublattice is reduced by the same magnitude. Thus, the total magnetization remains constant. The properties of an antiferromagnet in a *strong* external magnetic field being aligned parallel are discussed on p. 100.

The Hamiltonian in the approximation of the Model of Weiss can be expressed as:

$$\mathcal{H} = \sum_{ij} J_{ij} \mathbf{S}_i \cdot \mathbf{S}_j - g_J \mu_B \sum_i \mathbf{S}_i \cdot (\mathbf{B} + \mathbf{B}_{\mathrm{mf}}^{(i)}) \tag{5.79}$$

and the energy by:

$$E = \sum_{ij} J_{ij} \langle S_i \rangle \langle S_j \rangle - g_J \mu_B \sum_i \langle S_i \rangle (\mathbf{B} + \mathbf{B}_{\mathrm{mf}}^{(i)}) \tag{5.80}$$

with $i, j = 1, 2$ being correlated to both sublattices. But, this is only valid at $T = 0$ K with a perfect alignment within each sublattice. In this situation we have:

$$(J_{11} + J_{22})/g_J \mu_B = v_1 \text{ and } J_{11} = J_{22} \tag{5.81}$$

$$(J_{12} + J_{21})/g_J \mu_B = v_2 \text{ and } J_{12} = J_{21} \tag{5.82}$$

This implies:

$$J_{11} = J_{22} = \frac{1}{2} v_1 g_J \mu_B \tag{5.83}$$

$$J_{12} = J_{21} = \frac{1}{2} v_2 g_J \mu_B \tag{5.84}$$

Due to:

$$\langle S_1 \rangle = M_1 \tag{5.85}$$

$$\langle S_2 \rangle = M_2 \tag{5.86}$$

the first term in (5.80) results in:

$$\frac{1}{g_J \mu_B} \sum_{ij} J_{ij} \langle S_i \rangle \langle S_j \rangle =$$

$$\frac{1}{2} v_1 M_1^2 + \frac{1}{2} v_1 M_2^2 + \frac{1}{2} v_2 M_1 M_2 + \frac{1}{2} v_2 M_2 M_1 \tag{5.87}$$

$$= \frac{1}{2} v_1 (M_1^2 + M_2^2) + v_2 M_1 M_2 \tag{5.88}$$

Using (5.63) and (5.64) the second term in (5.80) is given by:

$$\sum_i \langle S_i \rangle (B + B_{\text{mf}}^{(i)}) = (M_1 + M_2) B + v_1 (M_1^2 + M_2^2) + 2 v_2 M_1 M_2 \tag{5.89}$$

Thus, the total energy per $g \mu_B$ amounts to:

$$E_0 = \frac{1}{2} v_1 (M_1^2 + M_2^2) + v_2 M_1 M_2$$

$$- (M_1 + M_2) B + v_1 (M_1^2 + M_2^2) + 2 v_2 M_1 M_2 \tag{5.90}$$

$$= -(M_1 + M_2) B - \frac{1}{2} (v_1 M_1^2 + v_1 M_2^2 + 2 v_2 M_1 M_2) \tag{5.91}$$

Using this expression we are able to discuss the more interesting situation that the external magnetic field B is directed perpendicular to the magnetization of both sublattices M_1 and M_2. Two forces are acting against each other. Whereas the external field rotates the magnetization of the sublattices into the field direction the molecular field tries to stabilize the antiparallel alignment of both sublattices with respect to each other. This leads to a rotation of M_1 and M_2 by an angle α which is determined by the energy minimum (see Fig. 5.7).

Considering the same magnitude of magnetization in both sublattices $M_1 = |M_1| = |M_2| = M_2$ the energy is given by:

$$E_\perp = -2 M_1 B \sin \alpha - v_1 M_1^2 + v_2 M_1^2 \cos 2\alpha \tag{5.92}$$

The energy minimum is reached at an angle α which can be calculated by:

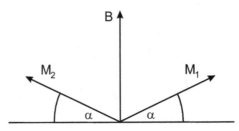

Fig. 5.7. An external magnetic field B induces a rotation of each sublattice magnetization M_1 and M_2 towards the field direction by the angle α whereas the molecular field tries to stabilize the antiferromagnetic alignment, i.e. $\alpha = 0$

$$0 = \frac{dE_\perp}{d\alpha} = -2M_1 B \cos \alpha - 2v_2 M_1^2 \sin 2\alpha \qquad (5.93)$$

Thus:

$$0 = 2M_1 \cos \alpha (B + 2v_2 M_1 \sin \alpha) \qquad (5.94)$$

using $\sin 2\alpha = 2 \sin \alpha \cos \alpha$. The solutions of this equation are given by:

$$-\frac{B}{2v_2 M_1} = \sin \alpha \quad \text{for } 0 \leq B \leq -2v_2 M_1 \qquad (5.95)$$

$$0 = \cos \alpha \quad \text{for } B > -2v_2 M_1 \qquad (5.96)$$

The projection of the total magnetization on the direction of the external magnetic field amounts to:

$$M_\perp = M_1 \sin \alpha + M_2 \sin \alpha = 2M_1 \sin \alpha \qquad (5.97)$$

For $0 \leq B \leq -2v_2 M_1$ we obtain $\sin \alpha = -B/(2v_2 M_1)$ and thus:

$$M_\perp = -\frac{2M_1 B}{2v_2 M_1} = -\frac{B}{v_2} \qquad (5.98)$$

For $B > -2v_2 M_1$ we get $\sin \alpha = 1$ because of $\cos \alpha = 0$ which results in:

$$M_\perp = 2M_1 \qquad (5.99)$$

This means that $M_\perp$ linearly increases with the external magnetic field until the magnetization of both sublattices points into the same direction. Now, we have a ferromagnetic system which is magnetically saturated with $M_S = 2M_1$.

Analogously, the magnetic susceptibility χ is divided into the component $\chi_\parallel$ being parallel to $\boldsymbol{M}$ and $\chi_\perp$ being perpendicular to $\boldsymbol{M}$ in order to determine the temperature dependence. The latter one is given by:

$$\chi_\perp \propto \frac{\partial M_\perp}{\partial B} = \frac{\partial}{\partial B}\left(-\frac{B}{v_2}\right) = \text{const.} \qquad (5.100)$$

Thus, $\chi_\perp$ is independent on temperature and amounts to:

$$\chi_\perp(T) = \chi_\perp(T_N) \tag{5.101}$$

As already mentioned above $\boldsymbol{M}$ remains constant when applying an external magnetic field parallel to $\boldsymbol{M}$. Due to this reason we obtain:

$$\chi_\|(T = 0) \propto \frac{\partial M_\|}{\partial B}(T = 0) = 0 \tag{5.102}$$

For higher temperatures this vanishing value changes to:

$$\chi_\|(T > 0) > 0 \tag{5.103}$$

because the external magnetic field works against the thermal fluctuations. Above T_N all spins are free to rotate thus loosing a preferred orientation. A distinction between $\chi_\|$ and $\chi_\perp$ becomes meaningless and we get paramagnetic behavior. For polycrystalline material the magnetic susceptibility is obtained by averaging:

$$\chi_{\text{poly}} = \frac{1}{3}\chi_\| + \frac{2}{3}\chi_\perp \tag{5.104}$$

The temperature dependence of the magnetic susceptibility and of the reciprocal magnetic susceptibility are shown in Fig. 5.8.

The arrangement of the sublattices can occur in many different ways because there are a lot of possibilities to place regularly the same number of antiparallel aligned magnetic moments, i.e. spin up and spin down electrons. The type of crystal lattice additionally influences this behavior. Possible types of antiferromagnets in the simple cubic form are shown in Fig. 5.9. Type A results in a layered structure which each layer being ferromagnetically ordered. It is called topological antiferromagnet. Type B leads to a chain-like arrangement of the spins. An antiferromagnetic arrangement of nearest neighbors often occurs in materials which couple via the superexchange interaction, e.g. MnO (details are discussed in Chap. 4.3).

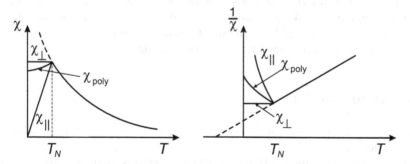

Fig. 5.8. Temperature dependence of the magnetic (**left**) and reciprocal magnetic susceptibility (**right**) for antiferromagnetic materials. $\chi_\|$ and $\chi_\perp$ are the magnetic susceptibilities measured by applying an external magnetic field parallel and perpendicular to the spin axis, respectively χ_{poly} is the susceptibility for polycrystalline material calculated by averaging (see (5.104))

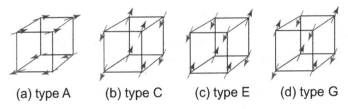

(a) type A   (b) type C   (c) type E   (d) type G

Fig. 5.9. Different types of antiferromagnetic order for simple cubic lattices

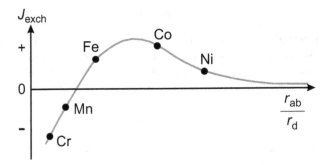

Fig. 5.10. The Bethe–Slater curve describes the relation of the exchange constant with the ratio of the interatomic distance r_{ab} to the radius of the d shell r_d

The exchange interaction between neighboring magnetic moments being described in the Heisenberg Hamiltonian can lead to a parallel or antiparallel alignment, i.e. to a ferromagnetic or antiferromagnetic arrangement. Ferromagnetism occurs for the exchange constant $J > 0$ whereas $J < 0$ results in antiferromagnetism. A correlation between the sign of the exchange constant and the ratio r_{ab}/r_d exists with r_{ab} being the interatomic distance and r_d the radius of the d shell. This behavior is graphically known as the Bethe–Slater curve (see Fig. 5.10). As indicated in the figure this curve allows to distinguish between ferromagnetic $3d$ elements like Fe, Co, and Ni exhibiting a parallel alignment and thus a positive exchange constant and antiferromagnetic elements like Mn and Cr with an antiparallel orientation of the magnetic moments and therefore a negative exchange constant.

5.3 Ferrimagnetism

Ferrimagnetism represents an intermediate position between ferro- and antiferromagnetism.

A spontaneous magnetization occurs below a critical temperature. At high temperatures the magnetic susceptibility exhibits a Curie–Weiss behavior with a negative paramagnetic Curie temperature.

The simplest characterization, but already satisfactory for the fundamental understanding, is given by the assumption of two magnetic sublattices with antiparallel orientation but different magnitude of each magnetization. Thus, the total magnetization does not vanish as in the antiferromagnetic case.

The molecular fields are described analogously to the antiferromagnetic case (see (5.63) and (5.64)). Due to symmetry reasons we have:

$$v_2 = w_1 < 0 \tag{5.105}$$

because of the antiferromagnetic coupling. But now, v_1 and w_2 are not identical:

$$v_1 \neq w_2 \tag{5.106}$$

Let us set:

$$v_2 = w_1 = -v \quad \text{with } v > 0 \tag{5.107}$$

and

$$v_1 = \alpha v \tag{5.108}$$

$$w_2 = \beta v \tag{5.109}$$

This procedure allows to discuss the behavior of a ferrimagnetic system as a function of the ratio of different molecular field constants instead of their absolute values. Thus, we obtain:

$$\boldsymbol{B}_{\text{mf}}^{(1)} = \alpha v \boldsymbol{M}_1 - v \boldsymbol{M}_2 \tag{5.110}$$

$$\boldsymbol{B}_{\text{mf}}^{(2)} = -v \boldsymbol{M}_1 + \beta v \boldsymbol{M}_2 \tag{5.111}$$

The magnetization of both sublattices is given by:

$$M_1 = n g_J \mu_B J \cdot B_J \left(\frac{g_J \mu_B J \mu_0}{kT} \cdot (B + \alpha v M_1 - v M_2) \right) \tag{5.112}$$

$$M_2 = n g_J \mu_B J \cdot B_J \left(\frac{g_J \mu_B J \mu_0}{kT} \cdot (B - v M_1 + \beta v M_2) \right) \tag{5.113}$$

Above the transition temperature we can again approximate (cf. (5.38)):

$$B_J(y) = \frac{J+1}{3J} y \tag{5.114}$$

which results in:

$$\boldsymbol{M}_1 = \frac{n g_J^2 \mu_B^2 J(J+1)\mu_0}{3kT} \cdot (B + \alpha v M_1 - v M_2) \tag{5.115}$$

$$= \frac{c_1}{T}(B + \alpha v M_1 - v M_2) \tag{5.116}$$

Analogously, the magnetization of the other sublattice amounts to:

$$\boldsymbol{M}_2 = \frac{c_2}{T}(B - v M_1 + \beta v M_2) \tag{5.117}$$

The solution of this linear set of equations is given by:

$$M_1 = \frac{c_1 T - c_1 c_2 \beta v - c_1 c_2 v}{T^2 - v(\alpha c_1 + \beta c_2)T + c_1 c_2 v^2(\alpha\beta - 1)} \cdot B \tag{5.118}$$

$$M_2 = \frac{c_2 T - c_1 c_2 \alpha v - c_1 c_2 v}{T^2 - v(\alpha c_1 + \beta c_2)T + c_1 c_2 v^2(\alpha\beta - 1)} \cdot B \tag{5.119}$$

The total magnetization $M = M_1 + M_2$ results in:

$$M = \frac{(c_1 + c_2)T - c_1 c_2 v(2 + \alpha + \beta)}{T^2 - v(\alpha c_1 + \beta c_2)T + c_1 c_2 v^2(\alpha\beta - 1)} \cdot B \tag{5.120}$$

Now we can calculate the inverse magnetic susceptibility:

$$\mu_0 \frac{1}{\chi} = \frac{B}{M} = \frac{T^2 - v(\alpha c_1 + \beta c_2)T + c_1 c_2 v^2(\alpha\beta - 1)}{(c_1 + c_2)T - c_1 c_2 v(2 + \alpha + \beta)} \tag{5.121}$$

Introducing suitable parameters θ, χ_0, and σ we obtain:

$$\mu_0 \frac{1}{\chi} = \frac{T}{c_1 + c_2} + \frac{1}{\chi_0} - \frac{\sigma}{T - \theta} \tag{5.122}$$

The temperature dependence of the inverse susceptibility is shown in Fig. 5.11. Whereas in the ferromagnetic and antiferromagnetic case the inverse susceptibility behaves as a linear function of the temperature this situation changes to a hyperbolic behavior for a ferrimagnetic system. Using (5.122) we can calculate the asymptotic behavior for $T \to \infty$:

$$\frac{1}{\chi} \propto \frac{T}{c_1 + c_2} + \frac{1}{\chi_0} \tag{5.123}$$

as well as the intersection point with the T-axis which determines the critical temperature T_C:

$$\frac{1}{\chi}(T_C) = 0 \tag{5.124}$$

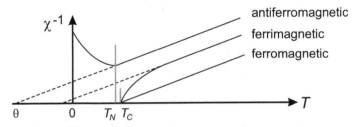

Fig. 5.11. Temperature dependence of the reciprocal magnetic susceptibility for ferromagnetic, antiferromagnetic, and ferrimagnetic material

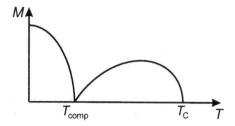

Fig. 5.12. In ferrimagnetic materials the situation can occur that the total magnetization M becomes zero at a temperature $T_\text{comp} < T_C$ with T_comp called compensation temperature

Thus:

$$T_C^2 - v(\alpha c_1 + \beta c_2)T_C + c_1 c_2 v^2(\alpha\beta - 1) = 0 \tag{5.125}$$

and we obtain:

$$T_C = \frac{v}{2}\left(\alpha c_1 + \beta c_2 \pm \sqrt{(\alpha c_1 - \beta c_2)^2 + 4c_1 c_2}\right) \tag{5.126}$$

The solution $T_C < 0$ results in a paramagnetic state down to $0\,\text{K}$. For $T_C > 0$ ferrimagnetism is present.

Below the transition temperature T_C each sublattice exhibits a spontaneous magnetization given by (5.112) and (5.113). The magnitude of the total magnetization M amounts to:

$$M = |M_1 - M_2| \tag{5.127}$$

$$= \frac{c_1 T - c_1 c_2 \beta v - c_1 c_2 v + c_1 c_2 v - c_2 T + c_1 c_2 \alpha v}{T^2 - v(\alpha c_1 + \beta c_2)T + c_1 c_2 v^2(\alpha\beta - 1)} \cdot B \tag{5.128}$$

$$= \frac{(c_1 - c_2)T - c_1 c_2 v(\beta - \alpha)}{T^2 - v(\alpha c_1 + \beta c_2)T + c_1 c_2 v^2(\alpha\beta - 1)} \cdot B \tag{5.129}$$

Due to the different magnitude of the magnetization in both sublattices the total magnetization often exhibits a complicated behavior as exemplarily depicted in Fig. 5.12. This becomes obvious for the following situation. Let $c_1/c_2 = 2/3$, $\beta = -1$, and $\alpha = 0$. Substituting in (5.129) leads to:

$$M = \frac{c_1}{2} \cdot \frac{3c_1 v - T}{T^2 - v(\alpha c_1 + \beta c_2)T + c_1 c_2 v^2(\alpha\beta - 1)} \cdot B \tag{5.130}$$

i.e. there is a temperature $T_\text{comp} < T_C$ which the total magnetization vanishes at: $M(T_\text{comp}) = 0$. T_comp is called compensation temperature.

5.4 Helical Order

Helical order often occurs in rare earth metals with hcp structure; they exhibit a layered crystalline structure with a stacking sequence of ABAB.... It is characterized by a parallel alignment of the spins within each layer, i.e. each

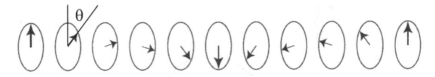

Fig. 5.13. Helical order: Each layer exhibits a ferromagnetic arrangement but the magnetization between adjacent layers is rotated by the angle θ

plane shows ferromagnetic behavior. But, a rotation of the magnetization occurs from layer to layer by an angle θ which is schematically shown in Fig. 5.13.

For the description of helical order we assume that a magnetic interaction is only present between adjacent layers exhibiting the strength J_1 and between next nearest layers with J_2. Consequently, the term $-\sum J_{ij}\boldsymbol{S}_i\boldsymbol{S}_j$ in the Hamiltonian leads to an energy of:

$$E = -2NS^2(J_1\cos\theta + J_2\cos 2\theta) \tag{5.131}$$

with N being the number of atoms per layer. The energy as a function of the rotation angle θ is minimum at $\partial E/\partial\theta = 0$ which leads to:

$$(J_1 + 4J_2\cos\theta)\sin\theta = 0 \tag{5.132}$$

This equation can be solved on the one hand by $\sin\theta = 0$, i.e. $\theta = 0$ or $\theta = \pi$. In this situation we have a ferromagnetic or an antiferromagnetic alignment, respectively, between adjacent layers. On the other hand the equation is solved by:

$$\cos\theta = -\frac{J_1}{4J_2} \tag{5.133}$$

which characterizes helical order or helimagnetism.

Let us discuss the behavior if helical arrangement is present. Due to $|\cos\theta| \leq 1$ we can deduce:

$$|J_1| \leq 4|J_2| \tag{5.134}$$

We see that helimagnetism only occurs if the interaction between next nearest layers is significantly larger than between adjacent planes.

The energies for ferro-, antiferro-, and helimagnetic arrangement amount to (see (5.131)):

$$E_{\mathrm{FM}} = -2NS^2(J_1 + J_2) \tag{5.135}$$

$$E_{\mathrm{AFM}} = -2NS^2(-J_1 + J_2) \tag{5.136}$$

$$E_{\mathrm{HM}} = -2NS^2\left(-\frac{J_1^2}{8J_2} - J_2\right) \tag{5.137}$$

The last equation is obtained using $\cos 2\theta = \cos^2\theta - \sin^2\theta = 2\cos^2\theta - 1$. For an energetic preference of helimagnetism two conditions must be fulfilled: (a) $E_{\mathrm{HM}} < E_{\mathrm{FM}}$ and (b) $E_{\mathrm{HM}} < E_{\mathrm{AFM}}$. From condition (a) we can conclude:

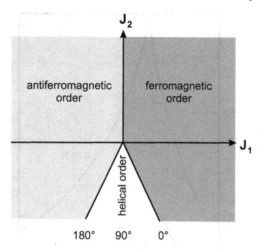

Fig. 5.14. Phase diagram for the model of planes being coupled by J_1 and J_2

$$-\frac{J_1^2}{8J_2} - J_2 > J_1 + J_2 \tag{5.138}$$

which implies:

$$2J_2 + J_1 + \frac{J_1^2}{8J_2} < 0 \tag{5.139}$$

Thus:

$$\frac{1}{8J_2}(16J_2^2 + 8J_1J_2 + J_1^2) < 0 \tag{5.140}$$

which leads to:

$$\frac{1}{8J_2}\left(J_2 + \frac{1}{4}J_1\right)^2 < 0 \tag{5.141}$$

We directly see that J_2 must be negative for the occurrence of helimagnetism. Condition (b) leads to:

$$\frac{1}{8J_2}\left(J_2 - \frac{1}{4}J_1\right)^2 < 0 \tag{5.142}$$

which results in the same conclusion.

Therefore, helical order requires an antiferromagnetic coupling between next nearest layers. The phase diagram shown in Fig. 5.14 summarizes our results.

5.5 Spin Glasses

Spin glasses are dilute alloys of magnetic ions in a non-magnetic matrix. The spins are frozen out below a critical temperature with a statistical distribution of their directions. This spin glass state only occurs in a limited concentration

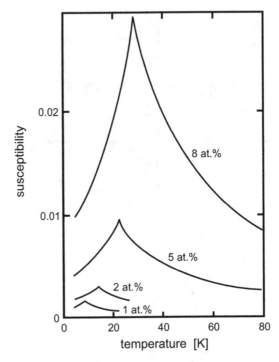

Fig. 5.15. Temperature dependence of the susceptibility for a different amount of Fe in an Fe-Au alloy. The measurements were carried out in a weak magnetic field. (Data taken from [5])

range of the magnetic ions. It must be high enough on the one hand to ensure an interaction via the RKKY coupling but low enough on the other hand in order to prevent the formation of clusters or the presence of a direct coupling of the magnetic moments.

Spin glasses exhibit a sharp maximum in the temperature dependence of the magnetic susceptibility (see Fig. 5.15). In this example magnetic Fe ions were dissolved in Au with increasing concentration c_{Fe} between 1 at% and 8 at%. The temperature T_0 which the susceptibility exhibits a maximum at becomes higher with increasing amount of Fe ions and depends on the concentration:

$$T_0 = a\, c_{Fe}^m \tag{5.143}$$

with m being constant.

Problems

5.1. Calculate the molecular field of Ni at $0\,\mathrm{K}$ in the theory of Weiss assuming a Curie temperature $T_C = 628$ K, $J = 1/2$, and a saturation magnetization per atom of 0.6 μ_B.

5.2. Consider an antiferromagnetic material exhibiting a susceptibility χ_0 at the Néel temperature T_N. Additionally, assume that the magnetization of the own lattice can be neglected for the molecular field in each sublattice. Determine the values of the susceptibility which would be measured when an external magnetic field is applied perpendicular to the spin axis at $T_1 = 0$, $T_2 = T_N/2$, and $T_3 = 2T_N$.

5.3. Determine the parameters θ, χ_0, and σ which allow to describe the inverse susceptibility of a ferrimagnetic system given in (5.122).

5.4. Helimagnetism
(a) Let us assume a negative coupling between next nearest layers, i.e. $J_2 < 0$, exhibiting a magnitude of $J_1\sqrt{3}/6$ with J_1 being the coupling strength between adjacent layers. Show that this material is helimagnetic.
(b) How large is the angle θ of the magnetization between adjacent layers?

6

Broken Symmetry

The occurrence of a spontaneously ordered state at low temperatures is a fundamental phenomenon in solid state physics. Examples are ferromagnetism, antiferromagnetism, and super conductivity.

It can be characterized by a temperature dependence of an important physical quantity with a significant difference above and below a critical temperature T^*. The description of each phase is carried out by an order parameter which vanishes for $T > T^*$ and exhibits a non-vanishing value for $T < T^*$. This means that the order parameter directly proves whether the system is in the ordered or disordered state.

In the case of magnetism this order parameter is given by the magnetization. Each ordered phase corresponds with the breaking of a symmetry which will be the first topic in this chapter. In the following we will deal with different models to describe the magnetization as a function of temperature. Subsequently, various properties like the magnetic susceptibility near the critical temperature which the phase transition occurs at will be discussed. Finally, magnetic excitations will be considered which become important at low temperatures.

6.1 Breaking of the Symmetry

We start our consideration on broken symmetry with the transition between the fluid and the solid state. The parameters which are responsible for this transition can be forces or pressure; mostly it is induced by a varying temperature.

A decreasing temperature of a fluid results in a slight contraction but the high degree of symmetry remains. Below a critical value (melting temperature) the transition to the solid state occurs which is related to the breaking of the symmetry. The fluid shows a total translational and rotational symmetry. Contrarily, the solid shown in Fig. 6.1 exhibits a fourfold rotational symmetry as well as a translational symmetry for a linear combination of lattice vectors.

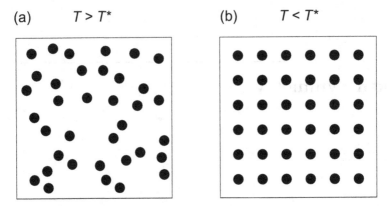

Fig. 6.1. Phase transition between the liquid (**a**) and solid state (**b**). Above the critical temperature T^* the system possesses a complete translational and rotational symmetry. Below T^* these symmetries have significantly been lowered

An analogous behavior occurs for ferromagnets. Above the critical temperature (Curie temperature T_C) the system possesses a complete rotational symmetry; all directions of classical spins or magnetic moments are equivalent (see Fig. 6.2). Below T_C a preferential alignment is present. A rotational symmetry only occurs around the direction of magnetization; this directly proves that the symmetry is broken.

An important aspect is the fact that the symmetry of these systems cannot be changed gradually. A specific type of symmetry can only be present or not. The consequences are that phase transitions are sharp and an unambiguous classification can be made between the ordered and disordered state.

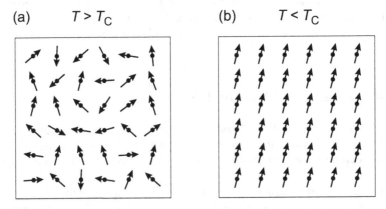

Fig. 6.2. Phase transition between the paramagnetic state for $T > T_C$ (**a**) and the ferromagnetic state for $T < T_C$ (**b**). In this situation the critical temperature is called Curie temperature

The ordered state occurs at low temperatures. This can be understood using thermodynamical considerations. The system tends to minimize the Helmholtz free energy $F = E - TS$ with E being the (internal) energy and S the entropy. At low temperatures the ordered ground state leads to a minimum free energy. At high temperatures F is minimized by a large value of the entropy S which is the disordered state.

6.2 Different Models of Magnetic Behavior

In the following we will discuss different models which describe the order parameter magnetization M as a function of temperature T.

Landau Theory

The free energy of a ferromagnetic system is described by a function of the order parameter using a power series in M. Both opposite magnetization states exhibit no energetic difference, i.e. they are energetically degenerated, which leads to vanishing terms of M to the power of odd values. Neglecting higher orders the free energy can then be written as:

$$F(M) = F_0 + a(T)M^2 + bM^4 \tag{6.1}$$

with F_0 being constant, b being positive and constant, $a(T) = a_0(T - T_C)$ and a_0 being positive. The ground state can be determined by minimizing the free energy F. As a necessary condition the first derivative of F with respect to M must be zero. The first derivatives are:

$$\frac{\partial F}{\partial M} = 2aM + 4bM^3 \tag{6.2}$$

$$\frac{\partial^2 F}{\partial M^2} = 2a + 12bM^2 \tag{6.3}$$

$$\frac{\partial^3 F}{\partial M^3} = 24bM \tag{6.4}$$

$$\frac{\partial^4 F}{\partial M^4} = 24b > 0 \tag{6.5}$$

This means:

$$0 = \frac{\partial F}{\partial M} \tag{6.6}$$

$$= 2M \left(a_0(T - T_C) + 2bM^2 \right) \tag{6.7}$$

The solutions are given by:

$$M = 0 \tag{6.8}$$

$$M = \pm \left(\frac{a_0(T_C - T)}{2b} \right)^{1/2} \tag{6.9}$$

Whereas the first solution is valid for the entire temperature range the second solution can only be fulfilled if $T < T_C$. The question arises whether the first solution $M = 0$ represents a stable state which means F reaches a minimum value. To answer this question we determine the first derivatives in different temperature regimes.

- $M = 0$ and $T > T_C$

$$\left.\frac{\partial^2 F}{\partial M^2}\right|_{M=0} = 2a_0(T - T_C) > 0 \tag{6.10}$$

Thus, the free energy becomes minimum. The system is stable above T_C if the magnetization vanishes.

- $M = 0$ and $T = T_C$

$$\left.\frac{\partial^2 F}{\partial M^2}\right|_{M=0} = 2a_0(T - T_C) = 0 \tag{6.11}$$

$$\left.\frac{\partial^3 F}{\partial M^3}\right|_{M=0} = 0 \tag{6.12}$$

$$\left.\frac{\partial^4 F}{\partial M^4}\right|_{M=0} = 24b > 0 \tag{6.13}$$

Again, the free energy becomes minimum and the system is stable directly at the Curie temperature for a vanishing magnetization.

- $M = 0$ and $T < T_C$

$$\left.\frac{\partial^2 F}{\partial M^2}\right|_{M=0} = 2a_0(T - T_C) < 0 \tag{6.14}$$

Now, the free energy becomes (locally) maximum. Therefore, the system is unstable below T_C if the magnetization vanishes. The magnetized state exhibiting the magnetization given in (6.9) possesses a lower free energy.

Thus, the ground state for different temperatures is given by:

$$M = \pm\sqrt{\frac{a_0(T_C - T)}{2b}} \quad \text{for} \quad T \leq T_C \tag{6.15}$$

$$M = 0 \quad \text{for} \quad T > T_C \tag{6.16}$$

The behavior of the free energy F as a function of the magnetization M is summarized in Fig. 6.3. The resulting temperature dependence of the magnetization in the region near the critical temperature is shown in Fig. 6.4.

This approach of a mean-field theory is characterized by the assumption that all magnetic moments are influenced by an identical averaged exchange field induced by all neighbors and is identical with the model of Weiss which was already discussed in Chap. 5.1. The advantage of this approach using mean

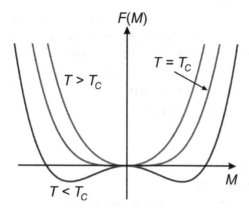

Fig. 6.3. Free energy $F(M)$ for temperatures below, at, and above the critical temperature T_C concerning (6.1)

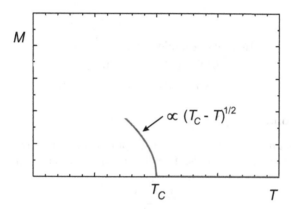

Fig. 6.4. Magnetization M as a function of temperature T near the critical temperature T_C concerning (6.15) and (6.16)

field theories is given by their simplicity. But, correlations and fluctuations are neglected which are important near T_C. Therefore, results near the Curie temperature are less confidential compared to that for low temperatures.

Heisenberg Model

This alternative approach is implemented in a microscopic model which only takes into account interactions between nearest neighbors. The corresponding part of the Hamiltonian is given by:

$$\mathcal{H} = -\sum_{ij} J_{ij} S_i \cdot S_j \tag{6.17}$$

with J_{ij} being the exchange integral. The summation is carried out only for nearest neighbor atoms i and j. For the following considerations it is important to distinguish between two different dimensionalities:

- D: dimensionality of the spin considered as a three-dimensional vector
- d: dimensionality of the crystal lattice

This allows to classify different models:

- $D = 1(z)$: Ising model
- $D = 2(xy)$: XY model
- $D = 3$: Heisenberg model
- $D = \infty$: spherical model

Thus, the part of the Hamiltonian $\mathcal{H}$ given in (6.17) can be written as:

$$\mathcal{H} = -\sum_{ij} J \left(\alpha (S_i^x S_j^x + S_i^y S_j^y) + \beta S_i^z S_j^z \right) \tag{6.18}$$

with

$$
\begin{array}{llll}
\text{Heisenberg model:} & \alpha = 1 & \beta = 1 & \text{(6.19)} \\
\text{Ising model:} & \alpha = 0 & \beta = 1 & \text{(6.20)} \\
\text{XY model:} & \alpha = 1 & \beta = 0 & \text{(6.21)}
\end{array}
$$

One-Dimensional Ising Model

This model assumes a chain with $N + 1$ spins which are located on a one-dimensional lattice and N bonds. The Hamiltonian amounts to:

$$\mathcal{H} = -2J \sum_{i=1}^{N} S_i^z S_{i+1}^z \tag{6.22}$$

with $J > 0$. In the ground state all spins are uniformly oriented in one direction, i.e. they are ferromagnetically ordered. The ground state energy can be determined using $S_i^z = 1/2$ to be

$$E_0 = -2JN \frac{1}{2} \cdot \frac{1}{2} = -\frac{1}{2} NJ \tag{6.23}$$

Let us now consider one defect which consists of an antiparallel alignment at a specific site of the chain, i.e. on the one side of the chain all spins are directed in one direction whereas on the other side of the defect all spins are aligned to the opposite direction.

The difference of the energy compared to the ground state amounts to J. The increase of the entropy is $S = k \ln N$. The free energy $F = E - TS$ of an infinitely long chain (i.e. $N \to \infty$) reaches $-\infty$ for the case that $T \neq 0\,\text{K}$. As a consequence defects can spontaneously be created. Thus, a long-range ordering cannot occur which results in a critical temperature identical to zero.

These considerations are not only valid for the one-dimensional Ising model but for most of the models in one dimension.

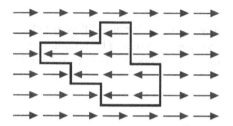

Fig. 6.5. Schematic representation of a defect in a two-dimensional Ising model

Two-Dimensional Ising Model

The spins are arranged on a two-dimensional lattice. A defect results in an increase of energy as well as of entropy both scaling with the length of the boundary of the defect as schematically shown in Fig. 6.5. This behavior is significantly different to that of the one-dimensional Ising model. The most important distinction consists of the possibility of an ordered state at finite temperatures. The exact solution for the two-dimensional Ising model was given by Onsager in 1944.

Three-Dimensional Ising Model

Without going in further detail, an ordered state also occurs for temperatures above $0\,\mathrm{K}$, i.e. $T_C > 0\,\mathrm{K}$.

6.3 Consequences

Important consequences of the broken symmetry are:

- Existence of phase transitions
  A sharp transition occurs at the critical temperature (e.g., fluid-solid, paramagnet-ferromagnet). The regime of the phase transition is called the critical regime.
- Rigidity
  After breaking the symmetry the system strongly tries to remain in the actual state because changes desire energy.
- Excitations
  Neglecting fluctuations due to quantum effects perfect order occurs at $T = 0\,\mathrm{K}$. With increasing temperature the degree of ordering is decreased due to excitations concerning the order parameter. For crystals lattice vibrations can be correlated with phonons. For ferromagnets spin waves are related to magnons.

- Defects
  The broken symmetry often results in a different behavior in regions being neighbored in a macroscopic system. The boundary represents a defect. In crystals the defect may be a dislocation or grain boundary. In ferromagnets it is a domain wall.

6.4 Phase Transitions

In the vicinity of the critical temperature important properties can be described using an exponential function (e.g., $M \propto (T^* - T)^x$). The parameter x is called "critical exponent". Continuous phase transitions are related to critical exponents which exclusively depend on the:

- dimensionality of the system d
- dimensionality of the order parameter D
- range of the forces being involved (long-short)

It is therefore sufficient to determine critical exponents for a specific universality class.

Mean-Field Theory

Despite neglecting correlations and fluctuations a correct description is given for $d \geq 4$. Accurate solutions can further be calculated for the following situations:

- In most cases for $d = 1$:
  No continuous phase transition occurs.
- $d \geq 4$:
  In all cases using mean-field theories
- $d = 2, D = 1$:
  This situation corresponds to the two-dimensional Ising model.
- $D = \infty$ for all values of d:
  Spherical model
- several cases for long-range interactions:
  The solution is given using mean-field theories.

Unfortunately, most of the real situations belong to $d = 3$ and short-range interactions. Thus, exact solutions are not possible and approximations must be carried out.

Critical Exponents for Ferromagnetic Systems

The most commonly used critical exponents describing ferromagnetic systems are:

- β characterizing the order parameter M for $T < T_C$:

$$M \propto (T_C - T)^\beta \tag{6.24}$$

- γ characterizing the susceptibility χ for $T > T_C$:

$$\chi \propto (T - T_C)^{-\gamma} \tag{6.25}$$

- α characterizing the specific heat c_H for $T > T_C$:

$$c_H \propto (T - T_C)^{-\alpha} \tag{6.26}$$

- δ characterizing the influence of an external magnetic field H on the magnetization M at $T = T_C$:

$$M \propto H^{1/\delta} \tag{6.27}$$

- ν characterizing the correlation length ξ for $T < T_C$:

$$\xi \propto (T_C - T)^{-\nu} \tag{6.28}$$

Calculation of the Critical Exponents

In the framework of the Landau theory it is possible to calculate the different critical exponents which we will carry out in the following.

- Critical exponent β
  As already calculated above (see (6.15)) the temperature dependence of the magnetization is given by:

$$M = \pm \left(\frac{a_0(T_C - T)}{2b} \right)^{1/2} \tag{6.29}$$

$$\propto (T_C - T)^\beta \tag{6.30}$$

Thus, $\beta = 1/2$.
- Critical exponent α
  The free energy F amounts to (see (6.1)):

$$F(M) = F_0 + a(T)M^2 + bM^4 \tag{6.31}$$

Inserting of (see (6.15)):

$$M^2 = \frac{a_0(T_C - T)}{2b} \tag{6.32}$$

results in

$$F = \begin{cases} F_0 - (a_0^2/4b)(T_C - T)^2 & \text{for} \quad T < T_C \\ F_0 & \text{for} \quad T \geq T_C \end{cases} \tag{6.33}$$

The specific heat is given by:

$$c_H = -T \frac{\partial^2 F}{\partial T^2} \tag{6.34}$$

For temperatures T near T_C but $T \geq T_C$ the specific heat amounts to:

$$c_H(T_C^+) = -T_C \left. \frac{\partial^2 F_0}{\partial T^2} \right|_{T=T_C} \tag{6.35}$$

which is independent on T. Thus, we obtain $\alpha = 0$. For temperatures T near T_C but $T < T_C$ we get:

$$c_H(T_C^-) = -T_C \left(\left. \frac{\partial^2 F_0}{\partial T^2} \right|_{T=T_C} - \frac{a_0^2}{2b} \right) = \frac{a_0^2 T_C}{2b} \tag{6.36}$$

which again results in $\alpha = 0$. Combining both results for temperatures near T_C, we find:

$$c_H(T_C^-) - c_H(T_C^+) = \frac{a_0^2 T_C}{2b} \tag{6.37}$$

This means that the specific heat c_H exhibits a discontinuity at the Curie temperature.

• Critical exponent γ

Applying an external magnetic field B the free energy exhibits an additional term and amounts to:

$$F = F_0 + aM^2 + bM^4 - BM \tag{6.38}$$

The minimum is given if the first derivative with respect to the magnetization is zero:

$$0 = \frac{\partial F}{\partial M} \tag{6.39}$$

$$= 2aM + 4bM^3 - B \tag{6.40}$$

Thus, we obtain:

$$B = 2aM + 4bM^3 \tag{6.41}$$

This allows to determine the susceptibility χ which is given by:

$$\mu_0 \frac{1}{\chi} = \frac{\partial B}{\partial M} \tag{6.42}$$

Using (6.41) we get:

$$\mu_0 \frac{1}{\chi} = 2a + 4bM^2 \tag{6.43}$$

Keeping in mind the magnitude of the magnetization as a function of temperature (see (6.15) and (6.16)) we find:

$$\mu_0 \frac{1}{\chi} = \begin{cases} -4a_0(T - T_C) & \text{for} \quad T < T_C \\ 2a_0(T - T_C) & \text{for} \quad T \geq T_C \end{cases} \tag{6.44}$$

Therefore:

$$\chi \propto (T - T_C)^{-1} \tag{6.45}$$

Thus, we get $\gamma = 1$.

- Critical exponent δ

Using (6.38) and the condition that the derivative $\partial F / \partial M$ must be zero we get:

$$B = 4bM^3 + 2aM \tag{6.46}$$

$$\propto M^3 \tag{6.47}$$

Therefore, we find:

$$M \propto B^{1/3} \propto H^{1/3} \tag{6.48}$$

and obtain $\delta = 3$.

- Critical exponent ν

Magnetic moments with distances $r \ll \xi$ being the correlation length are strongly correlated, i.e. the probability for a parallel orientation is about one. For large distances $r \gg \xi$ no correlations occur.

A spatial dependence of the magnetization M and an external magnetic field B can be introduced by means of spatially periodic fields. In this situation the free energy F is given by:

$$F = F_0 + aM^2 + bM^4 - BM \tag{6.49}$$

with

$$F_0 = -c(\nabla M)^2 \tag{6.50}$$

$$M = M_0 + \partial M \cdot e^{i\mathbf{k} \cdot \mathbf{r}} \tag{6.51}$$

$$B = \partial M \cdot e^{i\mathbf{k} \cdot \mathbf{r}} \tag{6.52}$$

$$a = a_0(T - T_C) \tag{6.53}$$

Using

$$\frac{\partial}{\partial M} \nabla M = \nabla \frac{\partial M}{\partial M} = M \tag{6.54}$$

we obtain:

$$\frac{\partial}{\partial M} (\nabla M)^2 = 2\nabla^2 M \tag{6.55}$$

which allows to calculate the first derivative of the free energy with respect to the magnetization which must be zero:

$$0 = \frac{\partial F}{\partial M} \tag{6.56}$$

$$= 2c\nabla^2 M + 2aM + 4bM^3 - B \tag{6.57}$$

$$= 2aM_0 + 2a\partial M \, e^{i\mathbf{k} \cdot \mathbf{r}}$$

$$+ 4b(M_0^3 + 3M_0^2 \partial M \, e^{i\mathbf{k} \cdot \mathbf{r}} + 3M_0(\partial M)^2 \, e^{2i\mathbf{k} \cdot \mathbf{r}} + (\partial M)^3 \, e^{3i\mathbf{k} \cdot \mathbf{r}})$$

$$- \partial B \, e^{i\mathbf{k} \cdot \mathbf{r}} + 2ck^2 \partial M \, e^{i\mathbf{k} \cdot \mathbf{r}} \tag{6.58}$$

Neglecting higher orders of the derivative we get:

$$0 = (2aM_0 + 4bM_0^3) - \partial B\, e^{i\boldsymbol{k}\cdot\boldsymbol{r}} + \partial M\, e^{i\boldsymbol{k}\cdot\boldsymbol{r}}(2a + 12bM_0^2 + 2ck^2) \quad (6.59)$$

The first bracket vanishes due to the equilibrium condition (see (6.2) and (6.6)) which results in:

$$\partial B = \partial M(2a + 12bM_0^2 + 2ck^2) \quad (6.60)$$

This calculation enables to determine the susceptibility $\chi_{\boldsymbol{k}}$:

$$\chi_{\boldsymbol{k}} = \mu_0 \frac{\partial M}{\partial B} = \frac{\mu_0}{2a + 12bM_0^2 + 2ck^2} \quad (6.61)$$

Above the Curie temperature $T \geq T_C$ the magnetization vanishes $M_0^2 = 0$. Using (6.44) we obtain:

$$\frac{1}{\chi_{\boldsymbol{k}}^+} = 2a_0(T - T_C) + 2ck^2 \quad (6.62)$$

$$= \frac{1}{\chi^+} + 2ck^2 \quad (6.63)$$

Thus:

$$\chi_{\boldsymbol{k}}^+ = \frac{\chi^+}{1 + k^2 \dfrac{c}{a_0(T - T_C)}} \quad (6.64)$$

$$= \frac{\chi^+}{1 + k^2\xi_+^2} \quad (6.65)$$

with ξ_+ being the correlation length above T_C and defined by:

$$\xi_+ = \left(\frac{c}{a_0(T - T_C)} \right)^{1/2} \quad (6.66)$$

For $T \leq T_C$ we analogously find:

$$\chi_{\boldsymbol{k}}^- = \frac{\chi^-}{1 + k^2\xi_-^2} \quad (6.67)$$

with

$$\xi_- = \left(\frac{c}{2a_0(T_C - T)} \right)^{1/2} \quad (6.68)$$

Therefore, we obtain:

$$\xi \propto (T - T_C)^{-1/2} \quad (6.69)$$

which results in $\nu = 1/2$

Scaling Laws

The scaling laws describe the relationship between different critical exponents which are valid for all exactly solvable models. The scaling laws can additionally be applied for different models that include approximations:

$$2 = \alpha + 2\beta + \gamma \tag{6.70}$$

$$\delta = 1 + \gamma/\beta \tag{6.71}$$

$$\alpha = 2 - d\nu \tag{6.72}$$

It should be noted that the last relationship is only valid within the Landau theory for $d = 4$. Additionally, we find that α, γ, and ν exhibit the same value below and above the Curie temperature which is by no means a triviality.

Consequently, only two independent critical exponents occur whereas the other ones are determined by the scaling laws. An overview on the values of critical exponents for different models is given in Table 6.1.

Table 6.1. Overview of critical exponents in different models

	α	β	γ	δ	ν
Landau theory	0	1/2	1	3	1/2
2d-Ising model	0	1/8	7/4	15	1
3d-Ising model	0.11	0.325	1.24	4.816	0.63
XY-model	−0.008	0.345	1.316	4.81	0.67
3d-Heisenberg model	−0.116	0.365	1.387	4.803	0.705
spherical model	−1	1/2	2	5	1

6.5 Magnetic Excitations

The description of, e.g., the magnetization M as a function of temperature by exponential laws and the calculation of critical exponents fails at lower temperatures $\approx T/T_C < 1/2$. Thus, another attempt must be used.

For low temperatures the description is carried out using low-energetic magnetic excitations. These spin waves are quantized by magnons. An analog are lattice vibrations in crystals which are quantized by phonons.

The important properties can be understood by the dispersion relation which describes the frequency dependence with the wave vector, i.e. $\omega(q)$ or the energy dependence with the momentum $\hbar\omega(\hbar q)$.

Magnons

The dispersion relation for magnons of an isotropic ferromagnet can be described using a semi-classical approach.

Let us assume a chain of equidistant (lattice constant a) "classical spin vectors" with length $|\boldsymbol{S}| = S$ and ferromagnetic coupling (i.e. $J > 0$) between adjacent spins $\boldsymbol{S}_j$ and $\boldsymbol{S}_{j+1}$.

The time dependence of the spins as classical angular momenta is determined by the actual torque which is due to the exchange field of neighboring spins:

$$\frac{\mathrm{d}\boldsymbol{S}_j}{\mathrm{d}t} = \boldsymbol{S}_j \times 2J(\boldsymbol{S}_{j-1} + \boldsymbol{S}_{j+1})/\hbar \tag{6.73}$$

The decomposition in Cartesian components leads to:

$$\frac{\mathrm{d}S_j^x}{\mathrm{d}t} = \frac{2J}{\hbar} \left(S_j^y \cdot (S_{j-1}^z + S_{j+1}^z) - S_j^z \cdot (S_{j-1}^y + S_{j+1}^y) \right) \tag{6.74}$$

$$\frac{\mathrm{d}S_j^y}{\mathrm{d}t} = \frac{2J}{\hbar} \left(S_j^z \cdot (S_{j-1}^x + S_{j+1}^x) - S_j^x \cdot (S_{j-1}^z + S_{j+1}^z) \right) \tag{6.75}$$

$$\frac{\mathrm{d}S_j^z}{\mathrm{d}t} = \frac{2J}{\hbar} \left(S_j^x \cdot (S_{j-1}^y + S_{j+1}^y) - S_j^y \cdot (S_{j-1}^x + S_{j+1}^x) \right) \tag{6.76}$$

The ground state is given if all spins are aligned in a given direction, e.g. along the z-axis. Thus, we have:

$$S_j^z = S \qquad S_j^x = S_j^y = 0 \tag{6.77}$$

An excited state can be characterized by a small deviation:

$$S_j^z \approx S \qquad S_j^x = \varepsilon' S \qquad S_j^y = \varepsilon'' S \tag{6.78}$$

with ε' and $\varepsilon'' \ll 1$. Inserting leads to:

$$\frac{\mathrm{d}S_j^x}{\mathrm{d}t} = \frac{2JS}{\hbar} \left(2S_j^y - S_{j-1}^y - S_{j+1}^y \right) \tag{6.79}$$

$$\frac{\mathrm{d}S_j^y}{\mathrm{d}t} = -\frac{2JS}{\hbar} \left(2S_j^x - S_{j-1}^x - S_{j+1}^x \right) \tag{6.80}$$

$$\frac{\mathrm{d}S_j^z}{\mathrm{d}t} \approx 0 \tag{6.81}$$

We use plane waves as solutions:

$$S_j^x = A\, \mathrm{e}^{i(qja - \omega t)} = AE_t \tag{6.82}$$

$$S_j^y = B\, \mathrm{e}^{i(qja - \omega t)} = BE_t \tag{6.83}$$

Inserting into (6.79) leads to:

$$-i\omega AE_t = 2JS(2BE_t - BE_t\, \mathrm{e}^{-iqa} - BE_t\, \mathrm{e}^{iqa})/\hbar \tag{6.84}$$

$$= 2JSBE_t(2 - (\mathrm{e}^{iqa} + \mathrm{e}^{-iqa}))/\hbar \tag{6.85}$$

$$= 4JSBE_t(1 - \cos qa)/\hbar \tag{6.86}$$

Thus:

$$-i\hbar\omega A = 4JS(1 - \cos qa)B \qquad (6.87)$$

Analogously, we obtain:

$$-i\hbar\omega B = -4JS(1 - \cos qa)A \qquad (6.88)$$

The non-trivial solution is given by:

$$A = iB \qquad (6.89)$$

which means that the oscillations in x- and y-direction exhibit a phase-shift of 90°. The spin wave dispersion is therefore given by:

$$\hbar\omega = 4JS(1 - \cos qa) \qquad (6.90)$$

which is shown in Figs. 6.6 and 6.7. As no gap occurs at $\hbar\omega = 0$ already smallest excitation energies can create spin waves. The same result is obtained using quantum mechanical considerations. It should be noted that magnons are bosons ($1\hbar$) because every magnon represents a delocalized switched spin.

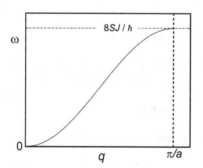

Fig. 6.6. Dispersion relation of a magnon in a one-dimensional chain

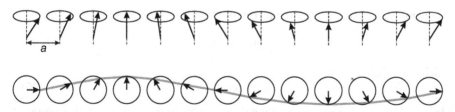

Fig. 6.7. Spin wave of a one-atomic chain in side view (*top*) and top view (*bottom*)

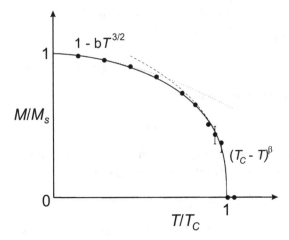

Fig. 6.8. Relative spontaneous magnetization of a ferromagnet as a function of the reduced temperature. For low temperatures the behavior can be expressed by the Bloch–$T^{3/2}$ law given in (6.102) using the spin wave model. Near the Curie temperature M/M_S is given by $(T_C - T)^\beta$ with β being a critical exponent

Magnetization Near $T = 0\,\mathrm{K}$

In the following we want to discuss the magnetization near $T = 0\,\mathrm{K}$. For low temperatures q is small which allows the approximation:

$$\cos qa = 1 - q^2 a^2/2 \tag{6.91}$$

Inserting into (6.90) results in:

$$\hbar\omega = 2JSq^2a^2 \tag{6.92}$$

Thus, we obtain:

$$\omega \propto q^2 \tag{6.93}$$

The density of spin wave states in three dimensions can be described by:

$$g(q) \propto q^2 \tag{6.94}$$

which leads to:

$$g(\omega) \propto \omega^{1/2} \tag{6.95}$$

The number of magnons n can be calculated by integration over all frequencies of the density of spin wave states and under consideration of the Bose distribution of magnons:

$$n = \int_0^\infty \frac{g(\omega)}{e^{\hbar\omega/kT} - 1}\, d\omega \tag{6.96}$$

Using (6.95) and setting:

$$x = \hbar\omega/kT \tag{6.97}$$

we get:

$$n = b \int_0^\infty \left(\frac{kT}{\hbar}\right)^{1/2} x^{1/2} \frac{kT}{\hbar} \frac{1}{e^x - 1}\, dx \tag{6.98}$$

$$= b \left(\frac{kT}{\hbar}\right)^{3/2} \int_0^\infty \frac{x^{1/2}}{e^x - 1}\, dx \tag{6.99}$$

$$= c\, T^{3/2} \tag{6.100}$$

Each magnon reduces the magnetization by $S = 1$, i.e.

$$M(0) - M(T) \propto n(T) = c\, T^{3/2} \tag{6.101}$$

Thus:

$$\frac{M(T)}{M(0)} = 1 - a\, c\, T^{3/2} \tag{6.102}$$

which is known as the so-called Bloch–$T^{3/2}$ law.

Summarizing, we found that the magnetization can be described near $T = 0\,\mathrm{K}$ by the Bloch–$T^{3/2}$ law and near the Curie temperature T_C by the scaling law $(T - T_C)^\beta$ (see Fig. 6.8).

7

Magnetic Anisotropy Effects

The considerations in the last chapters were related to isotropic systems, i.e. all physical properties are identical for different directions. Especially, we have discussed the energy as a function of the *magnitude* of the magnetization $|M|$ but we have neglected the energy dependence on the *direction* of M. In this chapter we will deal with that magnetic effects which depend on crystallographic directions.

The Heisenberg Hamiltonian is completely isotropic and its energy levels do not depend on the direction in space which the crystal is magnetized in. If there is no other energy term the magnetization would always vanish in zero applied field. However, real magnetic materials are not isotropic.

One illustrating example is given by spin waves which are quantized by magnons (see Chap. 6.5). The number of magnons diverges in one and two dimensions, i.e. the magnetization vanishes for $T > 0\,\mathrm{K}$ in the isotropic Heisenberg model (Mermin–Wagner–Berezinskii theorem). Thus, spin fluctuations can be exited with infinitely small energy which destroys any long range order. In the anisotropic case magnetic moments cannot be rotated to any direction with infinitely small energy. This additional amount of energy is due to an anisotropy energy which is responsible for the occurrence of ferromagnetism in two-dimensional systems at $T > 0\,\mathrm{K}$.

7.1 Overview of Magnetic Anisotropies

The most important different magnetic anisotropies which are discussed below in detail are:

- Magneto crystalline anisotropy
  The magnetization is oriented along specific crystalline axes.
- Shape anisotropy
  The magnetization is affected by the macroscopic shape of the solid.

- Induced magnetic anisotropy
  Specific magnetization directions can be stabilized by tempering the sample in an external magnetic field.
- Stress anisotropy (magnetostriction)
  Magnetization leads to a spontaneous deformation.
- Surface and interface anisotropy
  Surfaces and interfaces often exhibit different magnetic properties compared to the bulk due to their asymmetric environment.

7.2 Magneto Crystalline Anisotropy

The most important type of anisotropy is the magneto crystalline anisotropy which is caused by the spin orbit interaction of the electrons. The electron orbitals are linked to the crystallographic structure. Due to their interaction with the spins they make the latter prefer to align along well-defined crystallographic axes. Therefore, there are directions in space which a magnetic material is easier to magnetize in than in other ones (easy axes or easy magnetization axes). The spin-orbit interaction can be evaluated from basic principles. However, it is easier to use phenomenological expressions (power series expansions that take into account the crystal symmetry) and take the coefficients from experiment.

The magneto crystalline energy is usually small compared to the exchange energy. But the direction of the magnetization is only determined by the anisotropy because the exchange interaction just tries to align the magnetic moments parallel, no matter in which direction.

The magnetization direction $\boldsymbol{m} = \boldsymbol{M}/|\boldsymbol{M}|$ relative to the coordinate axes can be given by the direction cosine α_i as $\boldsymbol{m} = (\alpha_1, \alpha_2, \alpha_3)$ (see Fig. 7.1) with

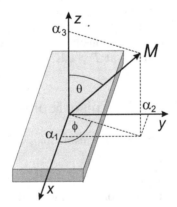

Fig. 7.1. Definition of the direction cosine

$$\alpha_1 = \sin\theta\cos\phi \tag{7.1}$$
$$\alpha_2 = \sin\theta\sin\phi \tag{7.2}$$
$$\alpha_3 = \cos\theta \tag{7.3}$$

These relations fulfill the condition:

$$\alpha_1^2 + \alpha_2^2 + \alpha_3^2 = 1 \tag{7.4}$$

which is often used below.

The magneto crystalline energy per volume E_{crys} can be described by a power series expansion of the components of the magnetization:

$$E_{crys} = E_0 + \sum_i b_i\alpha_i + \sum_{ij} b_{ij}\alpha_i\alpha_j + \sum_{ijk} b_{ijk}\alpha_i\alpha_j\alpha_k$$
$$+ \sum_{ijkl} b_{ijkl}\alpha_i\alpha_j\alpha_k\alpha_l + \mathcal{O}(\alpha^5) \tag{7.5}$$

The terms $\mathcal{O}(\alpha^5)$ with at least the fifth order in α are very small and can usually be neglected.

There is no energy difference for oppositely magnetized systems. Therefore, the energy only depends on the alignment:

$$E(\boldsymbol{M}) = E(-\boldsymbol{M}) \tag{7.6}$$

i.e.

$$E(\alpha_i) = E(-\alpha_i) \tag{7.7}$$

Thus, no odd terms of α_i occur in the series expansion (remember the relationship for the free energy $F(M) = F_0 + aM^2 + bM^4$) and we can reduce the expansion to:

$$E_{crys} = E_0 + \sum_{ij} b_{ij}\alpha_i\alpha_j + \sum_{ijkl} b_{ijkl}\alpha_i\alpha_j\alpha_k\alpha_l \tag{7.8}$$

Let us discuss this general equation for different crystallographic systems.

Cubic Systems

Due to $E(\alpha_i) = E(-\alpha_i)$ all cross terms $\alpha_i\alpha_j$ vanish, i.e. $b_{ij} = 0$ for $i \neq j$. Indices $i = 1, 2, 3$ are undistinguishable in systems with cubic symmetry, i.e. $b_{11} = b_{22} = b_{33}$. Using these considerations the term in second order amounts to:

$$\sum_{ij} b_{ij}\alpha_i\alpha_j = b_{11}(\alpha_1^2 + \alpha_2^2 + \alpha_3^2) = b_{11} \tag{7.9}$$

taking into account (7.4). The term in forth order is given by:

$$\sum_{ijkl} b_{ijkl}\alpha_i\alpha_j\alpha_k\alpha_l = b_{1111}(\alpha_1^4 + \alpha_2^4 + \alpha_3^4) + 6b_{1122}(\alpha_1^2\alpha_2^2 + \alpha_1^2\alpha_3^2 + \alpha_2^2\alpha_3^2) \quad (7.10)$$

The term in sixth order amounts to:

$$\sum_{ijklmn} b_{ijklmn}\alpha_i\alpha_j\alpha_k\alpha_l\alpha_m\alpha_n =$$

$$b_{111111}(\alpha_1^6 + \alpha_2^6 + \alpha_3^6)$$
$$+15b_{111122}(\alpha_1^2\alpha_2^4 + \alpha_1^4\alpha_2^2 + \alpha_1^2\alpha_3^4 + \alpha_1^4\alpha_3^2 + \alpha_2^2\alpha_3^4 + \alpha_2^4\alpha_3^2)$$
$$+90b_{112233}\alpha_1^2\alpha_2^2\alpha_3^2 \quad (7.11)$$

For a rearrangement of these equations we use the following relationships gained by the normalization condition given in (7.4):

$$1 = (\alpha_1^2 + \alpha_2^2 + \alpha_3^2)^2 \quad (7.12)$$
$$= \alpha_1^4 + \alpha_2^4 + \alpha_3^4 + 2(\alpha_1^2\alpha_2^2 + \alpha_1^2\alpha_3^2 + \alpha_2^2\alpha_3^2) \quad (7.13)$$
$$1 = (\alpha_1^2 + \alpha_2^2 + \alpha_3^2)^3 \quad (7.14)$$
$$= \alpha_1^6 + \alpha_2^6 + \alpha_3^6 + 6\alpha_1^2\alpha_2^2\alpha_3^2$$
$$+3(\alpha_1^2\alpha_2^4 + \alpha_1^4\alpha_2^2 + \alpha_1^2\alpha_3^4 + \alpha_1^4\alpha_3^2 + \alpha_2^2\alpha_3^4 + \alpha_2^4\alpha_3^2) \quad (7.15)$$

The multiplication of (7.4) with, e.g., $\alpha_1^2\alpha_2^2$ results in:

$$\alpha_1^2\alpha_2^2 = \alpha_1^4\alpha_2^2 + \alpha_1^2\alpha_2^4 + \alpha_1^2\alpha_2^2\alpha_3^2 \quad (7.16)$$

which leads to:

$$\alpha_1^4\alpha_2^2 + \alpha_1^2\alpha_2^4 = \alpha_1^2\alpha_2^2 - \alpha_1^2\alpha_2^2\alpha_3^2 \quad (7.17)$$

Now, the energy density can be expressed as:

$$E_{crys}^{cubic} = E_0 + b_{11} + b_{1111}(\alpha_1^4 + \alpha_2^4 + \alpha_3^4)$$
$$+6b_{1122}(\alpha_1^2\alpha_2^2 + \alpha_1^2\alpha_3^2 + \alpha_2^2\alpha_3^2)$$
$$+b_{111111}(\alpha_1^6 + \alpha_2^6 + \alpha_3^6) + 90b_{112233}\alpha_1^2\alpha_2^2\alpha_3^2$$
$$+15b_{111122}(\alpha_1^2\alpha_2^4 + \alpha_1^4\alpha_2^2 + \alpha_1^2\alpha_3^4 + \alpha_1^4\alpha_3^2 + \alpha_2^2\alpha_3^4 + \alpha_2^4\alpha_3^2)(7.18)$$
$$= K_0 + K_1(\alpha_1^2\alpha_2^2 + \alpha_1^2\alpha_3^2 + \alpha_2^2\alpha_3^2)$$
$$+K_2\alpha_1^2\alpha_2^2\alpha_3^2 + K_3(\alpha_1^2\alpha_2^2 + \alpha_1^2\alpha_3^2 + \alpha_2^2\alpha_3^2) + \cdots \quad (7.19)$$

with coefficients K_i (magneto crystalline anisotropy constants) which are functions of the coefficients $b_{...}$.

Tetragonal Systems

As in the case of cubic systems all cross terms $\alpha_i\alpha_j$ vanish, too. Due to the reduced symmetry only the indices 1 and 2 are indistinguishable. Under these conditions the term in second order amounts to:

$$\sum_{ij} b_{ij}\alpha_i\alpha_j = b_{11}\alpha_1^2 + b_{11}\alpha_2^2 + b_{33}\alpha_3^2 \tag{7.20}$$

Using $\alpha_1^2 + \alpha_2^2 = 1 - \alpha_3^2$ (see (7.4)) we obtain:

$$\sum_{ij} b_{ij}\alpha_i\alpha_j = b_{11} + (b_{33} - b_{11})\alpha_3^2 \tag{7.21}$$

$$= a_0 + a_1\alpha_3^2 \tag{7.22}$$

with coefficients a_i which are a function of the coefficients b_{ii}. The term in forth order is given by:

$$\sum_{ijkl} b_{ijkl}\alpha_i\alpha_j\alpha_k\alpha_l = b_{1111}\alpha_1^4 + b_{2222}\alpha_2^4 + b_{3333}\alpha_3^4$$

$$+6b_{1122}\alpha_1^2\alpha_2^2 + 12b_{1133}(\alpha_1^2\alpha_3^2 + \alpha_2^2\alpha_3^2) \tag{7.23}$$

$$= b_{1111}(\alpha_1^4 + \alpha_2^4) + b_{3333}\alpha_3^4 + 6b_{1122}\alpha_1^2\alpha_2^2$$

$$+12b_{1133}\alpha_3^2(\alpha_1^2 + \alpha_2^2) \tag{7.24}$$

Thus, the energy density amounts to:

$$E_{crys}^{tetra} = K_0 + K_1\alpha_3^2 + K_2\alpha_3^4 + K_3(\alpha_1^4 + \alpha_2^4) + \cdots \tag{7.25}$$

with coefficients K_i being dependent on the coefficients $b_{....}$. Replacing the direction cosine α_i by the angles θ and ϕ (see (7.1)–(7.3)) we get:

$$E_{crys}^{tetra} = K_0 + K_1 \cos^2\theta + K_2 \cos^4\theta + K_3 \sin^4\theta(\sin^4\phi + \cos^4\phi) \tag{7.26}$$

$$= K_0' + K_1' \sin^2\theta + K_2' \sin^4\theta + K_3' \sin^4\theta \cos 4\phi \tag{7.27}$$

The last term in (7.27) reflects the fourfold symmetry of this crystallographic system.

Hexagonal Systems

Analogous considerations and calculations lead to an energy density for hexagonal systems of:

$$E_{crys}^{hex} = K_0 + K_1(\alpha_1^2 + \alpha_2^2) + K_2(\alpha_1^2 + \alpha_2^2)^2$$

$$+K_3(\alpha_1^2 + \alpha_2^2)^3 + K_4(\alpha_1^2 - \alpha_2^2)(\alpha_1^4 - 14\alpha_1^2\alpha_2^2 + \alpha_2^4) \tag{7.28}$$

$$= K_0 + K_1 \sin^2\theta + K_2 \sin^4\theta + K_3 \sin^6\theta$$

$$+K_4 \sin^6\theta \cos 6\phi \tag{7.29}$$

It should be noted that the notation of the coefficients is partly inconsistent in the literature.

For tetragonal and hexagonal systems the magneto crystalline energy is related to a cylindrical symmetry up to terms of the second or forth order, respectively. The energy is only dependent on the angle θ between the magnetization direction and the z-axis. Therefore, we find a uniaxial symmetry.

The azimuthal angle ϕ characterizes the anisotropy concerning the basal plane and thus the energy in order to rotate the magnetization in a plane perpendicular to the z-axis. The terms $\cos 4\phi$ and $\cos 6\phi$ reflect the four- and sixfold symmetry of the tetragonal and hexagonal basal planes, respectively.

Generally we find:

- The coefficients (crystal anisotropy constants) depend on material and temperature.
- The experience proves that the constants K_1 and K_2 are sufficient for a good agreement between experiment and calculation.
- Sign and ratio of the constants determine the easy magnetization axis or the preferred axis, i.e. crystallographic directions which the magnetization is aligned on without external magnetic field. It can be identified by calculating the minimum energy of E_{crys}.

Magneto Crystalline Anisotropy for Cubic Crystals

The energies related to the magneto crystalline anisotropy for different crystallographic directions can be obtained using the direction cosine (see Fig. 7.1):

- E_{100}
  The [100] direction is characterized by $\theta = 90°$ and $\phi = 0°$. This results in $\alpha_1 = 1$, $\alpha_2 = \alpha_3 = 0$.
- E_{110}
  The [110] direction is characterized by $\theta = 90°$ and $\phi = 45°$ which leads to $\alpha_1 = \alpha_2 = 1/\sqrt{2}$, $\alpha_3 = 0$.
- E_{111}
  For the [111] direction we see that $\theta = 54.7°$ (due to $\tan\theta = \sqrt{2}$) and $\phi = 45°$. Thus, we obtain in $\alpha_1 = \alpha_2 = \alpha_3 = 1/\sqrt{3}$.

The magneto crystalline energy density for cubic materials was given by (cf. (7.19)):

$$E_{crys}^{cubic} = K_0 + K_1(\alpha_1^2\alpha_2^2 + \alpha_1^2\alpha_3^2 + \alpha_2^2\alpha_3^2) + K_2\alpha_1^2\alpha_2^2\alpha_3^2 + \cdots \qquad (7.30)$$

The corresponding energies are obtained by inserting the direction cosine:

$$E_{100} = K_0 \qquad (7.31)$$

$$E_{110} = K_0 + \frac{1}{4}K_1 \qquad (7.32)$$

$$E_{111} = K_0 + \frac{1}{3}K_1 + \frac{1}{27}K_2 \qquad (7.33)$$

The values of the anisotropy constants K_i for the ferromagnetic elements Fe, Co, and Ni at low temperatures are given in Table 7.1.

Table 7.1. Magnitude of the magneto crystalline anisotropy constants K_1, K_2, and K_3 of Fe, Ni, and Co at $T = 4.2\,\mathrm{K}$

		bcc-Fe	fcc-Ni	hcp-Co
K_1	[J/m^3]	$5.48 \cdot 10^4$	$-12.63 \cdot 10^4$	$7.66 \cdot 10^5$
	[eV/atom]	$4.02 \cdot 10^{-6}$	$-8.63 \cdot 10^{-6}$	$5.33 \cdot 10^{-5}$
K_2	[J/m^3]	$1.96 \cdot 10^2$	$5.78 \cdot 10^4$	$1.05 \cdot 10^5$
	[eV/atom]	$1.44 \cdot 10^{-8}$	$3.95 \cdot 10^{-6}$	$7.31 \cdot 10^{-6}$
K_3	[J/m^3]	$0.9 \cdot 10^2$	$3.48 \cdot 10^3$	$-$
	[eV/atom]	$6.6 \cdot 10^{-9}$	$2.38 \cdot 10^{-7}$	$-$

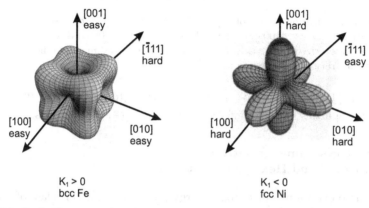

Fig. 7.2. Left: Energy surface for cubic symmetry with $K_1 > 0$ presenting the easy and hard magnetization axes. One example for this situation is bcc-Fe. **Right**: Energy surface for cubic symmetry with $K_1 < 0$ showing the easy and hard magnetization axes. One example is fcc-Ni

Considering only the most important anisotropy constant K_1 we find for a positive value $K_1 > 0$ that the [100]-direction is the easy magnetization axis and the [111]-direction is the hard (magnetization) axis (see left part of Fig. 7.2) due to the relationship $E_{100} < E_{110} < E_{111}$. One example is bcc-Fe. For a negative value $K_1 < 0$ the [111]-direction is the easy magnetization axis and the [100]-direction is the hard axis (see right part of Fig. 7.2) due to the relationship $E_{111} < E_{110} < E_{100}$. An example for this situation is fcc-Ni. This magnetic behavior is directly reflected in different magnetization curves (see Fig. 7.3).

The additional consideration of the anisotropy constant K_2 may significantly vary the magnetic properties. If K_1 and K_2 are of the same order of magnitude the [110]-direction can be the easy as well as the hard magnetization axis (see Table 7.2). The ratio of K_1/K_2 is about 250 for Fe. For Ni we find that this ratio is about 2, i.e. both constants nearly exhibit the same value. Therefore, the temperature dependence of the anisotropy constants must be taken into account (see Fig. 7.4). We observe that K_1 is negative for

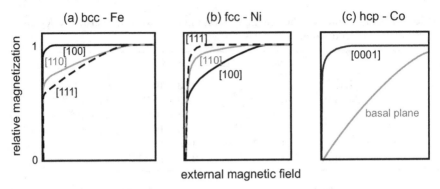

Fig. 7.3. Magnetization curves of (**a**) bcc-Fe, (**b**) fcc-Ni, and (**c**) hcp-Co. The easy magnetization axes are [100]- for Fe, [111]- for Ni, and [0001]-directions for Co. They are characterized by a small magnitude of an external magnetic field in order to achieve saturation magnetization

temperatures below about 400 K and changes sign above this value. Thus, the magnetic properties must be different in both temperature regimes.

Magneto Crystalline Anisotropy
for Tetragonal and Hexagonal Materials

The magneto crystalline anisotropy energy per volume for this class of materials was (cf. (7.27) and (7.29)):

$$E_{\text{crys}} = K_0 + K_1 \sin^2 \theta + K_2 \sin^4 \theta + \cdots \tag{7.34}$$

For positive values $K_1 > 0$ and $K_2 > 0$ we see from this equation that the minimum energy of E_{crys} is obtained for $\sin \theta = 0$, i.e. the c-axis [0001] is the easy magnetization axis (see left part of Fig. 7.5). If both constants K_1 as well as K_2 are negative the minimum is given for $\sin \theta = 1$. Now, the easy magnetization axis lies within the [0001]-plane and the [0001]-direction becomes the hard magnetization axis (see right part of Fig. 7.5). If one anisotropy constant

Table 7.2. Correlation of the easy, medium, and hard axis of magnetization with the magneto crystalline anisotropy constants K_1 and K_2 for cubic crystals (+: positive value; −: negative value)

K_1	K_2	Easy axis	Medium axis	Hard axis
+	$+\infty$ to $-9/4K_1$	100	110	111
+	$-9/4K_1$ to $-9K_1$	100	111	110
+	$-9K_1$ to $-\infty$	111	100	110
−	$-\infty$ to $9/4|K_1|$	111	110	100
−	$9/4|K_1|$ to $9|K_1|$	110	111	100
−	$9|K_1|$ to $+\infty$	110	100	111

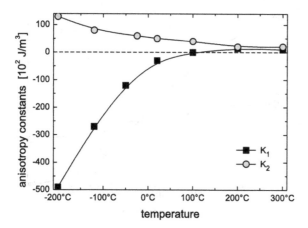

Fig. 7.4. Temperature dependence of the magneto crystalline anisotropy constants K_1 and K_2 of Ni (Data taken from [2])

is positive and one negative a continuous rotation occurs from the c-axis towards the [0001]-plane with $\sin^2 \theta = -K_1/2K_2$. One example for positive K_1 and K_2 is Co at low temperatures (see Table 7.1). Thus, the [0001]-direction is the easy magnetization axis. The rotation from the [0001]-direction towards the basal plane occurs between 500 K and 600 K due to the temperature dependence of K_1 and K_2 (see Fig. 7.6).

Influence of the Stoichiometry of Alloys

The anisotropy constants K_i depend on the relative amount of the constituents in magnetic alloys, too. This occurrence is exemplarily illustrated

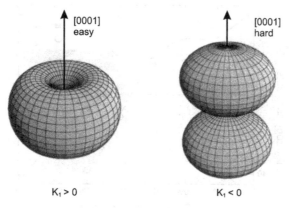

Fig. 7.5. Left: Energy surface for uniaxial symmetry with $K_1 > 0$. The easy magnetization axis is oriented along the [0001]-direction. **Right**: Energy surface for uniaxial symmetry with $K_1 < 0$. The [0001]-direction becomes the hard magnetization axis

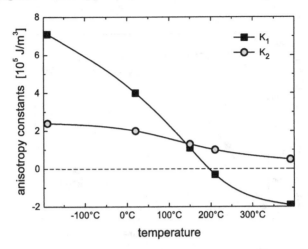

Fig. 7.6. Temperature dependence of the magneto crystalline anisotropy constants K_1 and K_2 of Co (Data taken from [2])

for $Fe_x Co_{1-x}$ in Fig. 7.7. With increasing amount of Co the anisotropy constant K_1 monotonously decreases. For low concentration it is positive and changes sign at about 45%. The behavior of the anisotropy constant K_2 is significantly more complex. For low and high amount of Co K_2 is positive whereas it is negative in the range between 35% and 70%.

The values of both anisotropy constants K_1 and K_2 as well as the magnitude of $-9K_1/4$ and $-9K_1$ which are important for the determination of the different magnetization directions are given in Table 7.3. This allows to determine the easy, medium, and hard magnetization axes as a function of the stoichiometry using Table 7.2 (see Table 7.4).

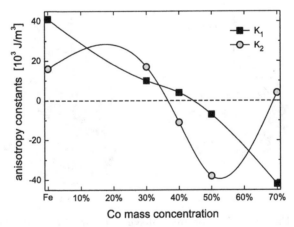

Fig. 7.7. Anisotropy constants K_1 and K_2 at room temperature of an FeCo alloy with different stoichiometry (Data taken from [6])

Table 7.3. K_1 and K_2 in 10^3 J/m³ at $T = 20°$C for an alloy consisting of Fe and Co with given atomic percentage

Co	Fe	K_1	K_2	$-\frac{9}{4}K_1$	$-9K_1$
0	100	42.1	15.0	−95	−379
30	70	10.2	16.2	−23	−92
40	60	4.5	−11.1	−10	−41
50	50	−6.8	−38.7	15	61
70	30	−43.3	5.3	97	390

Table 7.4. Sign (+: positive; −: negative) of the anisotropy constants K_1 and K_2 at $T = 20°$C for an alloy consisting of Fe and Co with given atomic percentage and correlation of the crystallographic directions with the different magnetization axes with regard to Table 7.2

Co	Fe	K_1	K_2	Easy	Medium	Hard
0	100	+	+	100	110	111
30	70	+	+	100	110	111
40	60	+	$\approx -\frac{9}{4}K_1$	100	110/111	111/110
50	50	−	−	111	110	100
70	30	−	$+, < \frac{9}{4}|K_1|$	111	110	100

The comparison with magnetization curves demonstrates the validity of this model (see Fig. 7.8). It is obvious that with increasing amount of Co the easy magnetization axis changes from [100]- to [111]-directions whereas the hard axis behaves conversely. The medium magnetization axis remains along the [110]-direction.

As discussed above the anisotropy constants additionally depend on the temperature which may alter the magnetization axes as shown in Table 7.5 for the example $Fe_{20}Co_{15}Ni_{65}$.

Table 7.5. Temperature dependence of the anisotropy constants K_1 and K_2 in 10^3 J/m³ for the alloy $Fe_{20}Co_{15}Ni_{65}$ and correlation of the crystallographic directions with the different magnetization axes

T	K_1	K_2	Easy	Medium	Hard
22°C	0.91	−11.2	111	100	110
201°C	−0.08	−1.8	111	110	100
398°C	−0.32	0.2	111	110	100

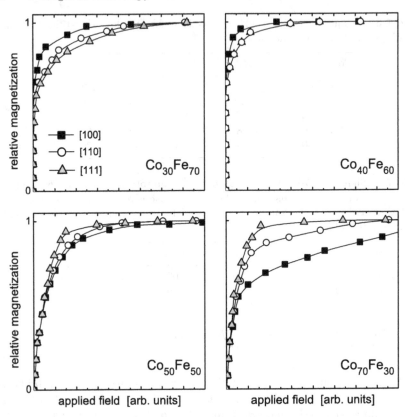

Fig. 7.8. Magnetization curves for an CoFe alloy with different stoichiometries. With increasing amount of Co the easy magnetization axis changes from [100]- to [111]-directions whereas the hard axis behaves conversely. The medium magnetization axis remains along the [110]-direction. (Data taken from [7])

Example for the Influence of the Magneto Crystalline Anisotropy

If an external magnetic field B is applied to an antiferromagnet which is perpendicular to the magnetization of the sublattices a continuous rotation of each sublattice magnetization towards the field direction occurs with increasing magnitude of the external field until saturation is reached (see Fig. 7.9(a)).

The situation becomes more complex if B is oriented parallel to each sublattice magnetization. Then the magnetization of the different sublattices behaves not so simple and significantly depends on the field strength. The sublattices remain parallel for small magnetic fields. A large external field induces a rotation of each sublattice magnetization being subsequently symmetrically rotated related to the direction of the external magnetic field. This change is called "spin-flop transition" (see Fig. 7.9(b)). In the antiferromagnetic case the angles of the sublattice magnetization with respect to the external magnetic field are $\theta = 0$ and $\phi = \pi$. For the spin-flop phase they amount to $\theta = \phi$.

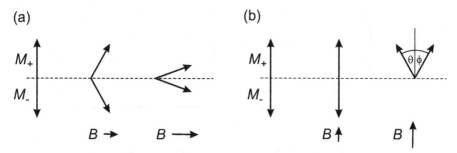

Fig. 7.9. (a) If the external magnetic field is oriented perpendicular to the sublattice magnetization a continuous rotation with increasing field strength occurs towards the field direction. (b) For a parallel alignment of the external field and the sublattice magnetization a critical field strength must be overcome in order to induce a rotation towards the direction of the external field

In the isotropic case the corresponding energy density for this system is given by:

$$E = -MB \cos\theta - MB\cos\phi + AM^2\cos(\theta + \phi) \tag{7.35}$$

with the last term describing the exchange interaction.

Considering the magneto crystalline anisotropy an additional term must be taken into account because the magnetization tries to be aligned along a low-index crystallographic direction:

$$E_{\text{crys}} = -\frac{1}{2}\Delta(\cos^2\theta + \cos^2\phi) \tag{7.36}$$

The energy density of the antiferromagnetic phase amounts to:

$$E = -AM^2 - \Delta \tag{7.37}$$

The spin-flop phase exhibits an energy density of:

$$E = -2MB\cos\theta + AM^2\cos 2\theta - \Delta\cos^2\theta \tag{7.38}$$

The energy minimum can be found by setting the first derivative to zero which results in:

$$\cos\theta = \frac{MB}{2AM^2 - \Delta} \tag{7.39}$$

Inserting into (7.38) leads to

$$E_{\text{spin-flop}} = -AM^2 - \frac{M^2B^2}{2AM^2 - \Delta} \tag{7.40}$$

Thus, below the critical field $B_{\text{spin-flop}}$ which is given by:

$$B_{\text{spin-flop}} = \sqrt{2A\Delta - (\Delta/M)^2} \tag{7.41}$$

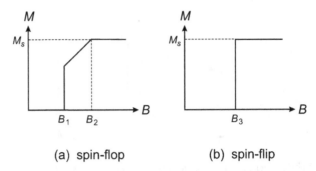

Fig. 7.10. (a) Dependence of the magnetization if a spin-flop transition occurs at the critical field B_1. At B_2 saturation is reached. (b) At a spin-flip transition the magnetization abruptly changes at the magnitude of the external magnetic field B_3 from zero to the saturation value M_S

the antiferromagnetic phase remains energetically favorable whereas above this value the system changes into the spin-flop phase.

Summarizing, the magnetization as a function of the external magnetic field is shown in Fig. 7.10a. If the external magnetic field overcomes the critical magnitude $B_{spin-flop} = B_1$ an abrupt change of the magnetization occurs. A further increase of the magnetic field strength results in a continuous rotation of each sublattice magnetization towards the direction of the external magnetic field. Thus, the total magnetization monotonously increases until saturation M_S is reached at the field strength B_2.

If the anisotropy effect plays a dominant role (i.e. Δ cannot be neglected) a spin-flip transition occurs. This situation is characterized by an abrupt and without a subsequent continuous rotation at the critical field strength B_3 (see Fig. 7.10b). Thus, the ferromagnetic phase is directly reached.

7.3 Shape Anisotropy

Polycrystalline samples without a preferred orientation of the grains do not possess any magneto crystalline anisotropy. But, an overall isotropic behavior concerning the energy being needed to magnetize it along an arbitrary direction is only given for a spherical shape. If the sample is not spherical then one or more specific directions occur which represent easy magnetization axes which are solely caused by the shape. This phenomenon is known as shape anisotropy. In order to get a deeper insight we have to deal with the stray and demagnetizing field of a sample.

The relationship $B = \mu_0(H + M)$ only holds inside an infinite system. A finite sample exhibits poles at its surfaces which leads to a stray field outside the sample. This occurrence of a stray field results in a demagnetizing field inside the sample.

The energy of a sample in its own stray field is given by the stray field energy E_{str}:

$$E_{str} = -\frac{1}{2} \int \mu_0 \boldsymbol{M} \cdot \boldsymbol{H}_{demag} \, dV \qquad (7.42)$$

with $\boldsymbol{H}_{demag}$ being the demagnetizing field inside the sample. The calculation is rather complicated for a general shape. It becomes more easy for symmetric objects which is shown in the following.

An ellipsoid possesses a constant demagnetizing field $\boldsymbol{H}_{demag}$ which is given by:

$$\boldsymbol{H}_{demag} = -\mathcal{N}\boldsymbol{M} \qquad (7.43)$$

with $\mathcal{N}$ being the demagnetizing tensor. Thus, the stray field energy amounts to:

$$E_{str} = 1/2 \cdot \mu_0 \int \boldsymbol{M}\mathcal{N}\boldsymbol{M} \, dV \qquad (7.44)$$

$$= 1/2 \cdot V\mu_0 \boldsymbol{M}\mathcal{N}\boldsymbol{M} \qquad (7.45)$$

with V being the volume of the sample. $\mathcal{N}$ is a diagonal tensor if the semiaxes a, b, and c of the ellipsoid represent the axes of the coordination system. Then, the trace is given by:

$$\mathrm{tr}\mathcal{N} = 1 \qquad (7.46)$$

An arbitrary direction of the magnetization with respect to the semiaxes can be characterized by the direction cosine α_a, α_b, and α_c. The tensor is given by:

$$\mathcal{N} = \begin{pmatrix} N_a & 0 & 0 \\ 0 & N_b & 0 \\ 0 & 0 & N_c \end{pmatrix} \qquad (7.47)$$

and the stray field energy per volume amounts to:

$$E_{str} = 1/2 \cdot \mu_0 M^2 (N_a \alpha_a^2 + N_b \alpha_b^2 + N_c \alpha_c^2) \qquad (7.48)$$

In the case of a sphere the tensor $\mathcal{N}$ amounts to:

$$\mathcal{N} = \begin{pmatrix} 1/3 & 0 & 0 \\ 0 & 1/3 & 0 \\ 0 & 0 & 1/3 \end{pmatrix} \qquad (7.49)$$

and the stray field energy density to:

$$E_{str} = 1/2 \cdot \mu_0 M^2 \cdot 1/3 (\alpha_a^2 + \alpha_b^2 + \alpha_c^2) \qquad (7.50)$$

$$= 1/6 \cdot \mu_0 M^2 \qquad (7.51)$$

due to the normalization condition of the direction cosine (cf. (7.4)). Thus, we find an isotropic behavior because all directions are energetically equivalent. This situation is only valid for a sphere.

In the following θ represents the angle between the magnetization M and the z-axis. For a spheroid the semiminor axes exhibit the same length $a = b$ but they are different to that of the semimajor axis being c. Thus, the diagonal elements of the demagnetizing tensor $\mathcal{N}$ are $N_a = N_b$ and $N_c = 1 - 2N_a$. For the stray field energy density we get:

$$E_{str} = 1/2 \cdot \mu_0 M^2 \left(N_a \sin^2 \theta \cos^2 \phi + N_a \sin^2 \theta \sin^2 \phi \right.$$
$$\left. +(1 - 2N_a) \cos^2 \theta \right) \tag{7.52}$$
$$= 1/2 \cdot \mu_0 M^2 (N_a + (1 - 3N_a) \cos^2 \theta) \tag{7.53}$$

For an infinitely long cylinder, we have $a = b$, $c = \infty$, and

$$\mathcal{N} = \begin{pmatrix} 1/2 & 0 & 0 \\ 0 & 1/2 & 0 \\ 0 & 0 & 0 \end{pmatrix} \tag{7.54}$$

In this case the stray field energy density amounts to:

$$E_{str} = 1/2 \cdot \mu_0 M^2 \cdot 1/2(\sin^2 \theta \cos^2 \phi + \sin^2 \theta \sin^2 \phi) \tag{7.55}$$
$$= 1/4 \cdot \mu_0 M^2 \sin^2 \theta \tag{7.56}$$

For an infinitely expanded and very thin plate, we have $a = b = \infty$, and

$$\mathcal{N} = \begin{pmatrix} 0 & 0 & 0 \\ 0 & 0 & 0 \\ 0 & 0 & 1 \end{pmatrix} \tag{7.57}$$

Now, the stray field energy density amounts to:

$$E_{str} = 1/2 \cdot \mu_0 M^2 \cos^2 \theta \tag{7.58}$$

This result is important for thin magnetic films and multilayers. Equation (7.58) can be written as:

$$E_{str} = K_0 + K_{shape}^V \sin^2 \theta \tag{7.59}$$

with $K_{shape}^V \propto -M^2 < 0$. The stray field energy reaches its minimum value at $\theta = 90°$. This means that the shape anisotropy favors a magnetization direction parallel to the surface, i.e. within the film plane.

A comparison of the anisotropy constants characterizing magneto crystalline and shape anisotropy (see Table 7.6 and cf. Table 7.1) shows that $K_{shape}^V > K_1$. We see that the shape anisotropy dominates the magneto crystalline anisotropy which results in an in-plane magnetization for thin film systems.

Table 7.6. Shape anisotropy constant K_{shape}^V of Fe, Ni, and Co

	bcc-Fe	fcc-Ni	hcp-Co
in [J/m³]	$1.92 \cdot 10^6$	$1.73 \cdot 10^5$	$1.34 \cdot 10^6$
in [eV/atom]	$1.41 \cdot 10^{-4}$	$1.28 \cdot 10^{-5}$	$9.31 \cdot 10^{-5}$

7.4 Induced Magnetic Anisotropy

For magnetic alloys exhibiting a cubic crystal structure a unidirectional magnetic anisotropy can often be achieved by tempering in an external magnetic field.

The prerequisites are a disordered distribution of the atoms in the crystal lattice (see Fig. 7.11(a)) and a high Curie temperature which allows rapid site-exchange processes in the magnetic state.

The external magnetic field orients the magnetization at high temperatures which must be below T_C. During cool-down or rapid thermal quenching the high-temperature state is frozen out under retention of the oriented magnetization direction. A magnetically induced anisotropic directional short-range order is created as schematically shown in Fig. 7.11(c). The binding energy of two neighbored atoms in a spontaneous magnetized crystal is given by:

$$E = a\,\ell \cdot (\cos^2 \phi - 1/3) \tag{7.60}$$

with ϕ being the angle between the magnetization $\boldsymbol{M}$ and the interatomic vector, $\ell = \ell(T;$ type of atoms), and a a constant.

In the following a binary alloy with two type of atoms A and B will be discussed. Three different kinds of bonds occur, between A and A, between B and B, and between A and B. Using (7.60) the energy is given by:

$$E = a \sum_i (N_{AAi}\ell_{AA} + N_{BBi}\ell_{BB} + N_{ABi}\ell_{AB})(\cos^2 \phi_i - 1/3) \tag{7.61}$$

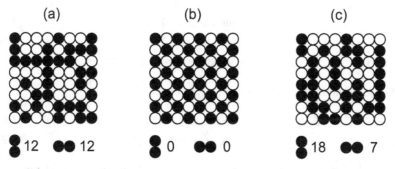

Fig. 7.11. (a) Random distribution, (b) perfect isotropic long range order, (c) anisotropic directional short range order characterized by neighbored atoms of one particular type (*filled circles*)

with

$$i : \text{direction of bond } i$$
$$\phi_i : \text{angle between } \boldsymbol{M} \text{ and the direction of bond } i$$
$$N_{XXi} : \text{number of } XX \text{ bonds in direction of bond } i$$

Using N_X as the number of atoms of type X we get:

$$2N_{AAi} + N_{ABi} = 2N_A \qquad (7.62)$$
$$2N_{BBi} + N_{ABi} = 2N_B \qquad (7.63)$$

which leads to:

$$N_{AAi} = N_{BBi} + \text{const.} \qquad (7.64)$$
$$N_{ABi} = \text{const.} - 2N_{BBi} \qquad (7.65)$$

Inserting into (7.61) yields for the induced magnetic anisotropy energy:

$$E_{\text{ind}} = aV \sum_i N_{BBi} \ell_0 \left(\cos^2 \phi_i - 1/3 \right) + \text{const.} \qquad (7.66)$$

with $\ell_0 = \ell_{AA} + \ell_{BB} - 2\ell_{AB}$. Now, this anisotropy energy is expressed as a function of direction cosine:

$$E_{\text{ind}} \cdot 1/aV = -F(\alpha_1^2 \beta_1^2 + \alpha_2^2 \beta_2^2 + \alpha_3^2 \beta_3^2)$$
$$-G(\alpha_1 \alpha_2 \beta_1 \beta_2 + \alpha_2 \alpha_3 \beta_2 \beta_3 + \alpha_1 \alpha_3 \beta_1 \beta_3) \qquad (7.67)$$

with α_i, β_i being the direction cosine of $\boldsymbol{M}$ during measurement and during annealing in an external magnetic field, respectively, and F and G being material constants. For specific types of crystals one or both of the constants can directly be given:

$$\text{isotropic} : G = 2F \qquad (7.68)$$
$$\text{simple cubic (sc)} : G = 0 \qquad (7.69)$$
$$\text{body-centered cubic (bcc)} : F = 0 \qquad (7.70)$$
$$\text{face-centered cubic (fcc)} : G = 4F \qquad (7.71)$$

For the isotropic system, for example, we get:

$$E_{\text{ind}} \cdot 1/aV = -F(\alpha_1 \beta_1 + \alpha_2 \beta_2 + \alpha_3 \beta_3)^2 \qquad (7.72)$$
$$= -K \cdot \cos^2(\theta - \theta_{\text{ann}}) \qquad (7.73)$$

with $F = K$ being the anisotropy constant describing the induced magnetic anisotropy leading to a uniaxial alignment and $(\theta - \theta_{\text{ann}})$ being the angle between the magnetization during the measurement and during annealing.

The rearrangement of pairs of atoms does not only occur during annealing in an external magnetic field but also by plastic deformation. Therefore, this type of induced anisotropy is called roll-magnetic anisotropy. Its occurrence is schematically explained for the example of an FeNi - alloy.

After a strong cold-rolling which reduces the thickness by about 98% a recrystallization occurs. In this fcc-like alloy a cube texture is present with the (100)-plane in the roll plane and the [100]-direction being parallel to the roll direction. A final cold-rolling subsequently reduces the thickness to about 50% with a remaining cube texture. Such a sheet exhibits a uniaxial magnetic anisotropy with its easy axis in the plane but perpendicular to the rolling direction (see Fig. 7.12). Magnetization parallel to the rolling direction takes exclusively place through domain rotation giving rise to a linear magnetization curve until saturation is reached.

The explanation is given in the following for the example of an A_3B-type superlattice which exhibits a crystal structure shown in Fig. 7.13. During rolling a plastic deformation takes place. One part of the crystal slips relative to another part along a gliding plane which is parallel to a (111)-plane for this fcc-like structure (see Fig. 7.14). One part of the crystal is displaced by one atomic distance which results in the creation of BB-type pairs that are not present in the undisturbed crystal. The distribution of these bonds is anisotropic producing a unidirectional anisotropy.

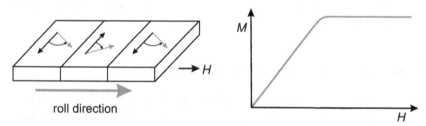

Fig. 7.12. Domain structure (**left**) and corresponding magnetization curve due to the rolling procedure (**right**)

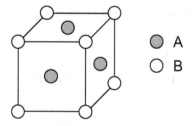

Fig. 7.13. Crystal structure of an A_3B-type superlattice

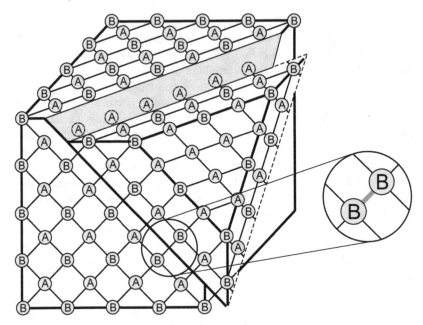

Fig. 7.14. Diagram indicating the appearance of BB pairs due to a single step slip along the (111)-plane (*gray shaded area*) in an A_3B-type superlattice

A further induced anisotropy is given by the exchange anisotropy. This type is discussed on p. 233 in more detail.

7.5 Stress Anisotropy (Magnetostriction)

Up to now we have assumed a rigid solid with a fixed lattice constant, i.e. the distance between atoms in the crystal was held constant. In the following an elastic degree of freedom is additionally allowed. This displacement influences the magnetic behavior of magnetic crystals. Vice versa, magnetic properties can alter elastic properties. The responsibility of this interplay is the magneto elastic interaction.

The magneto crystalline anisotropy energy per volume of cubic systems was given by (cf. (7.19)):

$$E_{\text{crys}} = K_1^{\text{cubic}} \cdot \mathcal{O}(\alpha^4) \tag{7.74}$$

whereas that of systems with a tetragonal crystal structure due to the reduced symmetry amounts to (cf. (7.25)):

$$E_{\text{crys}} = K_1^{\text{tetra}} \cdot \mathcal{O}(\alpha^2) \tag{7.75}$$

$$= K_1^{\text{tetra}} \sin^2 \theta \tag{7.76}$$

$$= K_1^{\text{tetra}}(1 - \alpha_3^2) \tag{7.77}$$

disregarding the constant part in each case. Because K_1^{tetra} belongs to a term of second order and K_1^{cubic} to a term of forth order it is possible roughly to estimate $K_1^{\text{tetra}} \approx 100 K_1^{\text{cubic}}$. We directly recognize that a small deformation, being characterized by the deformation parameter ε, which induces the transition from cubic to tetragonal symmetry is responsible for a distinct change of the anisotropy.

As a consequence the cubic unit cell spontaneously deforms to a tetragonal system below the Curie temperature due to a decrease of the energy E_{crys}. The distortion is limited due to a simultaneous increase of the elastic energy density $E_{\text{el}} = 1/2 \cdot C\varepsilon^2$. Thus, the total anisotropy energy density amounts to:

$$E = K_1(1 - \alpha_3^2)\varepsilon + 1/2 \cdot C\varepsilon^2 \tag{7.78}$$

Calculating the minimum energy allows to determine the relative change in length λ:

$$\lambda = \frac{\delta\ell}{\ell} = -\frac{K_1}{C}(1 - \alpha_3^2) \tag{7.79}$$

This spontaneous magnetostriction reaches typical values of $\lambda \approx 10^{-5}$.

In the following we will discuss magnetostriction using a phenomenological description by means of the constant λ. It is important that λ depends on the crystallographic direction. This means that the deformation parameter ε must be replaced by the deformation tensor $\mathcal{E}$ if the magneto crystalline anisotropy cannot be neglected.

The elastic energy density E_{el} can be characterized using the elements of the deformation tensor ε_{ij} and the elastic constants c_{ij}. For cubic systems we obtain:

$$E_{\text{el}}^{\text{cubic}} = \frac{1}{2}c_{11}(\varepsilon_{11}^2 + \varepsilon_{22}^2 + \varepsilon_{33}^2)$$
$$+ c_{12}(\varepsilon_{11}\varepsilon_{22} + \varepsilon_{11}\varepsilon_{33} + \varepsilon_{22}\varepsilon_{33}) + \frac{1}{2}c_{44}(\varepsilon_{12}^2 + \varepsilon_{13}^2 + \varepsilon_{23}^2) \tag{7.80}$$

whereas for systems with a hexagonal symmetry we get:

$$E_{\text{el}}^{\text{hex}} = \frac{1}{2}c_{11}(\varepsilon_{11}^2 + \varepsilon_{22}^2) + \frac{1}{2}c_{33}\varepsilon_{33}^2 + c_{12}\varepsilon_{11}\varepsilon_{22}$$
$$+ c_{13}(\varepsilon_{11} + \varepsilon_{22})\varepsilon_{33} + \frac{1}{2}c_{44}(\varepsilon_{13}^2 + \varepsilon_{23}^2) + (c_{11} - c_{12})\varepsilon_{12}^2 \tag{7.81}$$

The elastic constants of Fe, Ni, and Co are listed in Table 7.7.

If the direction of the magnetostriction in the measurement is characterized by the direction cosine β_i then we find for cubic crystals:

$$\lambda = \frac{\delta\ell}{\ell} = \frac{3}{2} \cdot \lambda_{100} \left(\alpha_1^2\beta_1^2 + \alpha_2^2\beta_2^2 + \alpha_3^2\beta_3^2 - \frac{1}{3} \right)$$
$$+ 3\lambda_{111}(\alpha_1\alpha_2\beta_1\beta_2 + \alpha_1\alpha_3\beta_1\beta_3 + \alpha_2\alpha_3\beta_2\beta_3) \tag{7.82}$$

Table 7.7. Elastic constants c_{ij} in 10^{11} N/m^2 of Fe, Ni, and Co at room temperature

	bcc-Fe	fcc-Ni	hcp-Co
c_{11}	2.41	2.50	3.07
c_{12}	1.46	1.60	1.65
c_{13}			1.03
c_{33}			3.58
c_{44}	1.12	1.18	0.76

with λ_{100} and λ_{111} being the saturation values of the magnetostriction for the arrangement that the direction of the measurement and of the spontaneous magnetization are along the [100]- or [111]-direction, respectively. The correlation between magnetostriction, elastic, and magneto elastic constants are:

$$\lambda_{100} = -\frac{2}{3} \cdot \frac{B_1}{c_{11} - c_{12}} \tag{7.83}$$

$$\lambda_{111} = -\frac{1}{3} \cdot \frac{B_2}{c_{44}} \tag{7.84}$$

The magnitude of the magneto elastic constants are given in Table 7.8 and that of the magneto elastic constants in Table 7.9. The magnetostriction in the [110]-direction λ_{110} is not independent on λ_{100} and λ_{111} and can be expressed as:

$$\lambda_{110} = \frac{1}{4}\lambda_{100} + \frac{3}{4}\lambda_{111} \tag{7.85}$$

Assuming an isotropic magnetostriction, i.e. $\lambda_{100} = \lambda_{111}$, we directly obtain $\lambda_0 = \lambda_{100} = \lambda_{111} = \lambda_{110}$. In this situation (7.82) can be written as:

$$\lambda = \lambda_0 \left(\cos^2 \theta - \frac{1}{3} \right) \tag{7.86}$$

with $\theta = \arccos \sum_i \alpha_i \beta_i$ being the angle between the spontaneous magnetization and the direction the magnetostriction is measured in. Due to the

Table 7.8. Magnitude of the magneto elastic constants B_1, B_2, B_3, and B_4 of Fe, Ni, and Co at room temperature

		bcc-Fe	fcc-Ni	hcp-Co
B_1	[J/m^3]	$-3.44 \cdot 10^6$	$8.87 \cdot 10^6$	$-8.10 \cdot 10^6$
	[eV/atom]	$-2.53 \cdot 10^{-4}$	$6.05 \cdot 10^{-4}$	$-5.63 \cdot 10^{-4}$
B_2	[J/m^3]	$-7.62 \cdot 10^6$	$1.02 \cdot 10^7$	$-2.90 \cdot 10^7$
	[eV/atom]	$-5.56 \cdot 10^{-4}$	$6.97 \cdot 10^{-4}$	$-2.02 \cdot 10^{-3}$
B_3	[J/m^3]			$-2.82 \cdot 10^7$
	[eV/atom]			$-1.96 \cdot 10^{-3}$
B_4	[J/m^3]			$-2.94 \cdot 10^7$
	[eV/atom]			$-2.05 \cdot 10^{-3}$

Table 7.9. Magnitude of the magnetostriction constants of Fe, Ni, and Co at room temperature

	bcc-Fe	fcc-Ni	hcp-Co
λ_{100}	$24 \cdot 10^{-6}$	$-66 \cdot 10^{-6}$	
λ_{111}	$-23 \cdot 10^{-6}$	$-29 \cdot 10^{-6}$	
λ_A			$-50 \cdot 10^{-6}$
λ_B			$-107 \cdot 10^{-6}$
λ_C			$126 \cdot 10^{-6}$
λ_D			$-105 \cdot 10^{-6}$

isotropic behavior of the magnetostriction a dependence on crystallographic directions is no more given.

The magnetostriction for hexagonal systems is given by:

$$\lambda = \frac{\delta\ell}{\ell} = \lambda_A \left((\alpha_1\beta_1 + \alpha_2\beta_2)^2 - (\alpha_1\beta_1 + \alpha_2\beta_2)\alpha_3\beta_3\right)$$
$$+ \lambda_B \left((1 - \alpha_3^2)(1 - \beta_3^2) - (\alpha_1\beta_1 + \alpha_2\beta_2)^2\right)$$
$$+ \lambda_C \left((1 - \alpha_3^2)\beta_3^2 - (\alpha_1\beta_1 + \alpha_2\beta_2)\alpha_3\beta_3\right)$$
$$+ \lambda_D (\alpha_1\beta_1 + \alpha_2\beta_2)\,\alpha_3\beta_3 \tag{7.87}$$

with

$$\lambda_A + \lambda_B = \frac{2B_2 c_{13} - (2B_3 + B_1)c_{33}}{c_{33}(c_{11} + c_{12}) - 2c_{13}^2} \tag{7.88}$$

$$\lambda_A - \lambda_B = \frac{B_1}{c_{11} - c_{22}} \tag{7.89}$$

$$\lambda_C = \frac{B_2(c_{11} + c_{12}) - (2B_3 + B_1)c_{13}}{c_{33}(c_{11} + c_{12}) - 2c_{13}^2} \tag{7.90}$$

$$4\lambda_D - (\lambda_A + \lambda_B + \lambda_C) = \frac{B_4}{c_{44}} \tag{7.91}$$

Using the magnetostriction constants given in Table 7.9 we want to discuss the magnetostrictive properties of pure Fe and Ni as well as of the alloy consisting of Fe and Ni. Generally, a positive magnitude of λ means a dilatation and a negative one a contraction.

In the case of Ni we find that λ_{100} and λ_{111} are negative. Using (7.85) we see that λ_{110} is negative, too. Therefore, Ni contracts along all three directions if it is becomes magnetic.

For Fe λ_{100} is positive whereas λ_{111} is negative. This results in a rather complex behavior. Along the easy magnetization direction being [100] we find a simple dilatation which leads to a distortion from a cubic to a tetragonal crystal lattice.

The alloy consisting of Fe and Ni (known as Invar or Permalloy) exhibits magnetostrictive properties which significantly depend on the stoichiometry

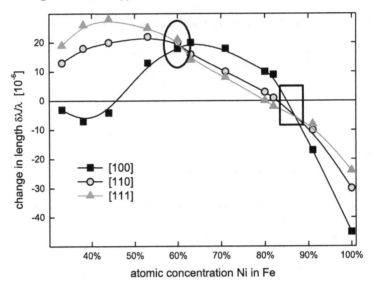

Fig. 7.15. Magnetostriction in single crystals of an FeNi alloy concerning different crystallographic directions (Data taken from [2])

(see Fig. 7.15). For $Fe_{40}Ni_{60}$ (see ellipse) one can see that $\lambda_{100} = \lambda_{111} = \lambda_{110}$. Thus, this particular alloy represents a cubic material with isotropic magnetostriction. The same behavior is found for $Fe_{15}Ni_{85}$ (see box); the measurements yield that $\lambda_{100} = \lambda_{111} = \lambda_{110} \approx 0$ for this composition. Additionally, the magnetostriction (nearly) vanishes. Due to that reason this specific $Fe_{15}Ni_{85}$ alloy is of technological importance.

7.6 Magnetic Surface and Interface Anisotropies

The considerations above were carried out for volume systems neglecting interfaces and surfaces. In the following low-dimensional systems are discussed concerning the anisotropy which is related to these interfaces.

Due to the broken symmetry at interfaces the anisotropy energy contains terms with lower order in α which are forbidden for three-dimensional systems. Therefore, each effective anisotropy constant K^{eff} is divided into two parts, one describing the volume and one the surface contribution:

$$K^{\mathrm{eff}} = K^{\mathrm{V}} + 2K^{\mathrm{S}}/d \qquad (7.92)$$

with K^{V} being the volume dependent magneto crystalline anisotropy constant and K^{S} the surface dependent magneto crystalline anisotropy constant. The factor of two is due to the creation of two surfaces. The second term exhibits an inverse dependence on the thickness d of the system. Thus, it is only important for thin films.

In order to illustrate the influence of the surface anisotropy we will discuss the so-called "spin reorientation transition". Rewriting (7.92) results in:

$$d \cdot K^{\text{eff}} = d \cdot K^{\text{V}} + 2K^{\text{S}} \tag{7.93}$$

Plotting this dependence as a $d \cdot K^{\text{eff}}(d)$ diagram allows to determine K^{V} as the slope of the resulting line and $2K^{\text{S}}$ as the zero-crossing which is exemplarily shown for a thin Co layer with variable thickness d on a Pd substrate (see Fig. 7.16). Due to the shape anisotropy K^{V} is negative. This can directly be seen by the negative slope which results in an in-plane magnetization. The zero-crossing occurs at a positive value K^{S}. This leads to a critical thickness d_c:

$$d_c = -\frac{2K^{\text{S}}}{K^{\text{V}}} \tag{7.94}$$

with

$$d < d_c : \text{perpendicular magnetization} \tag{7.95}$$
$$d > d_c : \text{in-plane magnetization} \tag{7.96}$$

due to the change of sign of K^{eff}. Thus, the volume contribution always dominates for thick films with a magnetization being within the film plane. The relative amount of the surface contribution increases with decreasing thickness followed by a spin reorientation transition towards the surface normal below d_c. This behavior is exemplarily illustrated in Fig. 7.17 representing hysteresis loops of a Au/Co(0001)/Au(111) system for different thicknesses of the ferromagnetic Co layer measured perpendicular (left part) and parallel

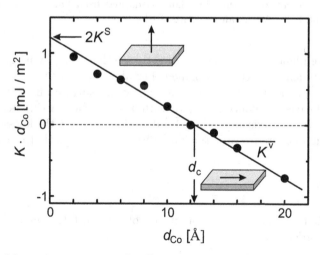

Fig. 7.16. Magnetic anisotropy of a Co thin film layer in a Co/Pd multilayer as a function of the Co thickness d_{Co}. The slope allows to determine K^{V}. The zero crossing amounts to $2K^{\text{S}}$. (Data taken from [8])

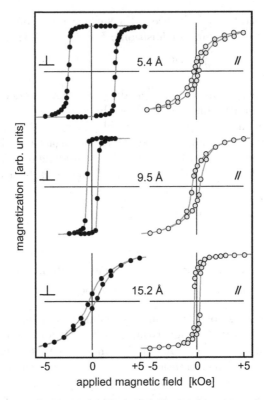

Fig. 7.17. Hysteresis loops of Au/Co(0001)/Au(111) systems for different thicknesses of the Co layer. **Left**: Measurement perpendicular to the film plane. **Right**: Measurement along the film plane. (Adapted from [9]; used with permission. Copyright 1988, American Institute of Physics)

to the surface plane (right part). We directly see that the easy magnetization axis is perpendicular to the film plane below about $d_c = 12$Å.

A further phenomenon which is caused by the surface anisotropy is the rotation of the easy magnetization axis within the surface. A thin bcc-Fe(110) film exhibits an easy axis along the $[1\bar{1}0]$-direction. Above a thickness of about 60 Å a rotation occurs towards the [001]-direction which is that of the bulk.

Problems

7.1. Show that the magneto crystalline anisotropy for tetragonal systems up to the forth order in α:

$$E_{\text{crys}}^{\text{tetra}} = K_0 + K_1\alpha_3^2 + K_2\alpha_3^4 + K_3(\alpha_1^4 + \alpha_2^4)$$

(see (7.25)) can be expressed as:

$$E_{\text{crys}}^{\text{tetra}} = K_0' + K_1' \sin^2 \theta + K_2' \sin^4 \theta + K_3' \sin^4 \theta \cos 4\phi$$

(see (7.27)) using the angles θ and ϕ (see (7.1)–(7.3)) by replacing the direction cosine α_i. Determine K_i' as a function of K_i.

7.2. Show that the magneto crystalline anisotropy for hexagonal systems up to the sixth order in α:

$$E_{\text{crys}}^{\text{hex}} = K_0 + K_1(\alpha_1^2 + \alpha_2^2) + K_2(\alpha_1^2 + \alpha_2^2)^2$$
$$+ K_3(\alpha_1^2 + \alpha_2^2)^3 + K_4(\alpha_1^2 - \alpha_2^2)(\alpha_1^4 - 14\alpha_1^2\alpha_2^2 + \alpha_2^4)$$

(see (7.28)) can be expressed as:

$$E_{\text{crys}}^{\text{hex}} = K_0 + K_1 \sin^2 \theta + K_2 \sin^4 \theta + K_3 \sin^6 \theta$$
$$+ K_4 \sin^6 \theta \cos 6\phi$$

(see (7.29)) using the angles θ and ϕ (see (7.1)–(7.3)) by replacing the cosine α_i.

7.3. Spin flop transition

(a) Show that below the critical field $B_{\text{spin-flop}}$ which is given by:

$$B_{\text{spin-flop}} = \sqrt{2A\Delta - (\Delta/M)^2}$$

(see (7.41)) the antiferromagnetic phase remains energetically favorable whereas above this value the system changes into the spin-flop phase.
(b) Show this behavior schematically by plotting the energy as a function of the external magnetic field $\boldsymbol{B}$.

7.4. After cooling a singly crystalline alloy exhibiting a body-centered cubic symmetry in an external magnetic field being applied along the [001], [110], or [111] direction the induced magnetic anisotropy is measured for a magnetization being within the $(1\bar{1}0)$ plane. Determine the ratio of the induced magnetic anisotropy constants for the three situations.

Magnetic Domain Structures

Ferromagnetic materials possess uniformly magnetized regions which exhibit a parallel orientation of all magnetic moments within this so-called magnetic domain on the one hand but on the other hand different directions of the magnetization in different domains. Thus, a demagnetized sample consists of domains each ferromagnetically ordered with a vanishing total magnetization. The boundary between neighbored domains are domain walls.

In this chapter general considerations on the behavior of magnetic domains and domain walls are made for macroscopic systems exhibiting a lower length scale of the order of microns. Properties concerning magnetic domains of low-dimensional systems will be discussed in Chaps. 12–15.

8.1 Magnetic Domains

In the year 1907 it was stated by P. Weiss that a ferromagnet possesses a number of small regions ("magnetic" domains). Each of them exhibits the saturation magnetization. It is important that the magnetization direction of single domains each along the easy axis are not necessarily parallel. Domains are separated by domain boundaries or domain walls.

These considerations allow to describe a lot of properties of different magnetic systems. Two examples are:

- In soft magnetic materials smallest external fields ($\approx 10^{-6}$ T) are sufficient to reach saturation magnetization ($\mu_0 M \approx 1$ T). The external field need not order all magnetic moments macroscopically (because in each domain they are already ordered) but has to align the domains. Thus, a movement of domain walls only occurs which requires low energy.
- It is possible that ferromagnetic materials exhibit a vanishing total magnetization $M = 0$ below the critical temperature without applying an external field. In this situation each domain still possesses a saturated magnetization but due to the different orientations the total magnetization amounts to zero.

The magnetization process consists of two independent steps:

- In small external magnetic fields
  The volume of regions which exhibit a magnetization direction being nearly the same as that of the external field grows at the expense of unfavorably oriented domains. The boundary movements can be reversible as well as irreversible.
- In strong external magnetic fields
  A rotation of M occurs towards the direction of the external magnetic field H.

Magnetization of an Ideal Crystal

In an ideal crystal which does not contain any defects the magnetization of a demagnetized ferromagnet starts with reversible domain wall movements. This process ends when all domain walls are annihilated or all walls are oriented perpendicularly to the external magnetic field. If this is the only process until saturation is reached we obtain a magnetization as a function of the external magnetic field which is shown in Fig. 8.1 labelled by "1". Subsequently, reversible rotation processes can occur. The magnetization curve which is obtained for pure rotation processes is shown in Fig. 8.1 labelled by "2". The combination of both processes results in a magnetization curve like "3" in Fig. 8.1.

Magnetization of a Real Crystal

A real magnetic system additionally possesses crystal defects which results in a domain wall potential. For small external magnetic fields reversible domain wall movements occur (Rayleigh regime). Higher fields (Barkhausen regime) lead to large jumps of the domain walls due to pinning at crystallographic

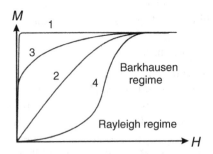

Fig. 8.1. *Curve 1*: Pure domain wall movements in an ideal crystal. *Curve 2*: Pure rotation processes in an ideal crystal. *Curve 3*: Combination of domain wall movements and rotation processes in an ideal crystal. *Curve 4*: Behavior of a real crystal

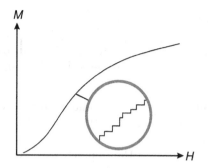

Fig. 8.2. The enlargement enables to observe the Barkhausen jumps

defects. These so-called Barkhausen jumps are directly observable in the hysteresis loop (see Fig. 8.2).

An important relationship is given by the dependence of the magnetization M as a function of an external magnetic field H. The graphic representation is known as the hysteresis curve or hysteresis loop (see Fig. 8.3) with M_S being the saturation magnetization, M_r the remanence, H_s the saturation field, and H_c the coercive field or coercivity. These values can be determined using the hysteresis loop by:

$$M_S = \max{(M)} \tag{8.1}$$
$$M_r = M(H = 0) \tag{8.2}$$
$$H_s = \min{(H'|M(H') = M_S)} \tag{8.3}$$
$$H_c = H(M = 0) \tag{8.4}$$

The area surrounded by the hysteresis loop is a direct measure of the magnetic hysteresis energy which has to be applied in order to reverse the magnetization.

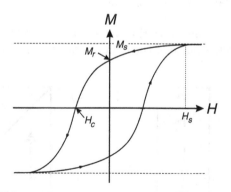

Fig. 8.3. Hysteresis loop schematically shown

8.2 Domain Walls

A classification of domain walls can be given by the angle of the magnetization between two neighbored domains with the wall as boundary.

- 180° wall
  A 180° domain wall represents the boundary between two domains with opposite magnetization (see Fig. 8.4(a)).
- 90° wall
  A 90° domain wall represents the boundary between two domains with magnetization being perpendicular to each other (see Fig. 8.4(b)).

The perpendicular direction of a domain wall corresponds to the bisecting line between the directions of the magnetization of neighbored domains. The occurrence of different types of domain walls depends on the crystallographic arrangement of the ferromagnet:

- Uniaxial ferromagnet
  An example of this type is given by Co which only exhibits 180° domain walls (see Fig. 8.5).
- Triaxial material
  The most prominent example is bcc-Fe with its easy magnetization axis along (100)-like directions. This material possesses 180° as well as 90° domain walls.
- Materials with four axes
  One example is given by fcc-Ni with an easy axis along (111)-like directions. The domain walls exhibit angles of 180°, 109°, and 71°.

A closer inspection of 180° domain walls reveals that they can be divided into two classes:

- Bloch wall
  The rotation of the magnetization occurs in a plane being parallel to the plane of the domain wall (see Fig. 8.6a).
- Néel wall
  The rotation of the magnetization vector takes place in a plane which is perpendicular to the plane of the domain wall (see Fig. 8.6b).

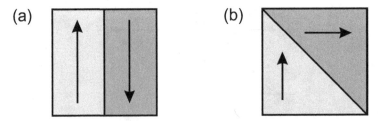

Fig. 8.4. (a) 180° and (b) 90° domain wall

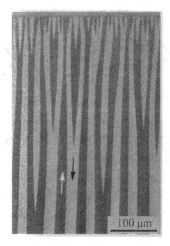

Fig. 8.5. Domains on the side plane of a Co crystal (From [10] (used with permission))

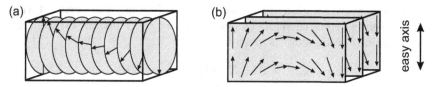

Fig. 8.6. Rotation of the magnetization in a (**a**) Bloch wall and (**b**) Néel wall

The Bloch wall assures a vanishing stray field. The Néel wall often occurs in ferromagnetic thin film systems with an in-plane magnetization. The magnetization remains within the film plane which is energetically favorable compared to the situation that the magnetization vector has to be perpendicular to the film plane (see Fig. 8.7) in order to reduce magnetic stray fields.

8.3 Domain Wall Width

The energy of two parallel spins is given by:

$$E = -2JS^2 \tag{8.5}$$

For a non-parallel arrangement of both spins $\boldsymbol{S}_1$ and $\boldsymbol{S}_2$, i.e. they exhibit an angle of $\varphi \neq 0$ with respect to each other, the energy amounts to:

$$E = -2J\boldsymbol{S}_1 \cdot \boldsymbol{S}_2 \tag{8.6}$$
$$= -2JS^2 \cos\varphi \tag{8.7}$$

which can be approximated for small angles φ by:

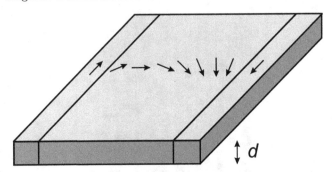

Fig. 8.7. The Néel wall is energetically favorable in thin film systems exhibiting only a small thickness d with an in-plane magnetization due to the avoidance of stray fields

$$E = -2JS^2 + JS^2\varphi^2 \tag{8.8}$$

using $\cos\varphi = 1 - \varphi^2/2$.

In a Bloch wall a rotation of 180° occurs across N spins. The mean rotation angle between two spins therefore amounts to $\varphi = \pi/N$. Thus, the total energy of every "spin rotation axis" is given by:

$$E = NJS^2\varphi^2 \tag{8.9}$$
$$= \pi^2 JS^2/N \tag{8.10}$$

If a is the lattice constant a Bloch wall exhibits $1/a^2$ spin rotation axes per m^2 leading to an exchange energy of:

$$E_{ex}^{BW} = JS^2 \cdot \frac{\pi^2}{Na^2} \tag{8.11}$$

Consequently, a domain wall tends to an infinite length ($N \to \infty$) due to a non-vanishing energy of twisted spins. But, the magneto crystalline anisotropy favors a short length. Thus, both energy contributions try to move the domain wall width into the opposite direction. The magneto crystalline anisotropy energy (cf. (7.34))

$$E_{crys} = K \sin^2\varphi \tag{8.12}$$

with $K > 0$ thus favors a parallel or antiparallel alignment of the spins in the two domains with the Bloch wall as boundary. The summation over all contributions of N spins can be approximated by

$$\sum_{i=1}^{N} K\sin^2\varphi_i \approx \frac{1}{2}NK \tag{8.13}$$

Assuming a planes we obtain

$$E^{BW}_{crys} = \frac{1}{2}aNK \tag{8.14}$$

Thus, the total energy of the Bloch wall is given by:

$$E^{BW} = JS^2\frac{\pi^2}{a^2}\cdot\frac{1}{N} + \frac{1}{2}aKN \tag{8.15}$$

The first term favors a large number N with spins involved in the domain wall whereas the second term favors a small number. The energy minimum can be determined by setting the first derivative to zero:

$$0 = \frac{dE}{dN} = \frac{1}{2}aK - JS^2\frac{\pi^2}{a^2}\cdot\frac{1}{N^2} \tag{8.16}$$

which results in:

$$N = \pi S\sqrt{\frac{2J}{a^3K}} \tag{8.17}$$

The constant

$$A = \frac{2JS^2}{a} \tag{8.18}$$

represents a measure of the stiffness of the magnetization vector against twisting as a consequence of the exchange forces which favor a parallel alignment. The domain wall width δ is defined as:

$$\delta = Na = \pi\sqrt{A/K} \tag{8.19}$$

We directly recognize that a high value of the stiffness favors a large domain wall width whereas a large magnitude of the anisotropy constant tries to reduce the width. For Fe we find a domain wall width of $\delta \approx 40\,\text{nm}$.

Domain walls exhibit a continuous rotation of the magnetization vector between two domains. Therefore, this definition of a domain wall width is

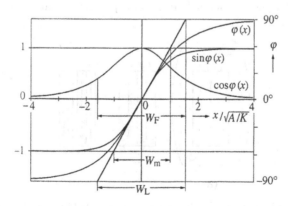

Fig. 8.8. Calculated wall profile of a 180° Bloch wall and different wall width definitions being given in the text (From [10] (used with permission))

unfortunately not uniform. The definition given by (8.19) is graphically represented by W_L (Lilley's domain wall width) in Fig. 8.8. A second definition deals with the slope of the magnetization component $\sin \varphi$ in the origin and leads to a width of $2\sqrt{A/K}$ represented by W_m in Fig. 8.8. A third definition of a domain wall width bases on the total wall flux and leads to $\int_{-\infty}^{\infty} \cos \varphi(x) \mathrm{d}x$ represented by W_F in Fig. 8.8.

8.4 Closure Domains

In a three-dimensional magnetic crystal we find magnetic domains and Bloch walls which exhibits no stray fields. The question arises what happens at the surface due to the reduction of the symmetry. Generally, the system tends to minimize its energy which is given by the sum of the Bloch wall energy E^{BW} and the stray field energy E_{str} concerning the magnetic part of the total energy. Therefore, the domain pattern becomes different at the surface compared to the bulk. These different and mostly additional domains are called closure domains.

Let us start with a sample which exhibits one single domain as the initial state (see Fig. 8.9(a)). It is obvious that $E^{\mathrm{BW}} = 0$ because no Bloch wall is present but E_{str} becomes rather large. In order to lower the energy Bloch walls are created (see Fig. 8.9(b) and (c)). The appearance and number of additional domains significantly depend on the anisotropy of the crystal.

Crystals Exhibiting a Strong Uniaxial Anisotropy

This type is often found for crystals with an hcp structure like Co. Only a small number of domain walls occurs within the crystal. Near the surface additional domains are created for a further reduction of the stray field as schematically shown in Fig. 8.9(c). This leads to a branching of the different domains. Nearby the surface a fine domain pattern is enforced to minimize

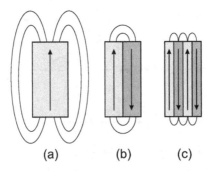

(a)   (b)   (c)

Fig. 8.9. The creation of Bloch walls reduces the stray field energy E_{str} but increases the wall energy E^{BW}

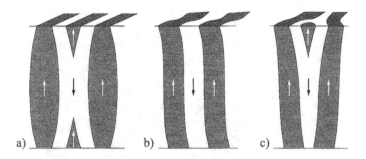

Fig. 8.10. Domain refinement and branching in two phase systems. (**a**) Planar two-dimensional branching. (**b, c**) Three-dimensional branching. (From [10] (used with permission))

the domain closure energy but in the bulk a wide pattern is favored to save domain wall energy. The branching process connects the wide and the narrow domains in a way that depends on crystal symmetry and in particular on the number of available easy directions. A distinction can be done between

- a two dimensional branching which can completely be described in a cross section drawing (see Fig. 8.10(a))
- a three dimensional branching which exploits geometrically the third dimension (see Fig. 8.10(b,c))

Further, we can distinguish between

- two-phase branching which has to achieve the domain refinement with two magnetization phases only (see Fig. 8.10)
- multi-phase branching which can use more than two magnetization directions in the branching process.

The branching phenomenon can be explained by the progressive domain refinement towards the surface by iterated generations of domains as schematically depicted in Fig. 8.11.

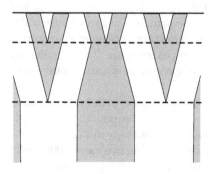

Fig. 8.11. Sketch of the iterated generation of domains towards the surface

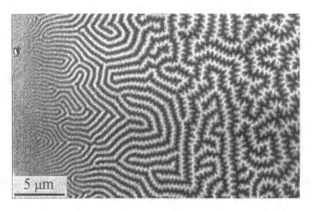

Fig. 8.12. Domain pattern on the basal plane of a wedge-shaped Co crystal-like thin film which exhibits a uniaxial anisotropy at room temperature. The easy axis is perpendicular to the surface. With increasing thickness from the **left** to the **right** the degree of branching increases. (From [10] (used with permission))

A cross section of the example Co was already shown in Fig. 8.5. The top view is given in Fig. 8.12 The image shows the domain distribution at the surface of a wedge-shaped Co thin film. On the left side a thin film is present which is too thin in order to create additional domains near the surface. The thicker film on the right side allows a dendritic occurrence of the domains which are due to the branching process.

Crystals Exhibiting a Weak Uniaxial Anisotropy

In order to avoid stray fields these systems tend to create domain walls and sometimes regions with a magnetization which are not aligned along an easy direction as schematically depicted in Fig. 8.13.

Crystals Exhibiting a Non-Uniaxial Anisotropy

The domain arrangement in bulk cubic crystals as multiaxial materials is primarily determined by the principle of flux closure which is due to stray field minimization. Almost as important is magnetic anisotropy in these materials. Most details of the domain patterns are therefore determined by the surface orientation relative to the easy magnetization directions. Several cases must be distinguished.

From the simplest case, a surface with two easy axes, to strongly misoriented surfaces with no easy axes the domain patterns become progressively more complicated. In positive anisotropy materials, i.e. $K_1 > 0$, such as Fe (see Table 7.1) the (100)-surface contains two easy [100]-directions (see Table 7.2). The (110)-surface of a negative anisotropy material, i.e. $K_1 < 0$, such as Ni (see Table 7.1) is analogous to the (100)-surface of Fe because it exhibits two

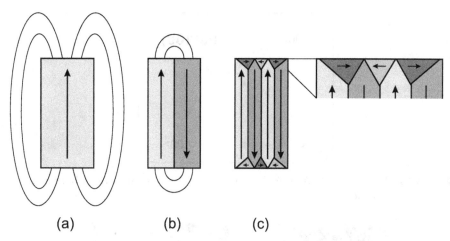

Fig. 8.13. Generation of domain walls (**b**) and closure domains (**c**) which exhibit a magnetization that is not parallel to an easy magnetization axis in order to reduce stray field energy

easy [111]-axes (see Table 7.2). Surfaces with only one easy magnetization axis are, for example, the (110)-surface for a positive anisotropy constant $K_1 > 0$ and the (112)-surface for $K_1 < 0$. The (111)-surface contains no easy axis for positive anisotropy. The same is true for (100)- and (111)-surfaces of materials with negative anisotropy. Certainly, domains in Ni-like materials are not the same as domains in Fe-like materials even if they are investigated on equivalent surfaces. They differ in the details of the allowed domain wall angles because the easy axes in Fe are all mutually perpendicular whereas this is not true for Ni-like substances (see p. 120). Domains in the two symmetry classes are highly analogous, however, so that we need not discuss them separately in each example.

- Two easy axes in the surface plane
  As already mentioned above this situation occurs for the Fe(100)-surface with two (100)-like directions in the bcc-(100) surface (see Fig. 8.14). For this magnetic crystallographic system the domain patterns which contains domains being separated by 180° and 90° domain walls are easy to interpret (see Fig. 8.15(a)). At every point of a sample usually one of the crystallographic easy axes is slightly preferred to the others due to small residual stress and possibly induced anisotropies. Figure 8.15(b) shows a pattern which was obtained at the same area as the pattern in Fig. 8.15(a) after another demagnetizing treatment. The details, in particular the magnetization directions, differ from those in Fig. 8.15(a) but the locally preferred axes agree in both patterns.
- One easy axis in the surface plane
  One example is given by Fe(110) with only one (100)-like axis within the bcc-(110) surface (see Fig. 8.16). The basic domain structure is rather

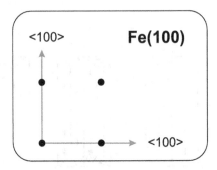

Fig. 8.14. The bcc-Fe(100) surface exhibits two (100)-like directions both representing easy magnetization axes

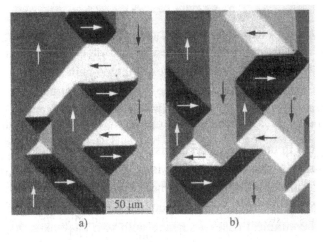

Fig. 8.15. Domains on a (100) surface of silicon-stabilized Fe. (**a**) The domains are all magnetized parallel to the surface yielding a particular clear pattern. (**b**) An alternate pattern from the same area after another demagnetization. (From [10] (used with permission))

simple. It consists of domains being magnetized parallel and antiparallel to the easy [100]-direction (see Fig. 8.17).

- Slightly misoriented surface
  This type of surface is characterized by an angle between the surface plane and the closest easy magnetization direction of about $\theta < 5°$. In this situation the basic domains correspond to those of the ideally oriented crystal. The different patterns are characterized by the introduction of shallow surface domains collecting the net flux which would otherwise emerge from the surface. Thus, the effective domain width at the surface is reduced. The flux is transported to a suitable surface of opposite polarity of the magnetic charges and distributed again. Because this system of compensating domains is superimposed on whatever basic domains would

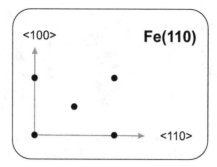

Fig. 8.16. The bcc-Fe(110) surface exhibits one (100)-like direction which represents an easy magnetization axis and a (110)-like direction being a hard magnetization axis

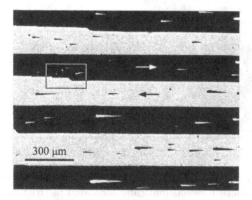

Fig. 8.17. Domains on a largely undisturbed (110)-oriented Fe crystal which contains small amounts of Si. The isolated lancets as well as the short kinks in the main walls are connected with internal transverse domains. (From [10] (used with permission))

be present without the misorientation these domains are known as supplementary domains. In the examples given below the magnetization is always assumed to follow strictly the easy magnetization directions.

– Two easy axes in the singular surface plane
  Typical additional domains belonging to a misoriented (100)-surface are shown in Fig. 8.18. There are two types of these so-called fir tree pattern. One is associated with 180° domain walls whereas the other one is connected with 90° domain walls. Both orientations of fir tree pattern may coexist depending on the overall basic domain pattern as demonstrated in Fig. 8.19.

– One easy axis in the singular surface plane
  All surface domains are magnetized along the only available easy axis but internally the additional easy directions due to the misorientation help in the flux distribution. In this so-called lancet pattern the flux

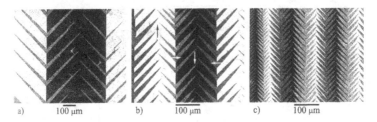

Fig. 8.18. Fir tree patterns on surfaces being misoriented by (**a**) 2°, (**b**) 3°, and (**c**) 4° with respect to (100). The diagrams exhibit different scales. (From [10] (used with permission))

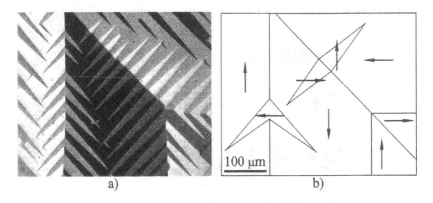

Fig. 8.19. Two kinds of fir tree pattern (**a**) associated with different wall types in the basic domain structure as schematically indicated in (**b**)(From [10] (used with permission))

is transported away from the surface either to the opposite surface or towards the neighboring domains. One example for the occurrence of lancet domains was already given in Fig. 8.17 which are obvious as the needle-like domains with the opposite magnetization within the large domains.

- Strongly misoriented surface

  The domain patterns encountered on strongly misoriented surfaces are most involved and certainly not completed understood. Surface domains on strongly misoriented surfaces are not at all representative for the under-lying bulk magnetization. The interior domains can only be inferred from subtle features and the dynamics of the surface pattern. As one example the surface of Fe(100) with $\theta = 7°$ is shown in Fig. 8.20. The striking fat walls are no real walls but traces of internal domains. The pattern mainly consists of basic domains along the one of the easy axes which are covered by shallow closure domains magnetized along the other axis. The cross section in (c) demonstrates how the charges on the surface are compensated

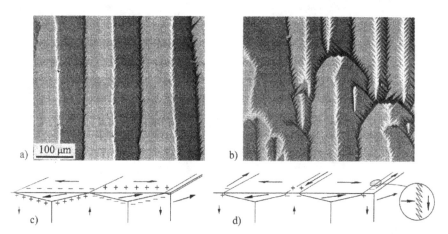

Fig. 8.20. Domain observation on an Fe(100)-surface deviating by about 7° from both easy axes. The pattern mainly consists of basic domains along one of the easy axes which are covered by shallow closure domains being magnetized along the other axis. The cross section (**c**) demonstrates how the magnetic charges on the surface are compensated by charges at the lower boundaries of the shallow domains. Wall energy is saved in the second step by slightly opening the basic domains (**d**). (From [10] (used with permission))

by charges at the lower boundaries of the shallow domains. Wall energy is saved in the second step by slightly opening the basic domains (d). The charges on the steeper wall segments formed in this opening process can be distributed by a small zigzag folding shown in the inset

9

Magnetization Dynamics

The fundamental question of the reaction of the magnetization M for changes of an external magnetic field H was already discussed in the chapters above for the situation of a differently constant or slowly varying external field like hysteretic behavior. Therefore, we have analyzed the static or nearly static situation.

This chapter deals with the consequences of rapid changes of H, i.e. the dynamic behavior. We will analyze the reaction of the magnetization

- if the external field is rapidly changed from one constant to another constant value,
- for an alternating magnetic field,
- and for a high-frequent alternating external field.

9.1 Magnetic After-Effect

The magnetic after-effect describes the situation if the magnetization does not follow the variation of an external magnetic field instantaneously, i.e. it exhibits a delayed reaction.

Let us assume that an external magnetic field $H = H_1$ is abruptly changed to $H = H_2$ at the time $t = 0$ (see Fig. 9.1).

The magnetization does not simultaneously follow the external field. The magnetization M is modified by a sudden jump of M_i at $t = 0$ with a subsequent slow change of M_n which is time dependent (see Fig. 9.2).

$M_n(t)$ additionally depends on the magnitude of M_i and the magnetic final state at H_2. If H_2 is, for example, in the regime where irreversible changes of the magnetization occur M_n is relatively large. If H_2 is in the regime of reversible rotation processes of the magnetization M_n is relatively small.

A simple situation is given for a logarithmic behavior of M_n with a relaxation time τ:

$$M_n(t) = M_{n0} \cdot \left(1 - e^{-t/\tau}\right) \tag{9.1}$$

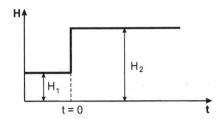

Fig. 9.1. Assumed change of the external magnetic field H as a function of time

with M_{n0} being the change of the magnetization without the amount of M_i. A semi-logarithmic plot of $M_n(t)$ as a function of t for Fe containing a small amount of carbon demonstrates agreement with experiment (see Fig. 9.3).

Due to $M = \chi H$ the susceptibility does not remain constant after the change of the external magnetic field caused by the magnetic after-effect:

$$\chi = \chi_i \ \text{ for } \ M \le M_1 + M_i \tag{9.2}$$
$$\chi = \chi_n \ \text{ for } \ M > M_1 + M_i \tag{9.3}$$

Without loss of generality we assume $H_1 = M_1 = 0$ in order to simplify the situation (see Fig. 9.4). Thus, we have:

$$M_i = \chi_i H \tag{9.4}$$
$$M_{n0} = \chi_n H \tag{9.5}$$

which allows to define the constant ξ:

$$\xi := \frac{\chi_n}{\chi_i} = \frac{M_{n0}}{M_i} \tag{9.6}$$

With:

$$M_\infty = M(t \to \infty) = M_i + M_{n0} \tag{9.7}$$
$$= (\chi_i + \chi_n)H \tag{9.8}$$

we get:

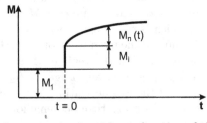

Fig. 9.2. Change of the magnetization M as a function of time assuming the variation of an external magnetic field as shown in Fig. 9.1

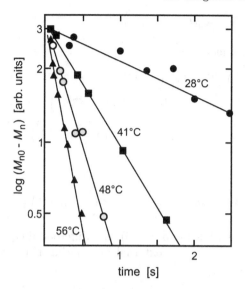

Fig. 9.3. Magnetic after-effect at different temperatures observed for Fe which contains small amounts of C (Data taken from [11])

$$M(t) = M_i + M_n(t) \tag{9.9}$$

$$= M_i + M_{n0} \cdot \left(1 - e^{-t/\tau}\right) \tag{9.10}$$

$$= M_\infty - M_{n0} + M_{n0} \cdot \left(1 - e^{-t/\tau}\right) \tag{9.11}$$

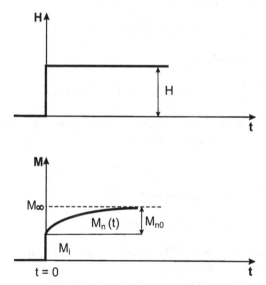

Fig. 9.4. Time dependence of the external magnetic field H and magnetization M assuming that $H_1 = M_1 = 0$ with respect to Figs. 9.1 and 9.2

$$= M_\infty - M_{n0} \cdot e^{-t/\tau} \tag{9.12}$$

Using (9.5) and (9.8) leads to:

$$M(t) = (\chi_i + \chi_n)H - \chi_n H e^{-t/\tau} \tag{9.13}$$

$$= \chi_i H + \chi_n H \left(1 - e^{-t/\tau}\right) \tag{9.14}$$

$$= \chi_i H \left(1 + \frac{\chi_n}{\chi_i}\left(1 - e^{-t/\tau}\right)\right) \tag{9.15}$$

With (9.6) we obtain:

$$M = M(t) = \chi_i H \left(1 + \xi\left(1 - e^{-t/\tau}\right)\right) \tag{9.16}$$

Let us rewrite this relationship of (9.16) as a differential equation which will be helpful for the discussion of external alternating magnetic fields. Equation (9.16) leads to:

$$M - \chi_i H = \chi_i H \xi \left(1 - e^{-t/\tau}\right) \tag{9.17}$$

which can be written as:

$$\frac{\mathrm{d}}{\mathrm{d}t}(M - \chi_i H) = \chi_i H \xi \frac{\mathrm{d}}{\mathrm{d}t}\left(1 - e^{-t/\tau}\right) \tag{9.18}$$

$$= \frac{1}{\tau}\chi_i H \xi e^{-t/\tau} \tag{9.19}$$

Equation (9.16) additionally leads to:

$$\chi_i H \xi e^{-t/\tau} = \chi_i H \xi + \chi_i H - M \tag{9.20}$$

Inserting into (9.19) results in:

$$\frac{\mathrm{d}}{\mathrm{d}t}(M - \chi_i H) = -\frac{1}{\tau}(M - \chi_i H(1 + \xi)) \tag{9.21}$$

This expression can be used to describe the influence of an alternating magnetic field H which is supposed to be:

$$H = H_0 e^{i\omega t} \tag{9.22}$$

on the magnetization which reacts with a delay:

$$M = M_0 e^{i(\omega t - \delta)} \tag{9.23}$$

with δ being a phase difference. The unknown quantities are δ and M_0 which will be determined in the following.

The derivatives of the external field and magnetization with respect to time are:

$$\frac{dH}{dt} = i\omega H = i\omega H_0\, e^{i\omega t} \tag{9.24}$$

$$\frac{dM}{dt} = i\omega M = i\omega M_0\, e^{i\omega t}\, e^{-i\delta} \tag{9.25}$$

Inserting into (9.21) results in:

$$i\omega M_0\, e^{-i\delta} - i\omega\chi_i H_0 = -\frac{1}{\tau} M_0\, e^{-i\delta} + \frac{1}{\tau}\chi_i H_0(1+\xi) \tag{9.26}$$

Using the relationship $e^{ix} = \cos x + i\sin x$ which can be rewritten as:

$$e^{-i\delta} = \cos\delta - i\sin\delta \tag{9.27}$$

results in the following set of equations after inserting into (9.26):

$$i\omega M_0(\cos\delta - i\sin\delta) - i\omega\chi_i H_0 =$$
$$-\frac{1}{\tau} M_0(\cos\delta - i\sin\delta) + \frac{1}{\tau}\chi_i H_0(1+\xi) \tag{9.28}$$
$$\omega M_0\sin\delta + i(\omega M_0\cos\delta - \omega\chi_i H_0) =$$
$$\frac{1}{\tau}\chi_i H_0(1+\xi) - \frac{1}{\tau} M_0\cos\delta + i(\frac{1}{\tau} M_0\sin\delta) \tag{9.29}$$

Thus, two conditions must be fulfilled:

$$\omega M_0\sin\delta = \frac{1}{\tau}\chi_i H_0(1+\xi) - \frac{1}{\tau} M_0\cos\delta \tag{9.30}$$

$$\frac{1}{\tau} M_0\sin\delta = \omega M_0\cos\delta - \omega\chi_i H_0 \tag{9.31}$$

From (9.31) we get:

$$\sin\delta = \omega\tau\cos\delta - \omega\tau\chi_i\frac{H_0}{M_0} \tag{9.32}$$

which is inserted into (9.30):

$$\omega^2\tau^2 M_0\cos\delta - \omega^2\tau^2\chi_i H_0 = \chi_i H_0(1+\xi) - M_0\cos\delta \tag{9.33}$$

Thus:

$$\cos\delta(M_0\omega^2\tau^2 + M_0) = \omega^2\tau^2\chi_i H_0 + \chi_i H_0(1+\xi) \tag{9.34}$$

which leads to:

$$\cos\delta = \frac{1 + \omega^2\tau^2 + \xi}{M_0(1 + \omega^2\tau^2)}\chi_i H_0 \tag{9.35}$$

This expression for $\cos\delta$ is inserted into (9.32):

$$\sin\delta = \omega\tau\frac{1 + \omega^2\tau^2 + \xi}{M_0(1 + \omega^2\tau^2)}\chi_i H_0 - \omega\tau\chi_i\frac{H_0}{M_0} \tag{9.36}$$

$$= \frac{\omega\tau\chi_i H_0(1 + \omega^2\tau^2 + \xi) - \omega\tau\chi_i H_0(1 + \omega^2\tau^2)}{M_0(1 + \omega^2\tau^2)} \tag{9.37}$$

$$= \frac{\omega\tau\chi_i H_0(1 + \omega^2\tau^2 + \xi - 1 - \omega^2\tau^2)}{M_0(1 + \omega^2\tau^2)} \tag{9.38}$$

$$= \frac{\omega\tau\xi}{M_0(1 + \omega^2\tau^2)} \chi_i H_0 \tag{9.39}$$

With (9.35) and (9.39) we obtain:

$$\tan\delta = \frac{\omega\tau\xi}{1 + \omega^2\tau^2 + \xi} \tag{9.40}$$

which allows to determine the phase angle δ. Now, we are able to get access to M_0 using (9.31):

$$M_0 = \frac{\omega\tau}{\omega\tau\cos\delta - \sin\delta} \chi_i H_0 \tag{9.41}$$

We directly see that the static behavior $M_0 = \chi_i H_0$ is only valid if $\delta = 0$. Due to the time delay of the magnetization which is equivalent with $\delta > 0$ it is obvious that the magnetization M exhibits a loss in the dynamic situation. Thus, δ is called loss angle and $\tan\delta$ loss factor.

One example for the loss factor $\tan\delta$ is shown in Fig. 9.5. The temperature dependence of the loss factor $\tan\delta$ is given at different frequencies for the same material with regard to Fig. 9.3. We see that the maximum of the loss factor occurs at different temperatures T which is a consequence of the temperature dependence of the relaxation time τ.

The dependence of the loss factor on the relaxation time was already given by (9.40). The loss factor decreases for large relaxations times due to τ^2 in the denominator. The loss factor also decreases for small relaxation times

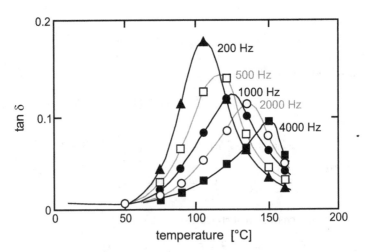

Fig. 9.5. Temperature dependence of the loss factor observed for Fe containing low carbon amounts. The numerical values are the frequencies of an alternating current field. (Data taken from [11])

because the numerator reaches zero and the denominator equals to $1 + \xi \neq 0$ for decreasing τ. Thus, there must be a maximum of the loss factor which can be determined by setting the first derivative to zero:

$$0 = \frac{d \tan \delta(\tau)}{d\tau} = \frac{\omega \xi(1 + \omega^2 \tau^2 + \xi) - \omega \tau \xi \cdot 2\omega^2 \tau}{(1 + \omega^2 \tau^2 + \xi)^2} \tag{9.42}$$

Thus:

$$1 + \omega^2 \tau^2 + \xi - 2\omega^2 \tau^2 = 0 \tag{9.43}$$

which results in:

$$\tau = \frac{\sqrt{1 + \xi}}{\omega} \tag{9.44}$$

This means that the relaxation time τ can experimentally be determined by that frequency which leads to a maximum of the loss factor at a given temperature.

9.2 Influence of High-Frequent Magnetic Fields

The reaction of the magnetic induction B (and thus also of the magnetization) on an external alternating magnetic field H with a time dependence given by (9.22) can be expressed as:

$$B = B_0 \, e^{i(\omega t - \delta)} \tag{9.45}$$

As a consequence the permeability μ becomes complex:

$$\mu = \frac{B}{H} \tag{9.46}$$

$$= \frac{B_0 \, e^{i(\omega t - \delta)}}{H_0 \, e^{i\omega t}} \tag{9.47}$$

$$= \frac{B_0}{H_0} \, e^{-i\delta} \tag{9.48}$$

Using (9.27) we get:

$$\mu = \frac{B_0}{H_0} \cos \delta - i \frac{B_0}{H_0} \sin \delta \tag{9.49}$$

Characterizing the real and negative imaginary part of the permeability by:

$$\mu' = \frac{B_0}{H_0} \cos \delta \tag{9.50}$$

$$\mu'' = \frac{B_0}{H_0} \sin \delta \tag{9.51}$$

we obtain:

$$\mu = \mu' - i\mu'' \tag{9.52}$$

with μ' being that component of B which is in-phase with H, i.e. the permeability μ equals to μ' for a vanishing loss. μ'' denotes the component of B which exhibits a phase shift of $90°$ compared to H. This means that energy is required if $\mu'' \neq 0$ in order to maintain this component. The ratio

$$\frac{\mu''}{\mu'} = \frac{(B_0/H_0)\sin\delta}{(B_0/H_0)\cos\delta} = \tan\delta \qquad (9.53)$$

corresponds to the loss factor $\tan\delta$. An increasing frequency ω results in different types of losses:

- Hysteresis loss
  The most important reason for this kind of loss is the movement of domain walls. It is therefore the dominant loss for low frequencies. For a small magnetization the loss factor depends on the amplitude of the external magnetic field. Thus, the distinction to other types of losses can be done by changing the amplitude of the external field. Hysteresis loss becomes less important with increasing frequency because the movement of domain walls is strongly damped at high frequencies and replaced by rotation of the magnetization.
- Eddy-current loss
  This kind of loss is proportional to ω^2 and therefore important in the high-frequency regime. A minimization can be carried out by reducing the dimension of the system in one or two directions perpendicularly to the direction of the magnetization, e.g. for thin sheets or in thin film systems which is important for high-speed data storage.
- Magnetic after-effect
  The loss factor $\tan\delta$ depends on the relaxation time τ and becomes maximum at the frequency $\omega = \sqrt{1+\xi}/\tau$ (see (9.44)). The relaxation time decreases for increasing temperature. Thus, the loss factor exhibits a maximum at a specific temperature which depends on the frequency ω.

At high frequencies a reduction of the permeability can occur correlated with an increase of the loss factor. This resonance behavior is due to the formation of electromagnetic standing waves inside the magnetic material if its dimension corresponds to an integer multiple of the wavelength in the material. This behavior results in a dimensional resonance which is exemplarily shown in Fig. 9.6 for a ferrite which is a magnetic alloy of the type MeO · Fe_2O_3 (Me: metal) with a very high electric resistance. The significant decrease of the permeability μ' is shifted to a higher frequency ν when the size of the ferrite is reduced.

A resonance behavior may also be induced due to the magneto crystalline anisotropy whose corresponding energy density is given by $E_{crys} = -MH\cos\theta$. The resonance frequency amounts to $\omega_0 = \gamma\mu_0 H_r$ with $\gamma = e/m$ being the electron gyromagnetic ratio and H_r the field at resonance. This

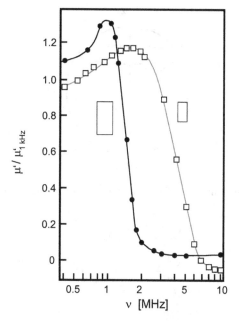

Fig. 9.6. Dimensional resonance observed for ferrite cores with different cross sections. The ordinate is normalized to the value at 1 kHz for μ'. (Data taken from [12])

means that the rotation of M about the easy magnetization axis for an external alternating magnetic field with the resonance frequency ω is in resonance with this field. Consequently, abrupt changes of μ' and μ'' occur.

9.3 Damping Processes

In Chap. 9.2 we have investigated the influence of an alternating external field on the *magnitude* of the magnetization. Now, we want to examine the influence of a constant external magnetic field on the *direction* of the magnetization.

Consideration without Damping Processes

Without loss of generality we can assume that a constant external magnetic field is oriented along the $-z$-direction, i.e. $\boldsymbol{H} = (0, 0, -H)$. Then, the time dependence of the magnetization is given by:

$$\frac{\mathrm{d}\boldsymbol{M}}{\mathrm{d}t} = -\gamma\mu_0(\boldsymbol{M} \times \boldsymbol{H}) \tag{9.54}$$

with γ being the gyromagnetic ratio. Writing (9.54) in Cartesian coordinates we get:

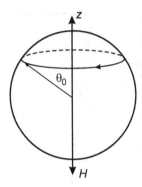

Fig. 9.7. Precession motion of the magnetization in the absence of damping

$$\frac{dM_x}{dt} = \gamma\mu_0 M_y H \tag{9.55}$$

$$\frac{dM_y}{dt} = -\gamma\mu_0 M_x H \tag{9.56}$$

$$\frac{dM_z}{dt} = 0 \tag{9.57}$$

The solution of this set of equations is given by:

$$M_x = M_S \sin\theta_0 \, e^{i\omega_0 t} \tag{9.58}$$

$$M_y = M_S \sin\theta_0 \, e^{i(\omega_0 t + \pi/2)} \tag{9.59}$$

$$M_z = M_S \cos\theta_0 \tag{9.60}$$

with M_S the saturation magnetization and $\omega_0 = \gamma\mu_0 H_r$. Thus, the magnetization exhibits a precessional motion with a constant angle θ_0 with respect to the z-axis (see Fig. 9.7). As an important consequence the external field is not able to rotate the magnetization towards its direction.

Consideration Including Damping Processes

Due to damping processes an additional term in (9.54) is necessary which results in the so-called Landau–Lifshitz equation:

$$\frac{d\boldsymbol{M}}{dt} = -\gamma\mu_0(\boldsymbol{M}\times\boldsymbol{H}) - \frac{4\pi\mu_0\lambda}{M^2}(\boldsymbol{M}\times(\boldsymbol{M}\times\boldsymbol{H})) \tag{9.61}$$

with λ being a measure of the damping ("relaxation frequency"). Using the relationship:

$$-\boldsymbol{M}\times(\boldsymbol{M}\times\boldsymbol{H}) = M^2\cdot\boldsymbol{H} - ((\boldsymbol{M}\cdot\boldsymbol{H})\boldsymbol{M}) \tag{9.62}$$

we get for the Landau–Lifshitz equation:

$$\frac{\mathrm{d}\boldsymbol{M}}{\mathrm{d}t} = -\gamma\mu_0(\boldsymbol{M}\times\boldsymbol{H}) + 4\pi\mu_0\lambda\left(\boldsymbol{H} - \frac{(\boldsymbol{M}\cdot\boldsymbol{H})\boldsymbol{M}}{M^2}\right) \qquad (9.63)$$

A closer look to the individual terms shows that $(\boldsymbol{M}\cdot\boldsymbol{H})\boldsymbol{M}/M^2$ corresponds to that component of $\boldsymbol{H}$ which is parallel to $\boldsymbol{M}$. Consequently, $\boldsymbol{H} - (\boldsymbol{M}\cdot\boldsymbol{H})\boldsymbol{M}/M^2$ is the component of $\boldsymbol{H}$ which is perpendicular to $\boldsymbol{M}$. The additional term therefore describes a torque acting on $\boldsymbol{M}$. This torque leads to a rotation of the magnetization towards the direction of the external magnetic field. Due to the damping process a precessional switching can therefore occur. Additionally, the precessional motion will decay unless a source of external energy is present in order to maintain it.

An implicit assumption which led to the Landau–Lifshitz equation (9.63) represents the ratio between the inertial term $-\gamma\mu_0(\boldsymbol{M}\times\boldsymbol{H})$ and the damping term $4\pi\mu_0\lambda(\boldsymbol{H} - (\boldsymbol{M}\cdot\boldsymbol{H})\boldsymbol{M}/M^2)$. Equation (9.63) is only valid if the inertial term is significantly larger than the damping term. By introducing the damping parameter α:

$$\alpha = \frac{4\pi\lambda}{\gamma M} \qquad (9.64)$$

we can express the assumption by $\alpha^2 \ll 1$. The accurate equation without restriction concerning the ratio of both terms is known as the Landau–Lifshitz–Gilbert equation and given by:

$$\frac{\mathrm{d}\boldsymbol{M}}{\mathrm{d}t} = -\gamma\mu_0\left(\boldsymbol{M}\times\left(\boldsymbol{H} - \frac{\alpha}{\gamma\mu_0 M}\cdot\frac{\mathrm{d}\boldsymbol{M}}{\mathrm{d}t}\right)\right) \qquad (9.65)$$

The equations above are obtained by neglecting terms of higher order in α^2. Writing (9.65) in Cartesian coordinates we get:

$$\frac{\mathrm{d}M_x}{\mathrm{d}t} = \omega_0 M_y + \alpha\frac{M_y}{M_S}\cdot\frac{\mathrm{d}M_z}{\mathrm{d}t} - \alpha\frac{M_z}{M_S}\cdot\frac{\mathrm{d}M_y}{\mathrm{d}t} \qquad (9.66)$$

$$\frac{\mathrm{d}M_y}{\mathrm{d}t} = -\omega_0 M_x + \alpha\frac{M_z}{M_S}\cdot\frac{\mathrm{d}M_x}{\mathrm{d}t} - \alpha\frac{M_x}{M_S}\cdot\frac{\mathrm{d}M_z}{\mathrm{d}t} \qquad (9.67)$$

$$\frac{\mathrm{d}M_z}{\mathrm{d}t} = \alpha\frac{M_x}{M_S}\cdot\frac{\mathrm{d}M_y}{\mathrm{d}t} - \alpha\frac{M_y}{M_S}\cdot\frac{\mathrm{d}M_x}{\mathrm{d}t} \qquad (9.68)$$

Ordering concerning the derivatives results in:

$$\frac{\mathrm{d}M_x}{\mathrm{d}t} = \frac{\omega_0}{1+\alpha^2}M_y + \frac{\omega_0\alpha}{1+\alpha^2}\frac{M_x M_z}{M_S} \qquad (9.69)$$

$$\frac{\mathrm{d}M_y}{\mathrm{d}t} = -\frac{\omega_0}{1+\alpha^2}M_x + \frac{\omega_0\alpha}{1+\alpha^2}\frac{M_y M_z}{M_S} \qquad (9.70)$$

$$\frac{\mathrm{d}M_z}{\mathrm{d}t} = -\frac{\omega_0\alpha}{1+\alpha^2}M_S + \frac{\omega_0\alpha}{1+\alpha^2}\frac{M_z^2}{M_S} \qquad (9.71)$$

The solution of this set of equations amounts to:

$$M_x = M_S \sin\theta \, e^{i\omega t} \tag{9.72}$$

$$M_y = M_S \sin\theta \, e^{i(\omega t + \pi/2)} \tag{9.73}$$

$$M_z = M_S \cos\theta \tag{9.74}$$

with θ_0 the initial angle of the magnetization vector, ω the angular frequency, and τ the time constant which behave like:

$$\tan\frac{\theta}{2} = \tan\frac{\theta_0}{2} \cdot e^{-t/\tau} \tag{9.75}$$

$$\omega = \frac{\omega_0}{1+\alpha^2} = \frac{\omega_0}{1 + \dfrac{1}{(\omega_0\tau_0)^2}} \tag{9.76}$$

$$\tau = \tau_0 \cdot (1+\alpha^2) = \tau_0 \left(1 + \frac{1}{(\omega_0\tau_0)^2}\right) \tag{9.77}$$

using

$$\omega_0 = \gamma\mu_0 H_r \tag{9.78}$$

$$\tau_0 = \frac{1}{\alpha\omega_0} = \frac{M_S}{4\pi\lambda\mu_0 H_r} \tag{9.79}$$

The precessional motion of the magnetization vector can easily be described for situations of a very small or a very large damping:

- Very small damping
  This case is characterized by $0 < \alpha^2 \ll 1$. Due to $\tau_0 = 1/(\alpha\omega_0)$ we get $\omega = 1/(\alpha\tau)$ and therefore $1/\omega \ll \tau$. This means that a large number of precessional circulations occurs until the magnetization is oriented along the direction of external magnetic field $\boldsymbol{H}$, i.e. until $\boldsymbol{M}$ points to the $-z$-direction (see left part of Fig. 9.8).
- Very large damping
  In this situation, we have $\alpha^2 \gg 1$ and therefore $1/\omega \gg \tau$. Consequently, the magnetization rotates with significantly less numbers of circulations towards the $-z$-direction. The extremal case of less than one circulation (creep behavior) is shown in the right part of Fig. 9.8.

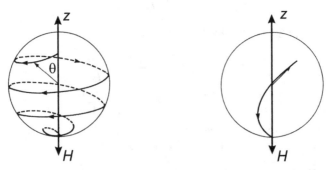

Fig. 9.8. Motion of the magnetization for small (**left**) and large damping (**right**)

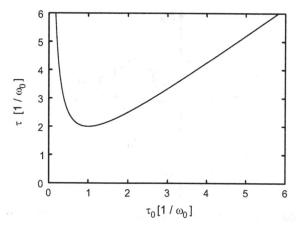

Fig. 9.9. Relaxation time for switching the magnetization as a function of the relaxation frequency

Therefore, the switching time of the magnetization increases for small as well as large values of the relaxation frequency. It is given by:

$$\tau(\tau_0) = \tau_0 \left(1 + \alpha^2\right) = \tau_0 + \frac{1}{\tau_0 \omega_0^2} \tag{9.80}$$

due to $\alpha = 1/(\tau_0 \omega_0)$ (see Fig. 9.9). The minimum can be determined by setting the first derivative to zero:

$$0 = \frac{d\tau}{d\tau_0} = 1 - \frac{1}{\tau_0^2 \omega_0^2} \tag{9.81}$$

which yields:

$$\tau_0 = \frac{1}{\omega_0} \quad \text{or} \quad \lambda = \frac{\gamma M_S}{4\pi} \tag{9.82}$$

This condition describes the critical damping with a minimum value of $\tau_{\min}$ which is given by:

$$\tau_{\min} = \tau\left(\tau_0 = \frac{1}{\omega_0}\right) = \frac{2}{\omega_0} = \frac{2}{\gamma \mu_0 H_r} \tag{9.83}$$

9.4 Ferromagnetic Resonance

A perfect alignment of all magnetic moments results in a precession when applying a static external magnetic field H which is referred to as uniform mode or Kittel mode. Without damping the precession frequency amounts to $\omega_0 = \gamma \mu_0 H_r$ at resonance. The occurrence of a damping mechanism leads to reversal of the magnetization towards the direction of H within several ns.

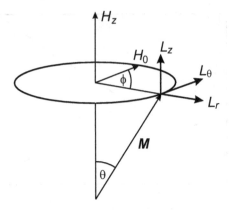

Fig. 9.10. Torque components exerted on the magnetization M by the rotational field H_0

In order to maintain the precessional rotation an additional high-frequent alternating field with amplitude H_0 and angular frequency ω about the z-axis is necessary which must exhibit a non-vanishing angle ϕ with respect to the component of the magnetization perpendicular to the z-axis. The applied torque L on the magnetization M in cylindrical coordinates is given by (see Fig. 9.10):

$$L_r = -M_S H_0 \cos \theta \sin \phi \tag{9.84}$$

$$L_\theta = M_S H_0 \cos \theta \cos \phi \tag{9.85}$$

$$L_z = M_S H_0 \sin \theta \sin \phi \tag{9.86}$$

Applying a static magnetic field H_z along the z-axis the corresponding Landau–Lifshitz equation is expressed by the set of equations:

$$\frac{dM_r}{dt} = \gamma \mu_0 M_S H_0 \cos \theta \sin \phi - 4\pi \lambda \mu_0 H_z \sin \theta \cos \theta \tag{9.87}$$

$$\frac{dM_\theta}{dt} = -\gamma \mu_0 M_S H_0 \cos \theta \cos \phi + \gamma \mu_0 M_S H_z \sin \theta \tag{9.88}$$

$$\frac{dM_z}{dt} = -\gamma \mu_0 M_S H_0 \sin \theta \sin \phi + 4\pi \lambda \mu_0 H_z \sin^2 \theta \tag{9.89}$$

In the stationary situation the following conditions are given:

$$\frac{dM_r}{dt} = 0 \tag{9.90}$$

$$\frac{dM_\theta}{dt} = M_S \omega \sin \theta \tag{9.91}$$

$$\frac{dM_z}{dt} = 0 \tag{9.92}$$

Comparing both sets of equation leads to:

$$\gamma\mu_0 M_S H_0 \sin\phi - 4\pi\lambda\mu_0 H_z \sin\theta = 0 \tag{9.93}$$

$$H_0 \cos\theta \cos\phi - H_z \sin\theta = -\frac{\omega}{\gamma\mu_0} \sin\theta \tag{9.94}$$

$$= -H_r \sin\theta \tag{9.95}$$

with $H_r = \omega_0/\gamma\mu_0$ being the resonance field. Using (9.93) we obtain

$$\sin\theta = \frac{\gamma M_S}{4\pi\lambda} \cdot \frac{H_0}{H_z} \sin\phi \tag{9.96}$$

$$= \frac{1}{\alpha} \cdot \frac{H_0}{H_z} \sin\phi \tag{9.97}$$

With the assumption of $\theta \ll \pi$ which leads to $\cos\theta = 1$, (9.95) results in:

$$H_0 \cos\phi - H_z \sin\theta = -H_r \sin\theta \tag{9.98}$$

Inserting the expression for $\sin\theta$ into (9.97) we obtain:

$$H_0 \cos\phi - \frac{1}{\alpha} H_0 \sin\phi = -\frac{1}{\alpha} \cdot \frac{H_0 H_r}{H_z} \sin\phi \tag{9.99}$$

and thus

$$\alpha - \tan\phi = -\frac{H_r}{H_z} \tan\phi \tag{9.100}$$

which leads to

$$\tan\phi = \alpha \frac{H_z}{H_z - H_r} \tag{9.101}$$

For the determination of properties near the resonance $H_z = H_r$ we have to distinguish whether H_z is larger or smaller than the resonance field. If H_z is smaller than and becomes equal to H_r, $\tan\phi \to -\infty$ and $\phi \to -\pi/2$. For a further increase of H_z, $\tan\phi$ changes to $+\infty$ and ϕ to $+\pi/2$.

Analogously to the complex permeability the susceptibility also becomes complex with $\chi = \chi' + i\chi''$. The real and imaginary part are given by:

$$\chi' = \frac{M_S}{H_0} \sin\theta \cos\phi \tag{9.102}$$

$$\chi'' = \frac{M_S}{H_0} \sin\theta \sin\phi \tag{9.103}$$

Inserting $\sin\theta$ as given in (9.97) we obtain:

$$\chi' = \frac{M_S}{\alpha H_z} \sin\phi \cos\phi = \frac{M_S}{2\alpha H_z} \sin 2\phi \tag{9.104}$$

$$\chi'' = \frac{M_S}{\alpha H_z} \sin^2\phi \tag{9.105}$$

This means that a change of sign for χ' and a maximum value for χ'' occur at the resonance condition $H_z = H_r$ (see Fig. 9.11). It is obvious that the smaller

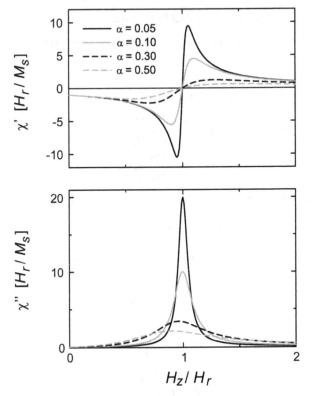

Fig. 9.11. Dependence of the real (**upper part**) and imaginary parts (**lower part**) of the susceptibility (χ' and χ'', respectively) on the intensity of the field H_z near the resonance field H_r for different values of α

α is the behavior of χ' and χ'' is more pronounced. An increase of α or the relaxation frequency λ results in a broadening of the absorption behavior which is a measure of χ''. The full width w at half maximum (FWHM) is given for $\phi = \pi/4$ because of $\sin^2 \pi/4 = 1/2$. Using $\tan \pi/4 = 1$ we obtain

$$1 = \alpha \, \frac{H_z}{H_z - H_r} \tag{9.106}$$

Therefore, the half maximum value is given at:

$$H_z = \frac{1}{1 - \alpha} \, H_r \tag{9.107}$$

Thus, the width w amounts to:

$$w = 2 \left(\frac{1}{1 - \alpha} \, H_r - H_r \right) \tag{9.108}$$

$$= 2H_r \, \frac{\alpha}{1 - \alpha} \tag{9.109}$$

$$= \frac{2H_r \cdot 4\pi\lambda}{\gamma M_S \left(1 - 4\pi\lambda/\gamma M_S\right)} \qquad (9.110)$$

$$= \frac{8\pi\lambda}{\gamma M_S - 4\pi\lambda} H_r \qquad (9.111)$$

This relationship allows the determination of the relaxation frequency λ by the measurement of the width w at half maximum.

Problems

9.1. Prove (9.62):

$$-\boldsymbol{M} \times (\boldsymbol{M} \times \boldsymbol{H}) = M^2 \cdot \boldsymbol{H} - ((\boldsymbol{M} \cdot \boldsymbol{H})\boldsymbol{M})$$

9.2. A spherically shaped sample with saturation magnetization $M_S = 2\,\mathrm{T}$ is magnetized along the z-axis which corresponds to $\theta = 0$. The relaxation frequency amounts to $\lambda = 2 \times 10^8$ Hz. A constant external magnetic field $H = -200\ \mathrm{Am}^{-1}$ is applied along the $-z$-direction. Determine the time which is required to rotate the magnetization from $\theta = 30°$ to $\theta = 150°$. Find the number of precession rotations that occur in this time interval.

Magnetism in Reduced Dimensions – Atoms

In the chapters above we have considered only bulk-like behavior of materials with infinite extension in three dimensions. An exemplary consequence is the neglect of surfaces. In the following chapters magnetic properties are discussed for systems which exhibit a restriction to the micro- or nanometer regime in at least one dimension (thin films, wires, nanoparticles, clusters, and atoms (see Fig. 10.1)). Due to this restriction an influence on the magnetic behavior occurs by additional parameters like:

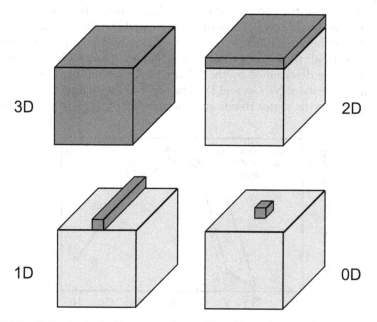

Fig. 10.1. Reducing a bulk system from three dimensions without any restriction (3D) to 2D leads to thin film systems, to 1D results in nanoscaled wires, and to 0D deals with atoms, clusters, and nanoparticles

- thickness of thin films
- possible magnetism of the underlying substrate
- crystallographic orientation between substrate and nanoscaled object
- interface between substrate and nanoscaled object
- additional interfaces or surface

In this chapter we will restrict our discussion to single atoms on surfaces or in matrices.

10.1 Single Atoms on a Surface

It is obvious that nanoscaled objects need a support (except free clusters and nanoparticles) which results in a significant influence due to the *existence of a substrate* and the *type of substrate itself* on the magnetism of the low-dimensional system.

For $3d$ metals only Fe, Co, and Ni are ferromagnetic. Some others like Cr are antiferromagnetic. This behavior is different for low-dimensional systems. Figure 10.2 shows the calculated local magnetic moments for $3d$ adatoms on a Pd(001) surface in comparison to corresponding results for the monolayers and the impurities in the bulk. It is seen that the moments of the $3d$ adatoms are well saturated and, except for V and Cr, are similar to the monolayers and the impurity moments. The impurity and adatom moments of Fe and Co are slightly larger than the corresponding monolayer ones. The moment of Ni is nearly the same for all cases. The more or less equal moments obtained for Fe, Co, and Ni for the three different environments are basically a consequence of the fact that in all these cases the majority band is practically filled so that the moments are determined by the valence electrons and increase by about $1\mu_B$ in the sequence of Ni, Co, and Fe. The effect of $sp-d$ interactions is relatively small since the major trends are determined by the $d-d$ hybridization.

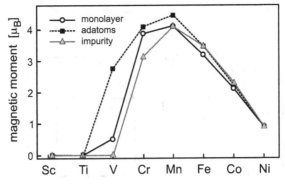

Fig. 10.2. Comparison between the calculated magnetic moments per atom of $3d$ transition metals as adatoms on Pd(001), $3d$ monolayers on Pd(001), and $3d$ impurities in bulk Pd. (Adapted from [13] (used with permission))

Dramatic environmental effects are, however, found for V and, to a somewhat smaller extent, for Cr. For a V impurity in bulk Pd the Stoner criterion is not satisfied since the $3d - 4d$ hybridization between the impurity $3d$ and the bulk $4d$ states shifts the virtual bound state above the Fermi energy so that the density of states at the Fermi energy E_F is rather small. This can also explain the rather small moment obtained for the V monolayer on Pd(001). On the other hand the hybridization of the V adatom is sufficiently reduced so that a rather large V moment of $2.8\mu_B$ is obtained. The adatom moments in Fig. 10.2 essentially follow Hund's rules of isolated atoms with the largest moment at the center of the series which is Mn.

Electronic and magnetic properties of ensembles consisting of one or a few atoms as free particles can be different compared to the situation that the atom(s) are in contact with matter:

- on a surface
- embedded in a matrix

The atom or particle can cause a charge transfer to or from the matter being in contact with which leads to a screening by the surrounding electron gas. The screening potential exhibits an oscillating behavior, resulting in the so-called Friedel oscillations. This behavior will be discussed in connection with Fig. 10.5.

For magnetic atoms or particles on non-magnetic surfaces or in non-magnetic matrices not only a charge transfer can occur. Additionally, a transfer of magnetic moments can be present. Again, the screening of charge results in Friedel oscillations whereas the screening of magnetic moments can be carried out in two different ways as schematically depicted in Fig. 10.3:

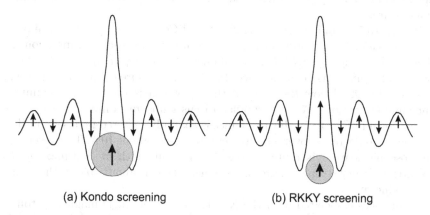

(a) Kondo screening   (b) RKKY screening

Fig. 10.3. Different types of screening which can occur for a magnetic atom on a non-magnetic surface or in a non-magnetic matrix: (**a**) Kondo screening; (**b**) RKKY screening

- Kondo screening
  The Kondo screening results in a compensation of the magnetic moment, i.e. in a reduction (see Fig. 10.3a). An example is given by Mn atoms in Cu. A single magnetic atom in a non-magnetic matrix is often referred to as a Kondo impurity.
- RKKY screening
  This kind of screening tends to result in an increase of the magnetic moment of the atom as schematically shown in Fig. 10.3b. One example is given by Co atoms in Pd.

Both kinds of screening lead to an oscillatory behavior of the magnetic moments with distance from the magnetic impurity. The RKKY screening can result in a ferromagnetic ground state of the atom whereas Kondo screening leads to a non-magnetic ground state.

The Kondo effect arises as a result of the exchange interaction between the localized magnetic moment of the impurity and the delocalized electrons in the metallic host. At sufficiently low temperatures T nearby conduction electrons begin to align their spins to screen the spin on the local moment. A many-body spin singlet state is thus formed consisting of the local magnetic impurity surrounded by a spin compensation cloud of conduction electrons such that no net moment remains at the Kondo site.

As the constituent electrons in the Kondo screening cloud come predominantly from the Fermi surface of the metallic host the Kondo effect may spectroscopically be observed as an acute perturbation or resonance in the renormalized density of states near the Fermi energy E_F.

This so-called Kondo resonance can by proven by means of photoelectron spectroscopy. One example is shown in Fig. 10.4 for different Ce compounds which exhibit a sharp peak directly at the Fermi edge well below the Konto temperature.

In Fig. 10.5 scanning tunnelling spectra of Co atoms on Cu(111) are presented in which the Kondo resonance is manifest as a sharp suppression in differential conductance dI/dU in the immediate vicinity of E_F ($U = 0$). This differential (tunnelling) conductance dI/dU is a measure of the local density of states (LDOS) below the apex of the tip which is used in the scanning tunnelling microscope. A more detailed discussion is given in Chap. 16.5.

As shown in Fig. 10.5a this resonance is spatially centered precisely on the Co atoms and decreases over a lateral length scale of about 10 Å. The observed spectroscopic feature is quite narrow (9 mV full-width at half-maximum, FWHM) and very strong (the resonance magnitude is about 40% of the zero-bias conductance).

Resonances associated with the Kondo effect over single magnetic atoms have been interpreted in terms of the Fano model of interfering discrete and continuum channels. The observed spectra can be fitted very well by a Fano line shape in the limit of small coupling to the discrete state; such fits reproduce the "dip" structure along with the observed asymmetry and the shift in the minimum from $U = 0$.

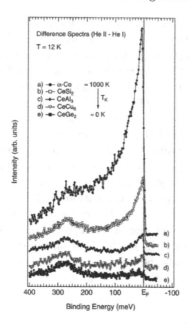

Fig. 10.4. Difference spectra of He II–He I taken with $h\nu = 40.8\,\mathrm{eV}$ (He II) and $h\nu = 21.2\,\mathrm{eV}$ (He I) for a series of heavy fermion Ce compounds which span a wide range of the Kondo temperature T_K. The original spectra were recorded with high energy resolution of about 5 meV and at 12 K. (Figure reprinted with permission from [14]. Copyright (1997) by the American Physical Society.)

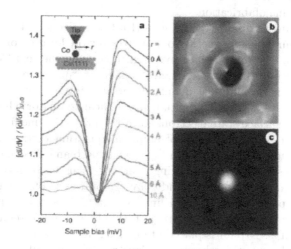

Fig. 10.5. Detection of the Kondo resonance localized around a single Co atom on Cu(111). (**a**) Tunnelling spectra acquired over the Co atom for increasing lateral displacement. (**b**) Topograph of an isolated Co atom. (**c**) dI/dU map of the same area. Dark to light corresponds to increasing conductance. (Reprinted by permission from Macmillan Publishers Ltd: Nature (see [15]), copyright (2000))

The Cu(111) surface exhibits a two-dimensional free electron gas. These electrons inhabit a surface state band which starts 450 mV below E_F. Co atoms placed on the surface are immersed in the two-dimensional electron sea. This fact is crucial to these results as it is precisely these electrons that form the projection medium for the quantum mirage which is described in Chap. 10.2. In the topograph of Fig. 10.5b the standing waves (equivalently, the energy-resolved Friedel oscillations) formed from the two-dimensional electrons scattering off a single Co atom are evident.

Using an STM it is possible to spatially map the Kondo screening cloud by simultaneously acquiring a dI/dU image while U is tuned to a small bias U_0 near the edge of the resonance ($|U_0|$ ranges from 5 to 10 mV). The presence of the Kondo resonance at a particular location on the surface removes spectral density between $U = 0$ and $U = U_0$ and causes the tip to approach the sample and thus increases the measured dI/dU signal. Figure 10.5c shows the dI/dU map associated with the single Co atom topograph in Fig. 10.5b. This visualization clearly reveals the location and extent of the Kondo resonance localized at the Co atom site.

10.2 Quantum Mirage

Image projection relies on classical wave mechanics. Well-known examples include the bending of light to create mirages in the atmosphere and the focusing of sound by whispering galleries. However, the observation of analogous phenomena in condensed matter systems is a more recent development facilitated by advances in nanofabrication.

Now, we want to discuss referring to [15] the projection of the electronic structure surrounding a magnetic Co atom to a remote location on the surface of a Cu crystal; electron partial waves scattered from the real Co atom are coherently refocused to form a spectral image or "quantum mirage". The focusing device is an elliptical quantum corral being assembled on the Cu surface. The corral acts as a quantum mechanical resonator while the two-dimensional Cu surface state electrons form the projection medium.

The discrete electronic states of elliptical quantum corrals are used to project the mirage. As illustrated in Fig. 10.6a and b an ellipse has the property that all classical paths emanating from one focus F_1 bounce specularly off the wall at arbitrary point P and converge on the second focus F_2. Furthermore, this path length (dashed line in Fig. 10.6a and b) remains fixed at $\overline{F_1PF_2} = 2a$ independent of the trajectory where a is the semimajor axis length. Hence, if a scatterer is placed at one focus all scattered waves will add with coherent phase at the other focus. Constructing a pair of resonators with the geometries sketched in Fig. 10.6a and b by means of an STM working at low temperature enables to position Co atoms to form the corral walls. STM topographs of the assembled resonators are shown in Fig. 10.6c and d where the electron-confining effects are evident. The mean spacing between

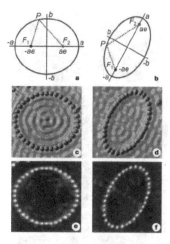

Fig. 10.6. Elliptical electron resonators. (a) Eccentricity $e = 0.5$; (b) $e = 0.786$. (c), (d) Corresponding topographs of the experimental realizations employing Co atoms to corral two-dimensional electrons on Cu(111). (e), (f) dI/dU maps acquired simultaneously with the corresponding topographs tuned to image the Kondo resonance. (Reprinted by permission from Macmillan Publishers Ltd: Nature (see [15]), copyright (2000))

the Co atoms comprising the resonator walls is roughly four atomic sites on the underlying Cu(111) lattice. This pair of ellipses shares the same $a = 71.3$ Å (hence the same focus-to-focus path length) but has significantly different eccentricity $e = \sqrt{1 - b^2/a^2}$ with b being the semiminor axis. The left and right columns of Fig. 10.6 pertain to $e = 1/2$ and $e = 0.786$ ellipses which corral 84 and 56 electrons, respectively.

The associated dI/dU maps of the empty resonators are displayed in Fig. 10.6e and f. These maps show the Kondo resonance localized around each Co wall atom but lack any significant Kondo signal within the ellipse. The faint interior features derive from the local density of states undulations of the ellipse eigenmodes closest to E_F. From these maps it is clear that the wall atoms themselves are not projecting the observed Kondo effect (Fig. 10.5a) into the confines of the resonator.

Next, the effects of placing a single Co atom at various locations inside each resonator is studied. The results are summarized in Fig. 10.7. For a given geometry with an interior atom a simultaneous topograph and dI/dU map is acquired; and then the respective dI/dU map is subtracted shown in Fig. 10.6e and f in order to remove the background electronic contribution of the standing wave LDOS and emphasize the Kondo component of the signal. The resulting dI/dU difference map was therefore tuned to spatially locate the Kondo signature of Co inside the resonator while the associated topograph revealed the actual atom locations. The striking results of this measurement shown in Fig. 10.7a–d reveal two interior positions which the Kondo resonance

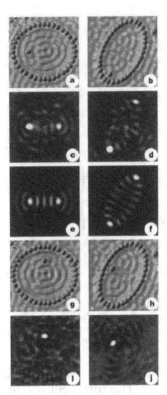

Fig. 10.7. Visualization of the quantum mirage. (**a**), (**b**) Topographs showing the $e = 0.5$ (**a**) and $e = 0.786$ (**b**) ellipse each with a Co atom at the left focus. (**c**), (**d**) Associated dI/dU difference maps showing the Kondo effect projected to the empty right focus resulting in a Co atom mirage. (**e**) and (**f**) Calculated eigenmodes at E_F (magnitude of the wave function is plotted). When the interior Co atom is moved off focus ((**g**) and (**h**), topographs), the mirage vanishes ((**i**) and (**j**), corresponding dI/dU difference maps). (Reprinted by permission from Macmillan Publishers Ltd: Nature (see [15]), copyright (2000))

dominates at for each ellipse: One signal is localized on the real Co atom at the left focus while the other signal is centered on the empty right focus. In effect, a phantom (albeit weaker) copy of the Co atom has been created by projecting the localized electronic structure from the occupied focus to the unoccupied focus thus creating the quantum mirage. However, also visible in the dI/dU difference maps of Fig. 10.7c and d is reproducible fine structure throughout the interior of the resonators. Further insight into these observations comes from treating the system as strictly quantized. Calculations reveal that the additional features in the conductance maps correspond very well to the spatial structure of the eigenstate closest to E_F (approximating the elliptical boundary as a hard-wall box). The magnitude of the corresponding

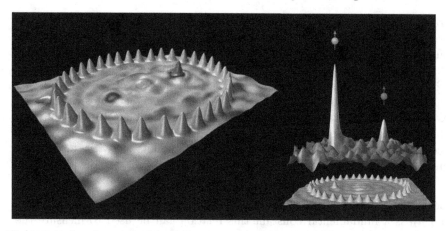

Fig. 10.8. Overlap of topographic and electronic properties for the situation that a Co atom is located in one focal point of the ellipse (From [16]. Image reproduced by permission of IBM Research. Unauthorized use not permitted.)

eigenstate for each ellipse is plotted in Fig. 10.7e and f. The quantum mirage is evidently projected predominantly through this state which acts as a conduit between foci for the Kondo signature. Modified geometries show that if the Co atom is placed at other positions within the ellipse besides one of the foci, even at high symmetry points as shown in Fig. 10.7g and h, a corresponding mirage is not observed (see Fig. 10.7i and j). The Kondo signature detected at the unoccupied focus (see Fig. 10.7c and d) exhibits about a third of the strength of that at the focal Co atom while possessing comparable spatial extent. A component of this signal arises from the simple perturbation of the eigenmodes pictured in Fig. 10.7e and f as their energies are shifted owing to an impurity inhabiting a region of high electron probability.

This means that specific electronic properties can be transferred which is shown in the left part of Fig. 10.8 representing an overlap of topographic and electronic properties. Due to the correlation of the Kondo resonance with magnetic behavior magnetic moments are also transferred to the second focal point *without* the existence of a magnetic atom at this site (see right part of Fig. 10.8).

10.3 Reversal of the Spin in a Single Atom

In an experiment using an STM the tunnelling current between the tip and the sample is carried by elastic and inelastic "channels". In an elastic tunnelling process the energy of the electron is conserved when it hops out of an occupied state of the negatively biased electrode into an empty state of the positive one. In contrast, an inelastic tunnelling process requires energy to be transferred between the tunnelling electron and the sample. Because this energy is quantized inelastic channels cannot contribute to the total tunnelling current if the bias potential is lower than the quantization energy. Above this threshold there will be a sudden conductance jump between tip and sample. This effect is the basis of inelastic electron tunnelling spectroscopy (IETS).

When highly diluted magnetic atoms in a non-magnetic host matrix are exposed to an external magnetic field B the electron potentials of spin up and spin down atoms become slightly different. The energy required to overcome the resulting energy gap in a spin-flip process amounts to twice the Zeeman energy

$$E_{\text{spin-flip}} = 2E_{\text{Zeeman}} = 2g\mu_B B \qquad (10.1)$$

Because B is an adjustable experimental parameter and the Bohr magneton μ_B is a fundamental constant this relation can be used to measure the Landé – g-factor which determines the spin and orbital contributions to the total magnetic moment.

In the investigation described in [17] and discussed below the reversal of the spin in a single atom was measured using IETS in order to determine the magnitude of g.

A topograph of a partially oxidized NiAl surface (see Fig. 10.9A) shows that the bare metal and the Al_2O_3 oxide regions are atomically flat. Contrast on the metal is caused by standing waves in surface state electrons. The cold sample was subsequently dosed with a small amount of Mn and the same area was imaged again (see Fig. 10.9B). Single Mn atoms are seen as protrusions with an apparent height of 0.13 nm on the bare metal surface and 0.16 nm on the oxide. The density of Mn atoms on the oxide is significantly smaller than on the metal presumably due to a lower sticking probability and motion along the oxide surface during adsorption. The upper set of spectra in Fig. 10.9C shows the marked magnetic-field dependence of the conductance when the tip is positioned over a Mn atom on the oxide. At $B = 7$ T the conductance is reduced near zero bias with symmetric steps up to a about 20% higher conductance at an energy of $\Delta = 0.8$ meV. These conductance steps are absent at $B = 0$. Furthermore, no conductance steps are observed when the tip is positioned over the bare oxide surface.

The characteristic signature of spin-flip IETS is a step up in the differential conductance dI/dU at a bias voltage corresponding to the Zeeman energy $E_{\text{Zeeman}} = \Delta = g\mu_B B$. Figure 10.10A shows that the conductance step shifts to higher energy with increased field. The measured Zeeman splitting is proportional to the magnetic field (Fig. 10.10B, black points). The data fit well with a straight line through the origin and a slope that corresponds to $g = 1.88$. A different Mn atom, this one within 1 nm of the edge of an oxide patch, shows a significantly different g value (gray points) of $g = 2.01$. The only difference between these two Mn atoms is the local environment. Thus, different lateral distances of the Mn atoms to the bare metal region significantly change electronic and magnetic properties.

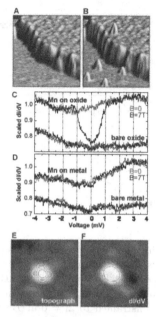

Fig. 10.9. Comparison of Mn atoms on oxide and on metal. (**A**) STM constant-current topograph of a NiAl(110) surface partially covered with Al_2O_3 (*upper right*). (**B**) Same area after dosing with Mn. (**C**) Conductance spectra on the Mn atom on oxide (*upper curves*) measured at $B = 7\,T$ (*black*) and $B = 0\,T$ (*gray*). The lower curves (shifted for clarity) were measured over the bare oxide surface. (**D**) Conductance spectra on a Mn atom on NiAl (*upper curves*) and on the bare NiAl surface (*lower curves*). (**E**) Topograph of the Mn atom on oxide. (**F**) Spatial map of dI/dU acquired concurrently; an increased signal (*light area*) maps the spatial extent of the spin-flip conductance step. (From [17]. Reprinted with permission from AAAS.)

10.4 Influence of the Substrate

A comparison of the magnetic moments of $3d$ adatoms on Pd, Ag, and Cu(001) surfaces is presented in Fig. 10.11. The lattice constant of Cu is about 10% smaller than the lattice constant of Ag which increases the $sp-d$ hybridization with the substrate considerably. As a result the magnetic moments of the adatoms on the Cu surface are always smaller than those on the Ag surface. Compared to the Ag and the Cu substrates the whole curve for the $3d$ adatoms on the Pd substrate is shifted to the right. In fact, we obtain that the Ti adatom is non-magnetic and the magnetic moments of V and Cr adatoms are slightly suppressed while the moments of the adatoms at the end of the d series are enhanced and larger than the moments on the Ag(Cu) surface. Exactly the same trend was observed for the $3d$ monolayers on the Pd(001)

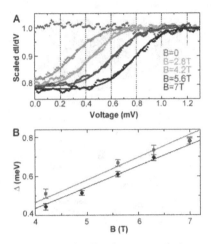

Fig. 10.10. Shift of the spin-flip conductance step with magnetic field. (**A**) Conductance spectra (points) for an isolated Mn atom on oxide at different magnetic fields. (**B**) Magnetic field dependence of the Zeeman energy Δ. Black points are extracted from fits in (A), and gray points were taken on a Mn atom near the edge of an oxide patch. Linear fits (see lines) constrained to $\Delta = 0$ at $B = 0$ yield g values of 1.88 and 2.01, respectively. (From [17]. Reprinted with permission from AAAS.)

surface and the $3d$ impurities in bulk Pd (see Fig. 10.2). It was shown that due to the hybridization with the $4d$ band of Pd the d states of the impurities and the monolayers at the end of the d series are shifted to higher energies leading to an increase of the LDOS at E_F and thus to higher moments. Due to the large extent of the $4d$ and the $5d$ wave functions the differences between the Pd(001) and the Ag(001) substrates should be more pronounced for the $4d$ and $5d$ adatoms. This is clearly shown in Fig. 10.12 where the results for the $4d$ and

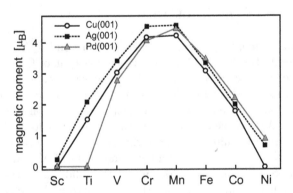

Fig. 10.11. Comparison of the calculated local magnetic moments of $3d$ adatoms on different substrates (Ag(001), Cu(001), Pd(001)). (Adapted from [13] (used with permission))

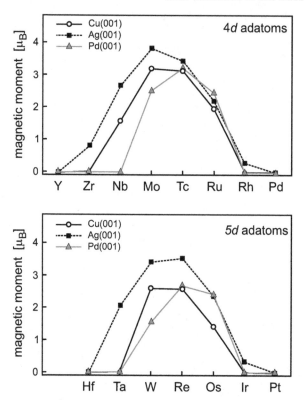

Fig. 10.12. Comparison of the calculated local magnetic moments of (**a**) 4*d* and (**b**) 5*d* adatoms on different substrates (Cu(001), Ag(001), Pd(001)). (Adapted from [13] (used with permission))

the 5*d* adatoms on the Pd, Ag, and the Cu surfaces are presented. Compared to the Ag surface the magnetic moments of Ru and Os being isoelectronic to Fe on the Pd surface are enhanced but the moments of Mo and W (isoelectronic to Cr) are suppressed while Nb and Ta (isoelectronic to V) are non-magnetic.

11

Magnetism in Reduced Dimensions – Clusters

In the following we want to increase the size, i.e. the number of atoms, of the magnetic "system" and thus discuss the behavior of magnetic clusters. In this context a cluster represents an ensemble of atoms which possesses varying properties with each additional atom ("every atom counts"). Contrarily, nanoparticles do not show significant changes of their (magnetic) properties if the number of atoms is changed by a few ones. The differentiation between clusters and nanoparticles can thus be given by:

- Clusters
  Every additional atom in a cluster significantly changes the properties. The behavior is not monotonous with increasing number of atoms (non-scalable regime).
- Nanoparticles
  Additional atoms only slightly change the physical properties. Different parameters like the Curie temperature exhibit a monotonous dependence on the number of atoms in the particle (scalable regime).

In this chapter we will restrict our discussion to clusters whereas in the next one the behavior of nanoparticles will be considered.

11.1 "Every Atom Counts"

To probe the magnetic properties of $3d$ ferromagnets x-ray absorption spectroscopy (XAS) at the $2p$ edges where electron transitions are predominantly into empty $3d$ states is well suited since the magnetic moments in this case are entirely carried by the $3d$ electrons. Using circularly polarized light the spin and orbital contributions m_S and m_L, respectively, to the magnetic moments of the cluster atoms can be separated.

The application of this experimental technique known as x-ray magnetic circular dichroism (XMCD) to $2p \rightarrow 3d$ absorption spectra requires the total number of $3d$ hole states n_h to be known in order to extract absolute values

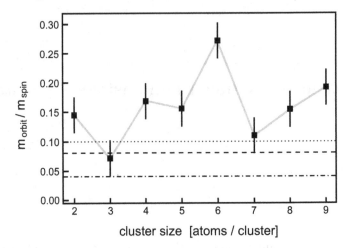

Fig. 11.1. Ratio of Fe_n orbital m_L to spin magnetic moments m_S vs. cluster size as compared to bulk iron (*dash-dotted line*), ultrathin films (*dashed line*), and iron nanoclusters with an average cluster size of about 400 (*dotted line*). (Adapted from [18] (used with permission))

of the spin and orbital magnetic moments. Since n_h is not known for small clusters coupled to a surface and could vary considerably as a function of cluster size magnetic moments will often be given as moments per $3d$ hole. The advantage of giving the ratio of orbital to spin contributions as a measure of the clusters' magnetic properties is that uncertainties in cluster magnetization and photon polarization which would contribute to an error in the absolute values will cancel each other upon dividing m_L by m_S.

Let us discuss the magnetic properties of Fe_n clusters ($n = 2 - 9$) on a magnetized Ni surface. The results of the analysis of Fe_n XMCD spectra are displayed in Fig. 11.1 and Fig. 11.2. Here, the ratios of m_L to m_S are plotted as a function of cluster size for Fe_2 through Fe_9. These ratios range from 0.11 to 0.27 and, in general, are larger than those observed for bulk iron (dash-dotted line in Fig. 11.1), iron ultrathin films (dashed line in Fig. 11.1), and iron nanoscaled cluster films (dotted line in Fig. 11.1). An increasing ratio of m_L to m_S is expected with decreasing dimensionality or coordination of a system; this trend is visible in Fig. 11.1 when going from bulk iron (dash-dotted line) to small iron clusters. In addition to this more general observation there is also a non-monotonous variation in the data displayed in Fig. 11.1 with the lowest ratio of m_L to m_S observed for Fe_3 and the highest ratio for Fe_6. This variation reflects the strong dependence of electronic and magnetic properties on cluster size and geometry in the small size regime.

In addition, the large changes observed in going from n to $n \pm 1$ atoms per cluster gives evidence that the sample preparation procedure yields single sized deposited clusters.

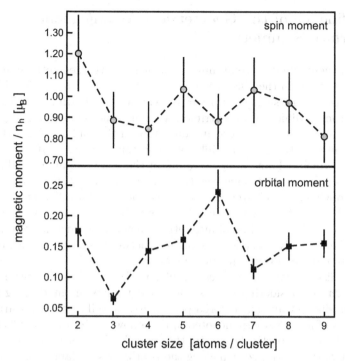

Fig. 11.2. Comparison of spin and orbital magnetic moments per $3d$ hole of Fe_n clusters on Ni/Cu(100). (Adapted from [18] (used with permission))

From calculations for small iron clusters on Ag(001) it is expected that the spin magnetic moment is enhanced over the bulk value and is constant within 10% for small clusters supported on a substrate. Furthermore, for iron atoms on Ag(100) strongly enhanced orbital moments are predicted by theory which was already discussed above (see Fig. 10.11). Applying these findings to Fe_n on Ni/Cu(001) we can conclude that the large values of m_L to m_S plotted in Fig. 11.1 are due to strongly enhanced orbital magnetic moments in Fe_n rather than due to reduced values of m_S. For a test of this conclusion Fe_n magnetic moments per $3d$ hole are evaluated which are shown in Fig. 11.2. Indeed, the upper panel in Fig. 11.2 shows that, with the exception of Fe_2, the spin magnetic moments of the Fe_n clusters are fairly large (approximately $1\mu_B/3d$ hole) and vary only within 15%–20%. On the other hand the lower panel in Fig. 11.2 proves that orbital moments per d hole are strongly enhanced compared to bulk and surface iron and that there is a much larger variation of these moments (by a factor of five from Fe_3 to Fe_6) which is responsible for the observed strong variation in the ratio which was presented in Fig. 11.1.

11.2 Influence of the Geometrical Arrangement and Surface Symmetry

As already mentioned above an important factor is the geometrical arrangement of the atoms in the cluster which is located on a surface. Five different types of clusters on a surface with $n = 1 - 5$ atoms are shown in Fig. 11.3: one single adatom, a dimer, a linear trimer, a square, and a cluster of five atoms with four atoms forming a cross and one atom in the middle of it. Additionally, different structures are possible for a constant number of atoms n in the cluster Fe_n which exhibit different values of the magnetic moment. Further, the possible arrangements significantly depend on the symmetry of the surface. It is obvious that clusters on a cubic surface exhibit different properties compared to clusters on a surface with a hexagonal symmetry. Thus, experimentally only averaged values of magnetic moments are generally determined. But, it is theoretically possible to distinguish between non-equivalent sites and to calculate individually the magnetic moments.

Let us discuss such properties for planar Fe, Co, and Ni clusters on a Ag(001) surface considering dimers and linear trimers oriented along the x-axis, square-like tetramers, centered pentamers, as well as a cluster arranged on positions of a 3×3 square denoted in the following simply as 3×3 cluster. In Fig. 11.4 for each particular cluster the equivalent atoms with respect to an orientation of the magnetization along the x- or y-axis are labeled by the same number. Note that for a magnetization aligned in the z-direction the atoms labeled by 2 and 3 in the pentamer and the 3×3 cluster become equivalent.

Calculations for different orientations of the magnetization prove that the spin moments are fairly insensitive to the direction of the magnetization while for the orbital moments remarkably large anisotropy effects apply. For a magnetization along the z-axis the calculated values of the spin and orbital moments for an adatom and selected clusters of Fe, Co, and Ni on Ag(100) are listed in Table 11.1. In there the position indices in a particular cluster refer to the corresponding numbers in Fig. 11.4 and the number of nearest neighbors of magnetic atoms (coordination number n_c) is also given.

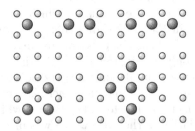

Fig. 11.3. Possible arrangements of clusters on the Ag fcc (001) surface. The *big dark spheres* represent transition metal adatoms and the *smaller white* ones represent surface Ag atoms. (Adapted from [19])

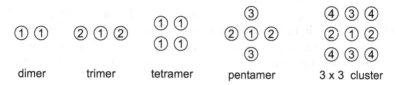

dimer   trimer   tetramer   pentamer   3 x 3 cluster

Fig. 11.4. Sketch of the planar clusters being considered. For an orientation of the magnetization along the x- or y-axis within the plane the equivalent atoms in a cluster are labeled by the same number. (Adapted from [20])

As compared to the corresponding monolayer values ($3.15\mu_B$ for Fe and $2.03\mu_B$ for Co) the spin moment of a single adatom of Fe ($3.39\mu_B$) and Co ($2.10\mu_B$) is slightly increased. In the case of Fe clusters the spin moments decrease monotonously with increasing coordination number. A slight deviation from that behavior can be seen for the 3×3 cluster where the atoms with $n_c = 2$ and 3 exhibit the same spin moment. For the central atom of the pentamer and, in particular, of the 3×3 cluster the monolayer value is practically achieved. The above results reflect a very short ranged magnetic correlation between the Fe atoms.

The general tendency of decreasing spin moments with increasing coordination number is also obvious for the Co clusters up to the pentamer case. For the 3×3 cluster, however, just the opposite trend applies. Establishing

Table 11.1. Calculated spin moments m_S and orbital moments m_L in units of μ_B for small clusters of Fe, Co, and Ni on Ag(100) with magnetization perpendicular to the surface. For each position in a particular cluster (see Fig. 11.4) n_c refers to the number of the neighboring magnetic (Fe, Co, Ni) atoms. (Data taken from [20])

Cluster	Position	n_c	m_S(Fe)	m_L(Fe)	m_S(Co)	m_L(Co)	m_S(Ni)	m_L(Ni)
adatom		0	3.39	0.88	2.10	1.19		
dimer	1	1	3.31	0.32	2.09	0.49		
trimer	1	2	3.29	0.25	2.07	0.45	0.77	0.21
	2	1	3.33	0.44	2.06	0.49	0.70	0.23
tetramer	1	2	3.26	0.18	2.08	0.32	0.76	0.28
pentamer	1	4	3.13	0.15	2.01	0.25	0.76	0.12
	2	1	3.35	0.37	2.10	0.59	0.71	0.33
	3	1	3.35	0.37	2.10	0.59	0.71	0.33
3×3 cluster	1	4	3.15	0.12	2.06	0.23	0.79	0.24
	2	3	3.23	0.16	2.04	0.30	0.71	0.20
	3	3	3.23	0.16	2.04	0.30	0.71	0.20
	4	2	3.23	0.33	2.00	0.29	0.63	0.19

a correlation between m_S and n_c for Co seems to be more ambiguous than for Fe because the changes of the spin moment are much smaller in this case. Nevertheless, it is tempting to say that in the formation of the magnetic moment of Co further off neighbors play a more significant role than in the case of Fe.

For clusters of Ni one can observe an opposite tendency as for Fe and Co: The spin moment enhances with increasing number of neighbors. This can be clearly seen from Table 11.1. Having in mind the calculated monolayer value of $0.71\mu_B$ the small cluster calculations indicate a fairly slow evolution of the spin moment of Ni with increasing cluster size implying that the magnetism of Ni is subject to correlation effects on a much longer scale than in Fe or Co.

Apparently, the orbital moments show a different, in fact, more complex behavior as the spin moments. For single adatoms of Fe and Co orbital moments are enhanced by a factor of about 6 and about 4.5, respectively, as compared to the monolayer values ($0.14\mu_B$ for Fe and $0.27\mu_B$ for Co). This is a direct consequence of the reduced crystal field splitting being relatively large in monolayers, and, in particular, in corresponding bulk systems.

For dimers of Fe and Co, the value of m_L drops to about 40% in magnitude as compared to a single adatom. The evolution of the orbital moment seems, however, to decrease explicitly only for the central atom of larger clusters. The (local) symmetry can be correlated with the magnetic anisotropy, i.e., with the quenching effect of the crystal field experienced by an atom. A single adatom and the central atom of the linear trimers, pentamers, and the 3×3 clusters exhibit well-defined rotational symmetry, namely, C_1, C_2, C_4, and C_4, respectively. The corresponding values of m_L, namely $0.88\mu_B$, $0.25\mu_B$, $0.15\mu_B$, and $0.12\mu_B$ for Fe, and $1.19\mu_B$, $0.49\mu_B$, $0.25\mu_B$, and $0.23\mu_B$ for Co, reflect the increasing rotational symmetry of the respective atoms. Although the outer magnetic atoms exhibit systematically larger orbital moments than the central ones even a qualitative correlation with the local environment (n_c) can hardly be stated. The orbital moment for the trimer of Ni is already close to the monolayer value ($0.19\mu_B$) but shows rather big fluctuations with respect to the size of the cluster and also to the positions of the individual atoms.

As already seen the magnitude of the magnetic moments depends on the coordination number. A further example of this behavior is now given for $4d$ clusters consisting of Ru on an Fe/Cu(100) surface (see Fig. 11.5). A negative moment means an antiferromagnetic coupling. One can see that small linear chains of Ru are ferromagnetic and ferromagnetically coupled to the Fe surface. At the same time in small plane islands of 9 atoms the transition from ferromagnetic to antiferromagnetic coupling is found. The central atom in the plane island of nine atoms has the same coordination as atoms in monolayers. One can see that the moment of this atom is very close to the moment of atoms in the Ru monolayer. Thus, the interaction between Ru atoms on fcc-Fe/Cu(100) leads to the transition from ferromagnetic to antiferromagnetic coupling.

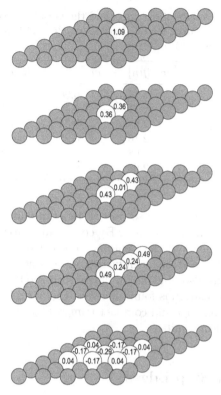

Fig. 11.5. Magnetic moment of Ru clusters on fcc Fe/Cu(100). The linear chains are orientated in the [110]-direction. All moments are given in Bohr magnetons. Negative values denote antiferromagnetic coupling. (From [21] (used with permission))

As already mentioned above the symmetry of the surface directly influences the magnetic properties of adsorbed clusters. It is obvious that the geometrical arrangement of small clusters is rather different on surfaces of cubic crystals (two- or fourfold) compared to that exhibiting a hexagonal symmetry (sixfold).

Thus, let us discuss this behavior on the magnetic properties of Co_n atoms for the situation that the cluster is adsorbed on a hexagonal surface realized by fcc-Pt(111). The spin and orbital moments as well as the anisotropy constant K per Co atom for particles with up to five atoms are given in Table 11.2. We see that m_S is very stable and nearly independent of the coordination number because the majority spin band is almost filled in all cases. In contrast to the spin moment the orbital moment strongly decreases with increasing Co coordination number from $0.60\mu_B$ for the single Co adatom to $0.22\mu_B$/atom for the tetramer.

All Co particles prefer a moment orientation perpendicular to the substrate. The values of K show the correct decrease with cluster size as well as the correct order of magnitude compared with experimental results. Dimer

Table 11.2. Calculated values of m_S, m_L, and the anisotropy constant in perpendicular direction K per Co atom for Co particles on Pt(111) being averaged over all Co sites (Data taken from [22])

n	m_S [μ_B]	m_L [μ_B]	K [meV]
1	2.14	0.60	+18.45
2	2.11	0.38	+4.11
3 (chain)	2.08	0.34	+3.69
3 (triangle)	2.10	0.25	+2.22
4	2.08	0.22	+0.75
5	2.08	0.27	+1.81

and trimer chain-like structures also possess large in-plane anisotropies with 6.91 and 6.19 meV/atom, respectively. By comparing particles with different shape, e.g., the tetramer and the pentamer the calculations show that the atomic coordination has a stronger influence on m_L and K than the absolute particle size which was also obvious for clusters on Ag(100) (see Table 11.1). But contrarily, no correlation is found between the particle point group symmetry (C_{3v} for the monomer and compact trimer, C_{2v} for the remaining particles) and changes of m_L and K.

11.3 Clusters at Step Edges

In the following let us mainly concentrate on the influence of step edges exemplarily on the (100) terrace thus describing in more detail the surfaces vicinal to the (100) surface. The two families of high-Miller-index surfaces which cut the crystal along the (100) × (111) and (100) × (110) step edges have Miller indices $(2m-1, 1, 1)$ and $(m, 1, 0)$, respectively. The terraces exhibited by these vicinal surfaces are m and $(m+1)$ atomic rows wide, respectively. As two examples the fcc-(711) and fcc-(410) vicinal surfaces (see Fig. 11.6) are chosen which exhibit terraces being four atomic rows wide. The calculation of $4d$ magnetic monatomic rows on fcc-(711) and fcc-(410) Ag vicinal surfaces are shown in Fig. 11.7. For both step orientations the rows were placed directly at step edges or in the middle of the terraces. A direct comparison between the magnetic moment profiles for the two different step orientations shows that the magnetic moments for the (100) × (111) rows are smaller than the ones for the (100) × (110) rows for all the elements except Rh. The main reason for this decrease in the magnetic moments is the $4d-4d$ hybridization acting between the atoms in each row which is much larger for the close-packed rows than for the open rows where neighboring row atoms are separated by a second neighbor distance.

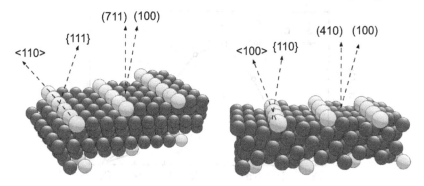

Fig. 11.6. Left: The fcc-(711) vicinal substrate considered to characterize the monatomic rows at the (100) × (111) step edge. **Right**: The fcc-(410) vicinal substrate considered to characterize the monatomic rows at the (100) × (110) step edge. (From [23] (used with permission))

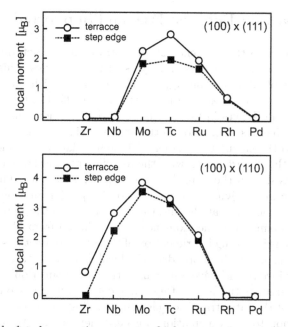

Fig. 11.7. Calculated magnetic moments of 4d monatomic rows (**top**) on the fcc-Ag(711) and (**bottom**) on the fcc-Ag(410) vicinal surfaces used to simulate the close-packed (100) × (111) and open (100) × (110) rows, respectively. The rows were placed both in terrace (*circles*) and step edge (*squares*) positions of the (100) Ag terraces. (From [23] (used with permission))

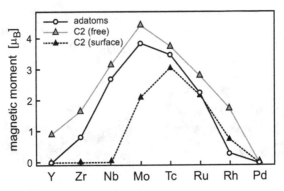

Fig. 11.8. Calculated local magnetic moments per adatom for $4d$ dimers (C2) on the Ag(001) surface and for free $4d$ dimers in free space (with Ag nearest neighbor distance). For comparison the local moments of single $4d$ adatoms are also given. (From [24] (used with permission))

11.4 Comparison Between Free and Supported Clusters

The influence due to the existence of a substrate can be determined by the comparison of supported clusters and layers with "free" ones. This is depicted for the example of $4d$ dimers on Ag(001) in Fig. 11.8. which shows the calculated local moments per adatom. In order to see the effect of the Ag substrate calculations for $4d$ dimers in free space are also performed which the molecular separation has been fixed for to the nearest neighbor distance in Ag. As a reference also the calculated moments of single adatoms are included. The comparison of the dimers on Ag with the free dimers shows a strong decrease of the moments due to the hybridization with the substrate. This is particularly dramatic for Nb where the moment of $3.16\mu_B$ for an atom of the free dimer is totally quenched at the surface. In addition to the reduction of the moments the maximum of the moment curve is shifted from Mo to Tc. Both effects arise from the strong hybridization of the $4d$ wave functions with sp-like valence electrons of Ag which broadens the local density of states and reduces the moments. Since the $4d$ wave functions of the early transition elements have an especially large spatial extent these moments are much strongly reduced than the ones of the later transition elements leading to the observed shift of the peak position to Tc. Calculations for $4d$ dimers in the Ag bulk show no or negligible magnetism. Thus, only at the surface the hybridization is weak enough so that the $4d$ magnetism survives.

Magnetism in Reduced Dimensions – Nanoparticles

As already mentioned in the introduction of Chap. 11 a nanoparticle represents an ensemble of atoms which does not change its behavior if the number of atoms is varied by only a small amount. In this chapter we will go into detail concerning the magnetic properties of nanoparticles.

12.1 Magnetic Domains of Nanoparticles

Assuming a soft magnetic material without any anisotropy a continuous vector field of the magnetization is expected. Neglecting any crystalline anisotropy a regular domain pattern can already be present only caused by the shape which is due to the minimization of the stray field energy. The conditions for the vanishing of charges of a magnetization vector field are:

- It has to be divergence-free, i.e. free of volume charges.
- It must be oriented parallel to the edges of the nanoparticle and to the surface in order to avoid surface charges.
- The magnitude of the magnetization vector must be constant.

For a rectangularly shaped element a continuous vector field does not simultaneously fulfill all conditions. Inside the element the stray field vanishes but the magnetization lies not parallel to the whole rim (see Fig. 12.1). All conditions are only fulfilled if the vector field exhibits linear discontinuities which are domain walls in reality. The magnetic state which is presented in Fig. 12.1(b) is known as Landau state. We see that an ordered domain pattern can already be created only due to the shape of magnetic elements.

An algorithm to construct the domain structure for curved shaped elements consists of the following characteristics:

- Take circles that touch the edge at two or more points and lie otherwise completely within the element. The centers of all circles represent a domain wall.

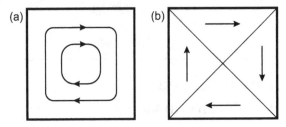

Fig. 12.1. (a) Instead of a continuous vector field, (b) a rectangular shaped soft magnetic element exhibits linear discontinuities. The magnetic state shown in (b) is called Landau state. (Reprinted from [25] with permission of the FZ Jülich)

- In each circle the magnetization vector is perpendicular to each touching radius (see Fig. 12.2a). The walls are then stray-field free.
- If the circle touches the edge in more than two points the center represents a crossing point of domain walls (dark circle in Fig. 12.2a).
- If the touching points fall together the domain wall ends in the center of the circle which becomes the central point of a domain with a continuous rotation of the magnetization (see Fig. 12.2b).
- If sharply shaped edges are present the domain wall proceeds to the corner (see Fig. 12.2c).

An unambiguousness of the domain pattern is not given as exemplarily shown for a circularly-shaped element. The algorithm leads to a single circle which touches the whole surface and the central point is the center of a domain which exhibits a continuous rotation of the magnetization (see Fig. 12.2d). But, different domain patterns are additionally observed (see Fig. 12.2e). These metastable configurations can also be constructed using the algorithm by a virtual cutting of the element and a subsequent applying of the algorithm (see Fig. 12.2f).

The realization of a specific domain pattern depends on the magnetic history and anisotropic effects. Thus, domain patterns are not identical whensoever a demagnetization was carried out before using alternating magnetic fields. This behavior is demonstrated using differently shaped permalloy elements (see Fig. 12.3). Here, those domain patterns are shown which occur with the highest probability. The elements exhibit different easy magnetization axes oriented along K_u characterizing a uniaxial anisotropy and different directions of the demagnetizing field $\boldsymbol{H}$.

Easy axes oriented along the long axis of the elements (see Fig. 12.3a and d) mostly result in domains being elongated in the direction of the easy axis with closure domains at the front side. This is not true for sharp elements and for ellipses with continuously changing domains (see Fig. 12.3b, c, e, f). If the easy magnetization axis is oriented along the short axis of the element domains occur which are rotated by 90° (see Fig. 12.3g-l). Despite of an identical

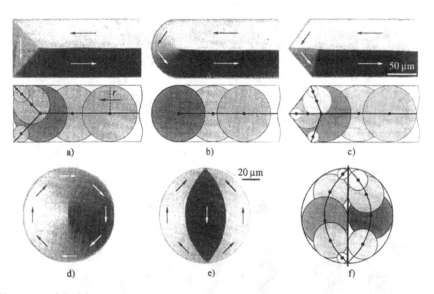

Fig. 12.2. (a)–(c) Differently shaped soft magnetic elements with their equilibrium domain structure (only one half of the elements is shown). The construction using the algorithm discussed in the text is additionally given. A virtual cut (f) allows to to construct metastable states (e) which possess more domain walls than the ground state (d). (Reprinted from [25] with permission of the FZ Jülich)

demagnetization procedure different domain patterns can occur (see forth and fifth row in Fig. 12.3).

Smaller magnetic particles possess simpler domain formations. Less domain walls are needed in order to minimize the stray field energy. The extremal case is represented by single domain particles (see Fig. 12.4). Micromagnetic simulations allow to calculate the different domain patterns (see Fig. 12.5). Large particles exhibit a lot of domain walls. Below a size of about 500 nm it is no more energetically favorable to form many domains. Now, the energy is minimized by the so-called C- and S-states. A further reduction of the size leads to single domain particles.

An additional important influence is given by the interaction of domain walls with the rim of the element, i.e. surface, of the magnetic particles (see Fig. 12.6). If an asymmetric Bloch wall does not touch the surface a reversible behavior is found (see the lower part of Fig. 12.6). An analogous behavior occurs for roundly shaped elements (see Fig. 12.7a → d → f → i). A contact of the domain wall with the surface is equivalent to its annihilation. Thus, an irreversible behavior occurs (see Fig. 12.7a → e → j → n). The formation of completely different backward pattern sequences results in different final zero-field states.

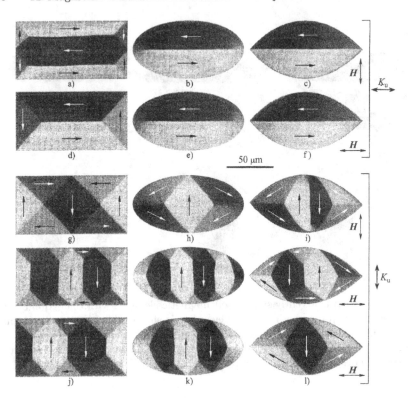

Fig. 12.3. Demagnetized states of various thick film elements. The particles differ in their shape and in their orientation of their easy axis relative to the particle axis. The resulting demagnetized states markedly depend on the alternating field axes used in the demagnetizing procedure. Interestingly, this is not true for elliptical and pointed shapes with a longitudinal easy axis (**b–e, c–f**). Different patterns can be formed under the same conditions as shown in the last two lines which apply to transverse easy axes and longitudinal alternating current fields. (From [10] (used with permission))

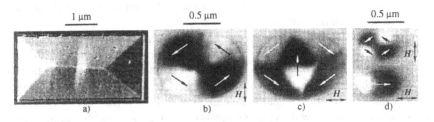

Fig. 12.4. (**a**) Rectangular shaped soft magnetic element. (**b**)–(**d**) Elliptical Co elements. The larger element shows either (**b**) a concentric state or (**c**) a three domain state. (**d**) The concentric state can also be observed in the smaller element after applying a field along the shorter axis. Otherwise a single domain state is observed which can be recognized by its black and white contrast. (Reprinted from [25] with permission of the FZ Jülich)

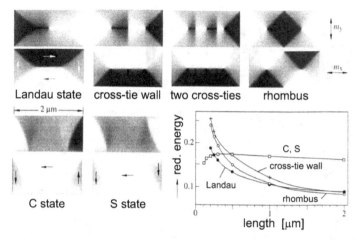

Fig. 12.5. Calculated various magnetization states of soft magnetic elements. The x- and y-components are gray coded. The diagram compares the reduced energy of different states as a function of the length for a constant length-to-width ratio of $2:1$. For elements with a length being smaller than $0.3\,\mu m$ the quasi single domain states (C and S) are the stable configurations. (Reprinted from [25] with permission of the FZ Jülich)

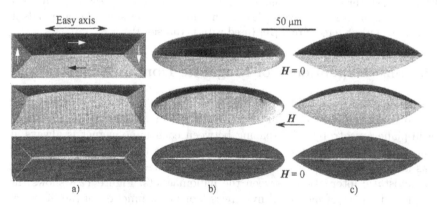

Fig. 12.6. Symmetrical demagnetized domain patterns (*top row*) of thick film elements. The domain walls are displaced reversibly in an applied field (*second row*) and return almost to their previous position after removing the field. This is documented in the third row by difference images between positive and negative remanent states. (From [10] (used with permission))

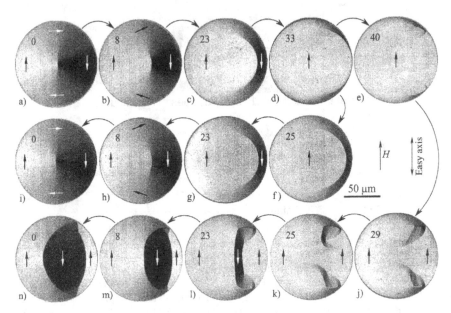

Fig. 12.7. Magnetization processes in circular thick film elements showing the interaction of a domain wall that is formed in an applied field (**a**)–(**d**) with the sample edge. If the repulsive interaction is not overcome the process stays reversible (**f**–(**i**). After a punch-through (**e**) a complete different sequence is observed (**j**)–(**n**). The field which is indicated in every element in units of A/cm starts (a) and ends in both cases (i,n) at zero. (From [10] (used with permission))

12.2 Size Dependence of Magnetic Domain Formation

For the detailed discussion of size dependent magnetic domain formation let us choose an array of circular disks consisting of an Ni–Fe–Mo alloy with an in-plane anisotropy. The spacing between each nanomagnet is always at least equal to the diameter of the nanomagnet. For the smallest structures it is as large as three times the diameter. This ensures that there is negligible magnetostatic interaction between the nanomagnets. Figure 12.8 shows the complete data set of measured hysteresis loops as a function of the diameter and thickness of the nanomagnets being normalized in height to remove the effects of thickness, array size, and array filling factor. The magnetic field was applied in the plane of the nanomagnets in the direction of the uniaxial anisotropy easy axis. One sees from Fig. 12.8 two classes of loops which are shown in detail in Fig. 12.9 with schematic annotation. The first class is typified by a loop of a nanomagnet with a diameter of 300 nm and a thickness of 10 nm (see Fig. 12.9a). As the applied field is reduced from minus saturation the nanomagnets retain full moment until a critical field slightly below zero. At this point nearly all magnetization is lost. The magnetization then progressively reappears as the field is increased from zero until positive satu-

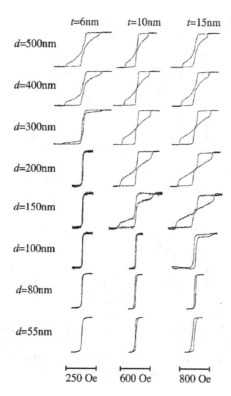

Fig. 12.8. Hysteresis loops measured as a function of diameter d and thickness t from circular nanomagnets. For each loop the horizontal axis is the applied field and the vertical axis is the magnetization. (Figure reprinted from [26] with permission. Copyright (1999) by the American Physical Society.)

ration is achieved. The sudden loss of magnetization close to zero field is very characteristic of the formation of a flux closing configuration; the simplest one is a magnetic vortex which the magnetization vector remains parallel to the nearest edge in at all points in the circular nanomagnet. A detailed discussion is given below.

In large structures this state lowers the energy of the system by reducing stray fields and hence lowering magnetostatic energy. Increasing the field then deforms the magnetic vortex by pushing its core away from the center of the nanomagnet until it becomes unstable and the vortex is eventually annihilated although a field of several hundred Oerstedt has not yet been reached.

The second class of loop is typified by a nanomagnet with a diameter of 100 nm and a thickness of 10 nm (see Fig. 12.9b). These loops retain a high remanence of about 80% and switch at a very low field of about 5 Oe. This is characteristic of single domain behavior: All nanomagnets within the array retain all of their magnetization to form an array of giant spins and magnetization reversal occurs by each giant spin rotating coherently.

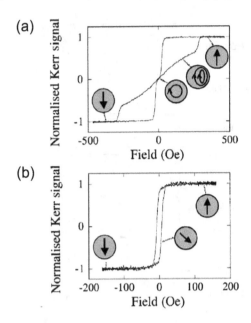

Fig. 12.9. Hysteresis loops measured from nanomagnets of diameter d and thickness t: (**a**) $d = 300$ nm, $t = 10$ nm; (**b**) $d = 100$ nm, $t = 10$ nm. The schematic annotation shows the magnetization within a circular nanomagnet, assuming a field oriented up the page. (From [26] (used with permission))

Now, we discuss the more complex magnetic behavior of *perpendicularly* magnetized particles on a ferromagnetic surface with an *in-plane* magnetization referring to [27].

Using an STM which is sensitive to the spin polarization of the tunnelling electrons (a short description of this experimental technique called spin polarized scanning tunnelling microscopy SPSTM is given on p. 288)) allows to determine the topography (see Fig. 12.10(a)) and the magnetic dI/dU signal (see Fig. 12.10(b)) of 1.28 monolayers (ML) Fe as grown on a W(110) substrate held at room temperature. Thus, Fig. 12.10(b) represents an image of the spin resolved density of states (SRDOS) at a given energy.

This preparation leads to Fe islands with a local coverage of 2 ML, i.e. double layer (DL) islands, surrounded by a closed and thermodynamically stable ML. As can be recognized in Fig. 12.10(a) the Fe DL islands which are about 10 nm wide are elongated along the [001] direction leading to a length of approximately 30 nm on average. This sample system can be characterized in terms of a thickness dependent anisotropy. While the ML preferentially keeps the magnetization within the film plane the DL islands exhibit a perpendicular anisotropy. Consequently, a magnetic probe tip which is sensitive to the perpendicular magnetization is expected to image the domain structure of the DL islands. Indeed, as can be recognized in Fig. 12.10(b) two different

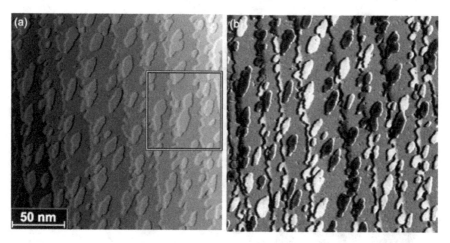

Fig. 12.10. (a) Topography and (b) magnetic dI/dU signal as measured with a Gd tip on Fe double-layer islands surrounded by a closed ML Fe/W(110) as prepared by evaporation of 1.28 ML Fe/W(110). Approximately an equal number of black and white double layer islands can be found representing opposite perpendicular magnetization directions. The region in the *rectangle* is shown in more detail in Fig. 12.11. (Reprinted from [27] with permission of IOP)

values of the dI/dU signal occur representing islands with their directions of magnetization aligned parallel or antiparallel to the tip magnetization. Approximately an equal number of black and white DL islands can be found. By changing the magnetization direction of adjacent Fe islands between up and down on a nanometer scale the stray field above the sample surface can dramatically be reduced.

Within the center of the box in Fig. 12.10(a) an island with a constriction can be seen. It may be caused by the nearby nucleation of two individual Fe DL islands in an early stage of the preparation process. As the growth proceeded the islands coalesced at the constriction thereby forming a single island. The region around this particular island is shown at higher magnification in Fig. 12.11(a)–(c). It can clearly be recognized that a high dI/dU signal is found in the upper part of the island while a much lower value is found in the other part. The domain wall is located just at the position where the constriction becomes narrowest which allows the domain wall energy to be minimized. Figure 12.11(d) shows a line section of the dI/dU signal measured along the line in (b). The y-scale indicates that at the position of the domain wall the dI/dU signal changes by a factor of three which corresponds to an effective spin polarization of the junction of about 50%. Examining the wall profile results in a domain wall width $w = 6.5\,\mathrm{nm}$.

A close inspection of Fig. 12.11(b) also reveals that some Fe DL islands are neither white nor black but instead exhibit an intermediate dI/dU signal. Four of these islands have been marked by circles. Obviously, these islands

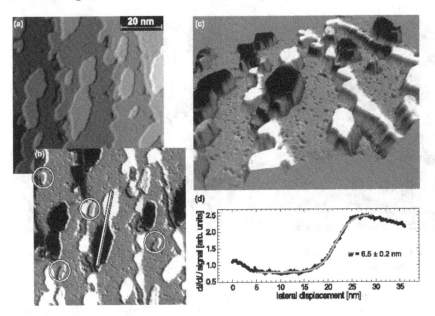

Fig. 12.11. (**a**) Topography and (**b**) magnetic dI/dU signal of the region in the rectangle in Fig. 12.10(a). In the center of the image an island with a constriction can be recognized. (**c**) Rendered perspective representation of the topographic data gray coded by the magnetic dI/dU signal. (**d**) Plot of dI/dU as measured on the constricted island along the line in (b). The fit (*gray line*) results in a domain wall width of 6.5 nm. (Reprinted from [27] with permission of IOP)

are rather small which suggests that a size dependent effect is responsible for the variation of the dI/dU signal.

In order to check this first impression the strength of the dI/dU signal of about 140 individual islands has been analyzed and plotted versus the width of each particular island along the $[1\bar{1}0]$ direction, i.e. the short island axis, in Fig. 12.12(a). The error bar represents the standard deviation over the island area. Three different size regimes can be recognized: (I) large islands exhibit either a high or a low value of the dI/dU signal. An intermediate signal strength has never been observed for islands with a width of more than approximately 3 nm. (II) For islands which exhibit a width between 1.8 and 3.2 nm a strong variation of the dI/dU signal is observed. (III) Finally, Fe DL islands with a width below 1.8 nm show an intermediate dI/dU signal.

Qualitatively, this observation can be understood on the basis of the sample's complicated nanostructure which is governed by the close proximity of regions with different anisotropies: While the closed ML exhibits an in-plane easy axis it is perpendicular for the DL. As long as the DL island is sufficiently large the local magnetization rotates by 90° from in-plane to out-of-plane at the boundary between the closed Fe ML and the DL islands. This situation is

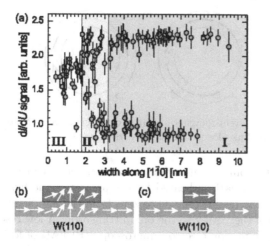

Fig. 12.12. (a) Plot of the dI/dU signal strength of about 140 individual Fe islands versus their geometrical width along the [1$\bar{1}$0] direction, i.e. the short island axis. (b) In large Fe double layer islands the magnetization rotates out of the easy plane of the ML into the perpendicular easy axis of the double layer thereby forming a 90° domain wall. (c) As the Fe double layer islands become too small it is energetically favorable to keep the magnetization in-plane as the domain wall costs too much energy. (Reprinted from [27] with permission of IOP)

schematically represented in Fig. 12.12(b). Since the corresponding material parameters of the Fe ML on W(110) allow the rotation to take place on a much narrower scale than in the DL the lateral reorientation transition mainly occurs in the ML region around the DL island. As the DL islands become smaller and smaller the energy which is gained by turning the magnetization into the easy magnetization direction of the DL decreases until it is smaller than the energy that has to be paid for the 90° domain wall that surrounds the DL island. Then, it is energetically favorable to keep the magnetization of the DL in-plane in spite of the fact that the local anisotropy suggests a perpendicular magnetization direction as shown in Fig. 12.12(c).

12.3 Ring Structures

For all ring structure geometries macroscopic measurements and micromagnetic simulations suggest the existence of two magnetic states: the "onion" state accessible reversibly from saturation and characterized by the presence of two opposite head-to-head walls and the flux-closure magnetic vortex state (see Fig. 12.13).

It is possible to use an appropriate minor loop field path to obtain rings in the onion state or in the magnetic vortex state at remanence. Measuring an array of such rings with a switching field distribution some of the rings

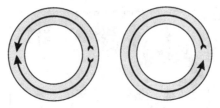

Fig. 12.13. Two different magnetic states can occur in ring structures. **Left:** "onion" state; **right:** flux-closure magnetic vortex state

can switch into the onion state while some will remain in the vortex state by following the field path indicated by the arrows in the hysteresis loop in Fig. 12.14a where the field is relaxed to zero from the middle of the switching field distribution of the vortex to the onion transition. A low resolution image by means of photoemission electron microscopy (PEEM) of four rings (outer diameter $D = 1100$ nm, inner diameter $d = 850$ nm, and thickness $t = 15$ nm of polycrystalline Co, taken at remanence) is shown in Fig. 12.14b. As expected some rings are still in the vortex state (two rings have a counter-clockwise and one ring has a clockwise circulation direction) while one ring has already switched into the onion state. The spin structure can also be calculated by micromagnetic simulations. Figure 12.15 shows the result for an onion state at remanence (a), a magnetic vortex state at applied field (b), and a reversed onion state with applied field after switching (c).

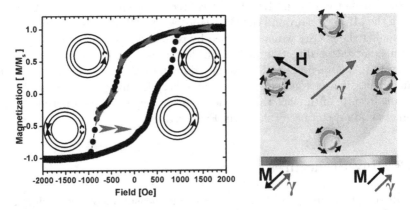

Fig. 12.14. (a) Hysteresis loop measured on an array of rings of Co. The magnetization configurations of the onion and the vortex states are shown schematically. The arrows indicate the field path used to obtain the rings in the states shown in (b). (b) Image of four polycrystalline Co rings. The top ring is in the clockwise vortex state, the bottom and right rings are in the counter-clockwise vortex state, and the left ring is in the onion state pointing along the direction of the applied field H. (From [28] (used with permission))

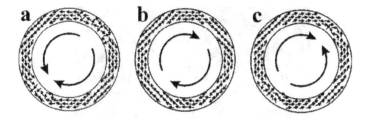

Fig. 12.15. Micromagnetic calculations of the onion state and vortex state in an applied field: (**a**) onion state at remanence, (**b**) vortex state, and (**c**) reversed onion state after switching. The small arrows represent the magnetic moments. (Reused with permission from [29]. Copyright 2002, American Institute of Physics)

12.4 Magnetic Vortices

A magnetic force microscopy (MFM) image of an array of 3×3 dots of permalloy with a diameter of $1\,\mu m$ and a thickness of $50\,nm$ is shown in Fig. 12.16. For a thin film of permalloy the magnetic easy axis typically has an in-plane orientation. If a permalloy dot has a single domain structure a pair of magnetic poles reflected by a dark and white contrast should be observed in an MFM image. In fact, the image shows a clearly contrasted spot at the center of each dot. It is suggested that each dot has a curling magnetic structure and the spots observed at the center of the dots correspond to the area where the magnetization is aligned parallel to the plane normal (see Fig. 12.17). However, the direction of the magnetization at the center seems to turn randomly, either up or down, as reflected by the different contrast of the center spots. This seems to be reasonable as up- and down-magnetization are energetically equivalent without an external applied field and do not depend on the vortex orientation (clockwise or counter-clockwise).

The question arises what is the diameter of this core. The investigation by MFM gives an upper limit of about $50\,nm$ (see Fig. 12.16) caused by the

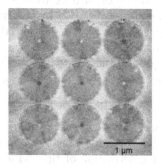

Fig. 12.16. MFM image of an array of permalloy dots which are $1\,\mu m$ in diameter and $50\,nm$ thick (From [30]. Reprinted with permission from AAAS.)

Fig. 12.17. Schematic of a magnetic vortex core. Far away from the vortex core the magnetization continuously curls around the center with the orientation in the surface plane. In the center of the core the magnetization is perpendicular to the plane (highlighted). (From [31]. Reprinted with permission from AAAS.)

intrinsic lateral resolution which is due to the detection of the stray field. An enhanced lateral resolution can be obtained using the technique of SPSTM which will be discussed in the following for another type of ferromagnetic nanoparticles referring to [31].

The magnetic ground state of high Fe islands on a W(110) surface of lateral and vertical size of about 200 nm and 10 nm, respectively, is expected to be a vortex. The dimensions of the particles are too large to form a single domain state because it would cost a relatively high stray field (or dipolar) energy. But they are also too small to form domains like those found in macroscopic pieces of magnetic material because the additional cost of domain wall energy cannot be compensated by the reduction of stray field energy. The magnetization continuously curls around the particle center drastically reducing the stray field energy and avoiding domain wall energy.

A constant current topograph of a single Fe island is shown in Fig. 12.18A. Though the dI/dU signal on top of such Fe islands is found to be spatially constant if measured with non-magnetic W tips a spatial pattern can be recognized in the dI/dU map (see Fig. 12.18B) measured with a tip coated with more than 100 ML of Cr. This variation is caused by spin polarized tunnelling between the magnetic sample and the polarized tip. The magnetization of the tip is parallel to the surface plane. Four different regions, referred to as domains, can be distinguished in Fig. 12.18B. Assuming positive polarization of tip and sample the observed pattern can be explained by a local sample magnetization that is parallel (bottom) and antiparallel (top) to the magnetization of the tip, respectively. An intermediate contrast in the left and the right domain shows that magnetization of tip and locally of the sample are almost orthogonal. A corresponding domain pattern exhibiting a flux-closure configuration is indicated by the arrows in Fig. 12.18B. However, because

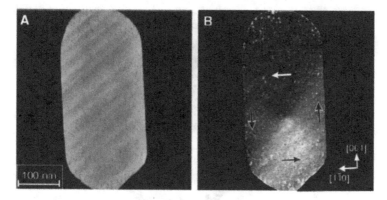

Fig. 12.18. (**A**) Topography and (**B**) map of the dI/dU signal of a single 8 nm high Fe island recorded with a Cr coated W tip. The vortex domain pattern can be recognized in (**B**). Arrows illustrate the orientation of the domains. Because the sign of the spin polarization and the magnetization of the tip is unknown the sense of vortex rotation could also be reversed. (From [31]. Reprinted with permission from AAAS.)

neither the absolute direction of magnetization of tip nor the sign of polarization of tip and sample is known the opposite sense of rotation would also be consistent with the data.

In order to gain a detailed insight into the magnetic behavior of the vortex core a zoom into the central region is carried out where the four "domains" touch and where the rotation of the magnetization into the surface normal is expected. Maps of the dI/dU signal measured with Cr-coated tips that are sensitive to the in-plane and out-of-plane component of the local sample magnetization are shown (see Fig. 12.19A and B, respectively). The dI/dU signal as measured along a circular path at a distance of 19 nm around the vortex core (circle in Fig. 12.19A) is plotted in Fig. 12.19C. The cosine-like modulation indicates that the in-plane component of the local sample magnetization continuously curls around the vortex core. Figure 12.19B which was measured with an out-of-plane sensitive tip on an identically prepared sample exhibits a small bright area approximately in the center of the island. Therefore, the dI/dU map of Fig. 12.19B confirms that the local magnetization in the vortex core is tilted normal to the surface (cf. Fig. 12.17). Figure 12.19D shows dI/dU line sections measured along the lines in (A) and (B) across the vortex core. It is predicted theoretically that the shape of a vortex core is determined by the minimum of the total energy which is dominated by the exchange and the magnetostatic or demagnetization energy. Compared with the latter, the magneto crystalline anisotropy energy, which is relevant for the width of bulk Bloch walls and the surface anisotropy are negligible as long as thin films made of soft magnetic materials like Fe are used. For the thin film limit (i.e., thickness = 0) the vortex width as defined by the slope of the

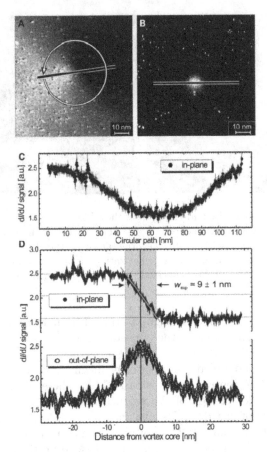

Fig. 12.19. Magnetic dI/dU maps as measured with an (**A**) in-plane and an (**B**) out-of-plane sensitive Cr tip. The curling in-plane magnetization around the vortex core is recognizable in (A) and the perpendicular magnetization of the vortex core is visible as a bright area in (B). (**C**) dI/dU signal around the vortex core at a distance of 19 nm [s. circle in (A)]. (**D**) dI/dU signal along the lines in (A) and (B). (From [31]. Reprinted with permission from AAAS.)

in-plane magnetization component in the vortex center is given by $w = 2\sqrt{A/K}$ (see p. 124) and amounts to 6.4 nm. This value is in reasonable agreement with the experimental result of $w = 9$ nm.

12.5 Single Domain Particles

Particles which exhibit only one single domain are called Stoner–Wohlfarth-particles and are therefore uniformly magnetized to saturation M_S. The description is given below in the Stoner–Wohlfarth model.

The variation of the magnetization in an external magnetic field depends on the anisotropy of this element characterized by the anisotropy constant K which can be due to the shape or to the crystalline structure. Without any anisotropy the magnetization M_S is always oriented parallel to the external magnetic field H.

If we assume a uniaxial anisotropy the corresponding anisotropy energy density term is given by:

$$E = K \sin^2 \theta - \mu_0 HM_S = K \sin^2 \theta - \mu_0 HM_S \cos(\gamma - \theta) \qquad (12.1)$$

with θ being the angle between M_S and the easy magnetization axis which shall be oriented along the x-axis (see Fig. 12.20). The plane which the magnetization can rotate in is defined by the easy axis and the vector of the external magnetic field H which exhibits the angle γ with respect to the x-axis. The decomposition of H in its Cartesian components leads to:

$$E = K \sin^2 \theta - \mu_0 M_S H_x \cos \theta - \mu_0 M_S H_y \sin \theta \qquad (12.2)$$

The shape or magneto crystalline anisotropy and the external magnetic field are acting in the opposite direction on the magnetization. Thus, the energy becomes minimum at a specific angle θ which can be determined by setting the first derivative to zero:

$$0 = \frac{dE}{d\theta} = 2K \sin \theta \cos \theta + \mu_0 M_S H_x \sin \theta - \mu_0 M_S H_y \cos \theta \qquad (12.3)$$

With:

$$\alpha = \frac{2K}{\mu_0 M_S} \qquad (12.4)$$

we obtain:

$$\alpha \sin \theta \cos \theta + H_x \sin \theta - H_y \cos \theta = 0 \qquad (12.5)$$

which can be rewritten as:

$$\frac{H_y}{\sin \theta} - \frac{H_x}{\cos \theta} = \alpha \qquad (12.6)$$

This equation is a quartic expression in $\cos \theta$. Therefore, two or four real solutions exist which depends on the magnitude of H. Two solutions result in

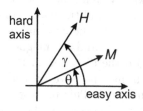

Fig. 12.20. Magnetization M of a single domain particle exhibiting a uniaxial anisotropy in an external magnetic field H

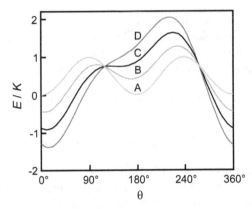

Fig. 12.21. Functional dependence of E/K on the angle θ between the easy magnetization axis and the direction of the magnetization for a constant value of $\gamma = 30°$ which represents the angle between the easy axis and the direction of the external magnetic field. The curves differ in the magnitude of the external field. *Curve A:* $H = 0.5K/\mu_0 M_S$; *curve B:* $H = 1.0K/\mu_0 M_S$; *curve C:* $H = 1.5K/\mu_0 M_S$; *curve D:* $H = 2.0K/\mu_0 M_S$

one minimum, i.e. the magnetization M_S exhibits one equilibrium orientation. Four solutions lead to two minima and thus to two equilibrium orientations of M_S. A graphical representation of both cases is given in Fig. 12.21 which shows the dependence of the energy E/K on the angle θ between the easy magnetization axis and the direction of the magnetization. The angle between the easy axis and the external magnetic field is set to $\gamma = 30°$ for all curves which differ in the magnitude of the external field.

Let us first discuss the situation that the external magnetic field is orientated parallel to the easy axis, i.e. $H_x = H$ and $H_y = 0$ which results in (cf. (12.5)) $\alpha \cdot \cos \theta + H = 0$. We obtain two real solutions for all values of $-1 \leq H/\alpha \leq +1$ and only one for $|H/\alpha| > 1$. Thus, if the applied field and the magnetization of the single domain particle are oriented along the positive value of the easy axis, i.e. $\gamma = 0$ and $\theta = 0$, the first derivative possesses a positive value in a stable position. Reducing the external field to zero and subsequently increasing it to negative values ($\gamma = 180°$) until $-1 \leq H/\alpha$ the magnetization has its saturation magnitude and is aligned along the positive direction of the easy axis. With an increase of the external field the first derivative of the energy decreases and the energy of the domain is approaching the local extremum. As soon as the external field reaches the value of α the first derivative changes its sign and the magnetization becomes unstable. It shows a sudden switch into the direction of the external field. Thus, the magnetization is switching from a positive to a negative value. Due to this behavior α is called switching field. This process leads to a completely irreversible magnetization as it can be seen in the right part of Fig. 12.23 for $\gamma = 0$.

Let us now assume that the external field is applied perpendicular to the easy axis, i.e. $\gamma = 90°$, $H_x = 0$, and $H_y = H$. In this situation (12.5) is given by $\alpha \cdot \sin\theta - H = 0$. Therefore, the component of the magnetization parallel to the applied field is a linear function of the external field and depicted in the right part of Fig. 12.23 for $\gamma = 90°$. We do not observe any hysteretic behavior; the magnetization is completely reversible. The saturation is reached if the magnitude of the external field exceeds the switching field. The first derivative of the energy changes sign while the magnetization turns to the direction of the applied field.

If the external field is applied towards an arbitrary direction to the easy axis the magnetization is partly reversible and partly irreversible. The critical angle which the magnetization switches at from one stable position to the other one can be determined using (12.5). One example is given in the right part of Fig. 12.23 for $\gamma = 45°$.

Within the plane defined in Fig. 12.20 two regions exist with one or two equilibrium orientations of M_S. The boundary between both regions where the equilibrium direction becomes discontinuous for a continuous change of the external field is characterized by the condition:

$$\frac{\partial^2 E}{\partial\theta^2} = 0 \tag{12.7}$$

This second derivative amounts to:

$$\frac{\partial^2 E}{\partial\theta^2} = H_x \cos\theta + H_y \sin\theta + \alpha(\cos^2\theta - \sin^2\theta) \tag{12.8}$$

Inserting of α (see (12.6)) leads to:

$$\frac{\partial^2 E}{\partial\theta^2} = \frac{H_x}{\cos^3\theta} + \frac{H_y}{\sin^3\theta} \tag{12.9}$$

Thus, at the boundary the following two conditions are fulfilled (see (12.6) and (12.9)):

$$\frac{H_y}{\sin\theta} - \frac{H_x}{\cos\theta} = \alpha \tag{12.10}$$

$$\frac{H_y}{\sin^3\theta} + \frac{H_x}{\cos^3\theta} = 0 \tag{12.11}$$

The solution of this set of equations is given by:

$$H_x = -\alpha \cos^3\theta \tag{12.12}$$

$$H_y = \alpha \sin^3\theta \tag{12.13}$$

Introducing reduced magnetic fields:

$$h_x = \frac{H_x}{\alpha} = -\cos^3\theta \tag{12.14}$$

$$h_y = \frac{H_y}{\alpha} = \sin^3\theta \tag{12.15}$$

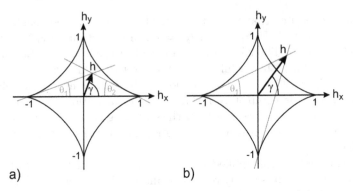

Fig. 12.22. Astroid curve of a Stoner–Wohlfarth particle with (a) two or (b) one equilibrium direction of the magnetization which depends on the magnitude of the reduced magnetic field h

we obtain:

$$h_x^{2/3} + h_y^{2/3} = 1 \qquad (12.16)$$

The graphical representation of the solution allows to determine the equilibrium direction of the magnetization M and is called Stoner–Wohlfarth astroid which is depicted in Fig. 12.22.

One of the possible tangent lines through the tip point of h represents the equilibrium direction of the magnetization. If h is within the astroid four tangent lines exist and there are two equilibrium directions of the magnetization which are given by that lines with the smallest angles with respect to the easy axis θ_1 and θ_2 (see Fig. 12.22a). But, if h is outside the astroid two tangent lines are possible and then only one equilibrium direction of M occurs which is realized by the line which exhibits the smaller angle with respect to the easy axis θ_1 (see Fig. 12.22b).

The astroid also allows to generate the hysteresis curve for an arbitrary direction of the external magnetic field. This is exemplarily shown in the left part of Fig. 12.23 assuming that the external field has a direction of A – B and its reduced magnitude varies between $-h$ and $+h$ characterized by the points 1 to 6.

A hysteresis loop of the Stoner–Wohlfarth particle is represented by starting its magnetization from the magnitude $-h$, given by point 1, then monotonously increasing the value along the path from point 2 to 5 to $+h$, given by point 6, and subsequently decreasing the magnitude back to $-h$. Moving the value of the external field along the path A – B the equilibrium direction of the magnetization is changing with the corresponding tangent lines to the right side of the astroid (solid lines) until point 5 is reached. At this point the magnetization flips and the tangent lines switch from the right to the left side of the astroid (dashed line). The stable orientation of the magnetization is characterized by the tangent line to the last point 6. At the way back along the path B – A the equilibrium direction belongs to the

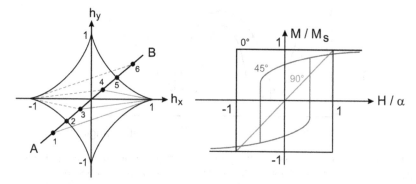

Fig. 12.23. Left: Variation of the reduced external magnetic field along the path **(A)**–**(B)** in relation to the astroid of a Stoner–Wohlfarth particle (details see text). **Right**: Hysteresis loop of a Stoner–Wohlfarth particle for different angles γ between the external magnetic field and the easy magnetization axis

tangent lines of the left side of the astroid (dashed lines). At point 2 the magnetization flips again and the tangent line switches from the left to the right side (solid line).

Therefore, the equilibrium direction of the magnetization is continuously changing if the external field is moving from the outside to the inside region of the astroid. An abrupt change of the direction of the magnetization occurs if the reduced external field crosses the astroid from the inside to the outside region. The points of the path A – B being inside the astroid exhibit two stable orientations of the magnetization which results in two different branches of the hysteresis loop (see right part of Fig. 12.23).

12.6 Superparamagnetism of Nanoparticles

As discussed above a sufficiently small ferromagnetic nanoparticle consists of a single domain. The direction of its magnetization M is determined by an external magnetic field H and by internal forces.

Let us assume that M can only rotate within a particular plane, e.g., being coplanar to the surface of a particle. Due to the magneto crystalline anisotropy the energy density E_{crys} depends on the rotation angle θ and characterizes the situation for a given direction of M. The difference between the maximum and minimum value of the energy density amounts to ΔE_{crys}. If the energy difference ΔE, i.e. $V \cdot \Delta E_{\text{crys}}$ with V being the volume of the particle, is very large in comparison with the thermal energy kT it is allowed to ignore thermal excitations for any reasonable measurement times. The static magnetization curves can simply be determined by minimizing the energy density at each H resulting in hysteresis because in certain field ranges there are two (e.g. upon uniaxial anisotropy shown) or more minima and transitions between them are neglected. If the energy difference ΔE is very small in

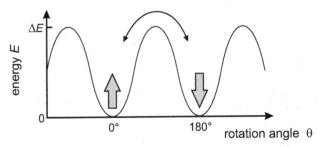

Fig. 12.24. Assuming a uniaxial magnetic anisotropy for a small nanoparticle the energy minimum is realized for both opposite magnetization states. A change between both states can only occur if the energy barrier ΔE is overcome which may be possible due to thermal excitations, i.e. if $kT > \Delta E$. Below the blocking temperature $T_B = \Delta E/k$ each magnetization state is stable

comparison with kT thermal excitation causes continual changes in the orientation of the magnetization for each individual particle which is schematically depicted in Fig. 12.24. In an ensemble of such particles it maintains a distribution of orientations. This behavior is like that of an ensemble of paramagnetic atoms (see Chap. 2.3). Thus, there is no hysteresis. This phenomenon is called "superparamagnetism".

Under intermediate conditions changes of the orientation occur with relaxation times being comparable with the time of a measurement. The fluctuation exhibits a period of:

$$\tau = \tau_0 \ e^{\Delta E/kT} \tag{12.17}$$

with ΔE being the energy difference between two opposite magnetization states. We directly see that the thermal fluctuations occur if the temperature T is larger than a critical temperature T_B. Thus, M vanishes. Below T_B the spin blocks are frozen out. Therefore, T_B is called blocking temperature and represents the superparamagnetic limit for a stable magnetization.

The fluctuation period becomes reduced if the temperature increases and if the volume of the particle decreases which is important concerning magnetic data storage devices.

As one example concerning superparamagnetism of small particles we will discuss, referring to [32], the magnetic behavior of Fe nanoparticles consisting of one atomic layer high patches on a Mo(110) surface. Figure 12.25(a) shows the correspondent constant-current STM image of the topography of 0.25 ML Fe deposited on Mo(110) at room temperature. Two atomically flat Mo(110) terraces are visible. They are decorated with Fe islands which are slightly elongated along the [001]-direction. Simultaneously with the topography maps of the differential conductance dI/dU were recorded using an out-of-plane sensitive Cr coated probe tip (see Fig. 12.25(b)). Although the spin averaged electronic properties of all Fe islands are identical one can recognize bright and dark islands in Fig. 12.25(b) representing two different values of the local

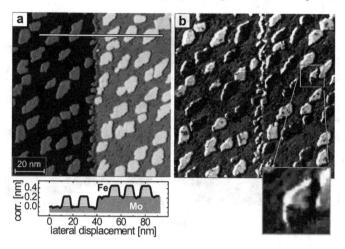

Fig. 12.25. (a) Topographic STM image and (b) the simultaneously measured out-of-plane sensitive magnetic dI/dU signal of two Mo(110) terraces decorated with Fe islands (overall coverage 0.25 ML). The line section (*lower panel*) reveals that the substrate's step edge and the islands are of monatomic height. During image recording one island switches from dark to bright (inset). (From [32] (used with permission))

dI/dU signal. This variation is caused by spin polarized vacuum tunnelling between the magnetic tip and islands which are magnetized perpendicularly either up or down.

Most of the islands exhibit a surface area $A \geq 40$ nm^2 and therefore possess a barrier large enough to inhibit superparamagnetic switching at $T = 13$ K being the temperature during measurement. Consequently, their magnetization direction remains constant resulting in the same dI/dU signal in successive scans. Few smaller islands, however, are found to be magnetically unstable on the time scale of the experiment, i.e. several minutes. Such an island with $A = 26$ nm^2 is shown at higher magnification in the inset of Fig. 12.25(b). The dI/dU signal changes between two subsequent scan lines from a low value (dark) at an early time of the scan (bottom part of the image) to a higher value (bright). This signal variation is caused by superparamagnetic switching.

From (12.17) it is expected that the switching rate exponentially increases with increasing temperature. This can be checked by successively scanning along the same line thereby periodically visiting four islands "a"–"d" (inset of Fig. 12.26). Measurements were performed at temperatures $T_1 = 13$ K and $T_2 = 19$ K for 5.5 min with a line repetition frequency of 3 Hz. At every passage the dI/dU signal of the islands was recorded (see Fig. 12.26). At $T_1 = 13$ K (left panel) the relatively large islands "a" ($A = 30$ nm^2), "d" ($A = 71$ nm^2), and "c" ($A = 28$ nm^2) do not switch. This magnetic stability is a result of their large anisotropy energy that prevents superparamagnetic switching at

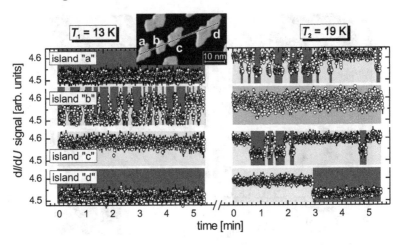

Fig. 12.26. Magnetic switching behavior of the Fe islands (**a**)–(**d**) (inset) measured at $T = 13$ K (*left panel*) and $T = 19$ K (*right panel*). The switching rate increases with increasing temperature thereby proving the thermal nature of the observed effect. At $T = 19$ K the switching rate of island b exceeds the line repetition rate resulting in an intermediate and blurred signal. (From [32] (used with permission))

this temperature. In contrast, island "b" ($A = 18$ nm^2) with its lower barrier reverses its magnetization direction 31 times within the observation time. In qualitative agreement with (12.17) the switching rate of any island increases as the temperature is increased to $T_2 = 19$ K (right panel). Islands "a", "c", and "d" reverse their magnetization direction 18, 10, and 1 times, respectively. Since the switching rate of island "b" exceeds the line repetition rate single switching events cannot be resolved and an intermediate and blurred dI/dU signal is measured above island "b".

With regard to magnetic data storage the critical size for the superparamagnetic limit can be estimated for room temperature. Typical parameters in (12.17) are $\tau_0 \approx 10^{-10}$ s, $kT = 25$ meV for $T = 300$ K, and $E \approx n \cdot 0.2$ meV with n being the number of atoms. Assuming a long-term stability of 10 years we get $\tau = 3 \cdot 10^8$ s. Equation (12.17) can be rewritten as:

$$E = kT \ln \frac{\tau}{\tau_0} \qquad (12.18)$$

Inserting the parameters we obtain $n = 5300$ atoms as the lower limit. Assuming a cube this number corresponds to an edge length of about 4 nm. For thin films with a thickness of five monolayers it corresponds to a size of 8 nm. Such cubes correspond to a storage density of about 10000 Gbit/inch^2 whereas today's hard disks exhibit a value of about 100 Gbit/inch^2.

12.7 Magnetism of Free Nanoparticles

The preparation of nanoparticles as free particles enables the characterization *without* the influence of a surface or a matrix. Their properties are between that of atoms and of the solid state due to the enhanced number of surface atoms with reduced coordination number. Assuming a cube-shaped nanoparticle the amount of surface atoms is given in Table 12.1 for different sizes. This means that particles containing 1000 atoms exhibit 50% surface atoms. And even particles with 100000 atoms possess 13% atoms at the surface.

The preparation of free nanoparticles can be carried out in a gas beam under vacuum conditions in order to avoid the influence of contaminants. The magnetic moments of free Fe_n nanoparticles with $n = 25 - 700$ at a temperature of $120\,K$ are given in Fig. 12.27. Up to about $n = 120$ the magnetic moment amounts to about $3\mu_B$ with relatively large oscillations and maxima near $n = 55$ and $n = 110$. The magnetic moment gradually decreases from $3\mu_B$ at $n = 240$ to $2.2\mu_B$ at $n = 520$. For larger sizes it remains nearly constant.

Several features can qualitatively be understood from element-specific properties of Fe. Atoms of this element possess 8 valence electrons; as already discussed on p. 37 approximately 7 electrons are in the $3d$ bands and one in the $4s$ band. In the case that the $3d$ spin up band is fully polarized it is occupied with 5 electrons (since it is completely below the Fermi level) leaving 2 electrons in the $3d$ spin down band. Consequently, each atom contributes $(5-2)\mu_B = 3\mu_B$ to the total moment. (i.e. intersect the Fermi level) the magnetic moment is reduced to less than $3\mu_B$. This is the case for bulk Fe where the magnetic moment amounts to about $2.2\mu_B$. Applying this knowledge to Fig. 12.27 it appears that the fully polarized majority spin band case applies for small nanoparticles ($n < 140$). Between $n \approx 140$ and $n \approx 550$ there seems to be a gradual transition from the fully polarized band to a more bulk-like situation. For larger n the magnetic moment corresponds approximately with the bulk value.

Table 12.1. Atoms in the bulk and at the surface for cube-shaped nanoparticles with different sizes

Shells	Atoms	Bulk atoms	Surface atoms	Ratio of surface atoms [%]
1	1	0	1	100
2	8	0	8	100
3	27	1	26	96
5	125	27	98	78
10	1000	512	488	49
20	8000	5832	2168	27
50	125000	110592	14408	12
100	1000000	941192	58808	6

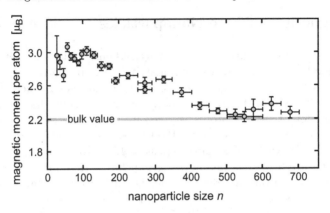

Fig. 12.27. Magnetic moments per atom of iron nanoparticles at $T = 120\,\mathrm{K}$. Horizontal bars indicate nanoparticle size ranges with n being the number of atoms in the nanoparticle. (Data taken from [33])

Figure 12.28 shows the magnetic moments of nickel and iron nanoparticles as a function of their sizes for two temperatures. We see a decrease of the magnetic moments with increasing size and with increasing temperature as well as the trend towards the bulk limit. For free Ni particles bulk-like behavior is given above 400 atoms at $T = 78\,\mathrm{K}$ (see Fig. 12.28a). The magnetic moments of Fe nanoparticles as a function of temperature are shown in Fig. 12.28b. At room temperature Fe particles with about 600 atoms exhibit a reduced magnetic moment compared with the bulk. This is due to a different crystalline structure of the particles (fcc) compared to the bulk (bcc).

As a first estimation of the structure of a free nanoparticle a truncated octahedron is shown in Fig. 12.29. This polyhedron is the equilibrium shape of fcc cobalt or nickel nanoparticles. The different positions are depicted in Fig. 12.29.

A more realistic morphology for cobalt nanoparticles is given in Fig. 12.30a. The surface of a perfect truncated octahedron is randomly covered with additional Co atoms. This particle contains 1489 atoms. Considered surface anisotropy constants are nearly one order of magnitude larger than shape anisotropy ones. Easy and hard axes for shape and surface anisotropies are reported on the switching field distribution in Fig. 12.30b. The aspect ratio is equal to 1.014. The shape and surface anisotropies add up. However, the resulting anisotropy is not large enough compared with experimental results. Therefore, surface atoms must be "organized" in order to increase the aspect ratio and anisotropy constants.

The most probable cluster geometries are displayed in Fig. 12.31a and Fig. 12.31b. The first particle contains 1357 atoms, (001)- and (111)-facets are filled with surface atoms; its aspect ratio is 1.09. The second particle contains 1405 atoms; (001)-, (111)-, and ($\bar{1}\bar{1}\bar{1}$)-facets are filled with surface atoms. Its aspect ratio is 1.15. Considered anisotropy constants are close to experimental

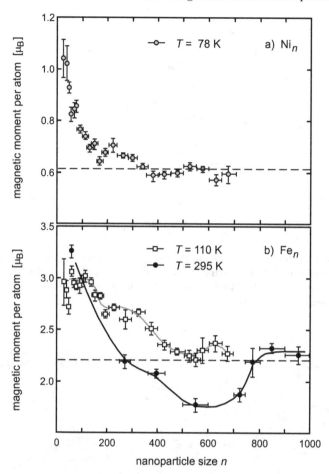

Fig. 12.28. Magnetic moment per atom in Bohr magneton units as a function of size. The lines are guides for the eyes. (**a**) Nickel results at 78 K. The values of the magnetic moment at 295 K are not reported as they would be superposed to the ones pictured. With increasing size the magnetic moment of the nanoparticles converges towards the bulk value (*dashed line*). (**b**) Iron results at 78 K and 295 K. Note the decrease from 130 up to 600 atoms below the bulk value (*dashed line*) and the progressive convergence as the size is further increased. (Adapted from [34] (used with permission))

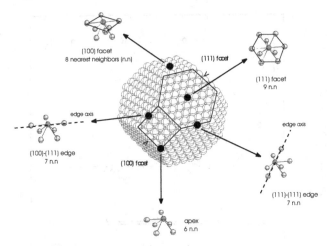

Fig. 12.29. The different atomic positions at the surface of a perfect truncated octahedron containing 1289 atoms. This polyhedron is the equilibrium shape of fcc clusters according to the Wulff theorem. It exhibits 8 (111)- and 6 (100)-facets. (From [35] (used with permission))

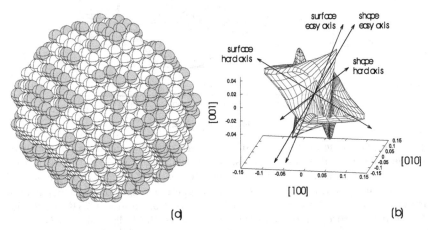

Fig. 12.30. (a) Perfect truncated octahedron (*bright atoms*) randomly covered with surface atoms (*dark atoms*). This particle contains 1489 atoms and the surface coverage is 42%. Aspect ratios are different from one which leads to second-order terms in the anisotropy energy. (b) Corresponding three-dimensional switching field distribution. This surface is very complex due to the mixing of second- and fourth-order anisotropy terms with different easy and hard magnetic axes. Easy and hard axes for shape and surface anisotropies are directly reported on the surface. (From [35] (used with permission))

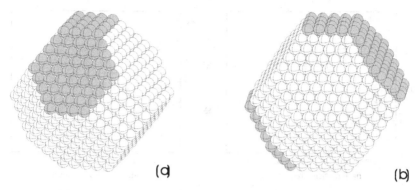

Fig. 12.31. (a) Perfect truncated octahedron (*bright atoms*) whose (001)- and (111)-facets are filled with surface atoms (*dark atoms*). It contains 1357 atoms and the surface coverage is 14.2%. Its aspect ratio is close to one and adding shape and surface anisotropies leads to anisotropy constants close to experimental ones. (b) Perfect truncated octahedron (*bright atoms*) whose (001)-, (111)-, and ($\bar{1}\bar{1}\bar{1}$)-facets are filled with surface atoms (*dark atoms*). It contains 1405 atoms and the surface coverage is 24.2%. Its aspect ratio is close to one and adding shape and surface anisotropies leads to anisotropy constants close to experimental ones. (From [35] (used with permission))

values. Moreover, these two last geometries are consistent with the growth mode of cobalt nanoparticles. Indeed, the growth of a truncated octahedron to one which is one atomic layer larger occurs by the filling of successive facets. The resulting nanoparticles are nearly spherical, i.e. their aspect ratio is close to one, but the surface anisotropy is strong enough to explain the large value of experimental anisotropy constants. The physical picture is the following: The interface anisotropy leads to easy and hard axes for the whole surface spins which are strongly exchange coupled with core spins. As a result magnetic anisotropy is completely driven by interface anisotropy. This assertion is only available for small nanoparticles with large surface-to-volume ratios. For the cobalt nanoparticles the mean value is 37%. However, surface contribution becomes negligible for larger particles.

12.8 Nanoparticles in Contact with Surfaces

After the discussion concerning free nanoparticles, i.e. without any interaction between the particle and the environment, given above we start our considerations assuming only weak interactions.

In order to realize experimentally a weak interaction between particle and surface chemically inert substrates like HOPG (highly oriented pyrolytic graphite) can be used.

Figure 12.32. shows the experimental values of m_L and m_S as a function of nanoparticle size for exposed Fe nanoparticles. The effect of the dipole contri-

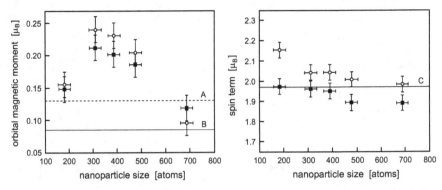

Fig. 12.32. Atomic orbital magnetic moment m_L (**left**) and spin term m_S (**right**) as a function of nanoparticle size for exposed Fe nanoparticles on HOPG. In both plots open circular and filled square symbols represent measurements taken for different geometric arrangements. Lines A and B in the plot for the orbital magnetic moment indicate the value measured in the 25 nm Fe film and the true typical value found in bulk Fe, respectively. Line C in the plot for the spin term indicates the value measured for the Fe thin film. (Reprinted from [36]. Copyright 2002, with permission from Elsevier)

bution to the spin term is evident across the size range being studied. It is also evident that its contribution increases in the smaller nanoparticles. The spin moment gradually increases as the size is reduced. For the 180 atom nanoparticles an enhancement relative to the bulk value of about 10% is measured. Initially, the orbital moment also increases with decreasing nanoparticle size reaching a maximum for 300 atom nanoparticles of about 2.5 times the true bulk value. Both increases can be attributed to the increasing proportion of surface atoms. A sharp drop in m_L is evident, however, for the 180 atom nanoparticles. As the nanoparticle size shrinks, apart from an increasing fraction of surface atoms, there will also be an increase in the proportion influenced by the crystalline field at the surface/nanoparticle interface; this will act to reduce m_L. This effect starts to dominate for nanoparticle sizes <180 atoms.

A stronger interaction between particle and solid state which can experimentally realized using metallic surfaces results in significantly different magnetic properties.

A more systematic overview on the size dependence of m_L/m_S is given in Fig. 12.33 where the results for exposed Fe nanoparticles on HOPG up to 2.3 nm are combined with data for Fe nanoparticles between 6 and 12 nm on Co/W(110). In the left part of Fig. 12.33 the data from very small Fe nanoparticles on Ni/Cu(100) are included as one data point (see Fig. 11.1) as the average of the results from Fe nanoparticles with 1–9 atoms. The experimental data for large Fe nanoparticles (filled circle) exhibit m_L/m_S values from 0.07 at a nanoparticle size of 12 nm up to 0.095 for Fe nanoparticles with 6 nm. All these values are clearly above the corresponding bulk value of

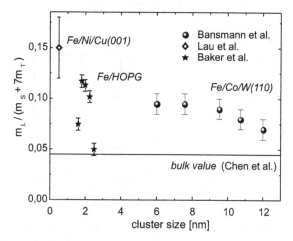

Fig. 12.33. Ratio of orbital to spin moment for Fe nanoparticles on various surfaces: Fe nanoparticles on Ni/Cu(100) (*left part*), Fe nanoparticles on HOPG (*middle*), and large Fe nanoparticles on Co/W(110) (*right part*). The bulk value is indicated by the *solid line*. (Reprinted from [37]. Copyright 2005, with permission from Elsevier)

0.043. The enhanced ratio is related to an increase of the orbital moment in the outer two shells which exhibit a large number of surface atoms compared to the total number of atoms in the nanoparticle. Assuming a spherical shape the ratio of surface-to-volume atoms amounts to 43% in case of a 6 nm Fe nanoparticle and still 23% for Fe nanoparticles with a size of 12 nm.

In the discussion above the large Fe nanoparticles have been deposited onto ferromagnetic Co(0001) films on W(110) where the clusters feel the presence of a strong exchange field inducing an in-plane magnetic anisotropy in the nanoparticles. Now, let us concentrate on the magnetic behavior of Fe nanoparticles on a non-magnetic support. For the realization of this situation 12 nm Fe nanoparticles have been deposited onto a clean W(110) surface and variable external magnetic fields have been applied in-plane and out-of-plane to the sample. The measured total magnetic

moments (i.e. the sum of orbital and spin moment) are the projections of the real magnetic moments on the direction of the magnetic field. Clearly, the nanoparticles can be much more easily magnetized in the surface plane than perpendicular to it (see Fig. 12.34). Although a magnetic saturation could not be reached with such small magnetic fields the in-plane magnetization nevertheless shows a total magnetic moment of about $1.7\mu_B$ at 14 mT. Additionally, the inset shows an in-plane hysteresis loop confirming a small but finite remanence that could not be observed in an out-of-plane magnetization. This pronounced in-plane anisotropy can be explained by two effects which both favor an in-plane magnetization: (i) A strong surface anisotropy at the Fe/W interface as known from Fe(110) films on W(110) and (ii) due to a wetting of the tungsten surface which leads to an oblate nanoparticle shape.

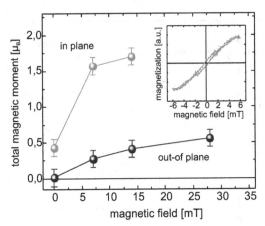

Fig. 12.34. Total magnetic moment of 12 nm Fe nanoparticles on W(110) plotted for in-plane and out-of-plane magnetization. The inset displays a hysteresis loop taken with an in-plane magnetization (Reprinted from [37]. Copyright 2005, with permission from Elsevier)

Thus, the shape anisotropy favors an in-plane magnetization compared to the perpendicular direction because the Fe particles are flattened being in contact with the surface whereas free particles are spherically shaped.

The results discussed above are related to particles on the surface being largely separated from each other. Now, let us discuss what happens if the number of nanoparticles on the surface increases. In Fig. 12.35 the ratio of

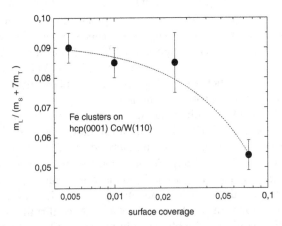

Fig. 12.35. Ratio of orbital to spin moment of 9 nm Fe nanoparticles on Co/W(110) as a function of the surface coverage. The x-axis denotes the area on the surface covered by Fe nanoparticles when assuming the mean size of the free particles. The dotted line serves to guide the eye. (Reprinted from [37]. Copyright 2005, with permission from Elsevier)

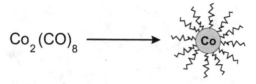

$$Co_2(CO)_8 \longrightarrow$$

Fig. 12.36. Wet chemically prepared Co nanoparticle surrounded by a ligand shell due to decomposition of the precursor material di-cobalt octa-carbonyl $Co_2(CO)_8$

orbital to spin moment $m_L/(m_S + 7m_T)$ (with $m_T = \langle T_z \rangle$ considering the dipole interaction) for 9 nm Fe nanoparticles is displayed as a function of the surface coverage. Under the assumption that the spin moment is bulk-like it is obvious that the percolation of individual particles to larger ones reduces the orbital moment similar to smaller Fe nanoparticles. At a coverage of less than one monolayer the orbital moment nearly reaches the bulk value of iron.

12.9 Wet Chemically Prepared Nanoparticles

A totally different procedure to produce magnetic nanoparticles is realized by wet chemical preparation. Suitable reactions of metal salts lead to nanoparticles which are surrounded by a ligand shell (see Fig. 12.36). The size of the particles can be tuned by different salts, solvents, concentration, temperature, and time of reaction. The size distribution may be rather small. Ordered structures are often formed due to the ligand shell (see Fig. 12.37). The distance between the particles can be adjusted using different types of ligands and by varying their length. Additionally, the ligand shell prevents an agglomeration and passivates the metallic nanoparticles against oxidation.

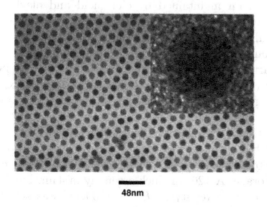

48nm

Fig. 12.37. TEM image of a 2D assembly of 9 nm Co nanoparticles. Inset: High resolution TEM image of a single particle. (Reused with permission from [38]. Copyright 1999, American Institute of Physics)

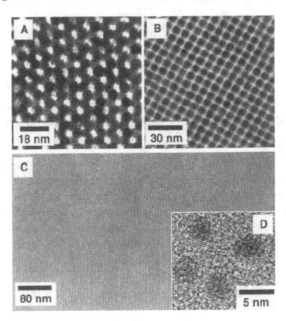

Fig. 12.38. (**A**) TEM micrograph of a 3D assembly of 6 nm as-synthesized $Fe_{50}Pt_{50}$ particles deposited from a hexane/octane dispersion onto a SiO-coated copper grid. (**B**) TEM micrograph of a 3D assembly of 6 nm $Fe_{50}Pt_{50}$ sample after replacing oleic acid/oleyl amine with hexanoic acid/hexylamine. (**C**) HRSEM image of a 180 nm thick, 4 nm $Fe_{52}Pt_{48}$ nanocrystal assembly. (**D**) High-resolution TEM image of 4 nm $Fe_{52}Pt_{48}$ nanocrystals. (From [39]. Reprinted with permission from AAAS.)

A TEM image (see Fig. 12.38A) shows a thin section of a hexagonally closed packed 3D array of 6 nm $Fe_{50}Pt_{50}$ particles with a nearest neighbor spacing of about 4 nm maintained by oleic acid and oleyl amine capping groups. Room temperature ligand exchange of long-chain capping groups for shorter ones allows the interparticle distance to be adjusted. Ligand exchange with hexanoic acid/hexylamine yields a cubic packed multilayer of 6 nm $Fe_{50}Pt_{50}$ particles with about 1 nm spacings (see Fig. 12.38B). The symmetry of the observed superlattices is influenced by several experimental parameters including the relative dimensions of the metal core and the organic capping as well as the annealing history of the sample.

Annealed FePt nanocrystal assemblies may be smooth ferromagnetic films that can support high-density magnetization reversal transitions (bits). Such an ordered ensemble consisting of hard magnetic FePt particles can be used for magnetic data storage. A 120 nm thick assembly of 4 nm $Fe_{48}Pt_{52}$ nanocrystals with an in-plane coercivity of $H_c = 1800$ Oe was selected for initial recording experiments. Atomic force microscopy studies of this sample indicate a 1 nm root-mean square variation in height over areas of 3 mm by 3 mm. The read-back sensor voltage signals (see Fig. 12.39) from written data tracks

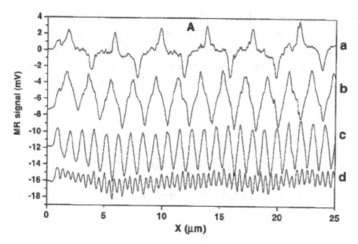

Fig. 12.39. Magnetoresistive (MR) read-back signals from written bit transitions in a 120 nm thick assembly of 4 nm diameter $Fe_{48}Pt_{52}$ nanocrystals. The individual line scans reveal magnetization reversal transitions at increasing linear densities. (From [39]. Reprinted with permission from AAAS.)

correspond to linear densities of 500, 1040, 2140, and 5000 flux changes per millimeter (fc/mm) (curves a to d, respectively). These write/read experiments demonstrate that this 4 nm $Fe_{48}Pt_{52}$ ferromagnetic nanocrystal assembly supports magnetization reversal transitions at moderate linear densities that can be read back non-destructively. Much higher recording densities beyond the highest currently achievable linear densities of about 20000 fc/mm can be expected if the thickness of these ferromagnetic assemblies can be reduced to about 4 nm.

The characteristics of magnetic nanoparticles make them interesting for biomedical applications, too. Magnetic particle aggregates stabilized by hydrophilic (water-soluble) polymers are small enough to enter cells or pass borders like the blood-brain barrier. Their surface can be functionalized for selective interaction and their magnetic properties make them controllable by an external magnetic field. The controllability includes magnetic separations as well as far more complicated methods such as guided drug delivery or hyperthermia. The superparamagnetism of the particles ensures that no further aggregation or even coagulation of the particles occurs during and after patient treatment with a magnetic field. Aggregation and coagulation to large particle clusters could have fatal effects, especially in small blood vessels.

Among the magnetic materials with suitable properties magnetite (Fe_3O_4) is the only one that has up to now been allowed for use in humans. It is the only material which is known to be biocompatible without relevant toxicity in the applied dosing range. Once injected into the blood stream the particles will stay in the body until they are washed out through the liver and kidney. The systems being used are magnetite nanoparticles with a size of about

10 nm being stabilized by hydrophilic polymeric shells like dextrane or car-bodextrane. By functionalizing the shell it is possible to attach, for example, drugs by ionic links that can be set free at a desired site after being directed there by an external magnetic field. Magnetite nanoparticles produce enough heat in an alternating magnetic field to be applied in hyperthermia treatment.

Problems

12.1. Prove that one of the possible tangent lines through the tip point of h in the astroid of a single domain particle represents the equilibrium direction of the magnetization for the situation that h is within the astroid (see Fig. 12.22a).

Magnetism in Reduced
Dimensions – Nanoscaled Wires

In the following chapter we want to consider the magnetic properties of nanoscaled wires. They are characterized by a large aspect ratio of their length to their width and height. We will distinguish between wires with a width in the sub-micrometer regime on the one hand and those which a built up by only single atoms on the other hand. It is obvious that such systems exhibit a pronounced anisotropic behavior due to its shape.

13.1 Wires Exhibiting a Width
in the Sub-Micrometer Regime

Let us assume that the height and the width of the wire is in the sub-micrometer regime whereas the length is significantly larger. This implies a high aspect ratio. One example of such a system is presented in Fig. 13.1. The sample is a thin Fe film of 13 nm thickness which has been transformed into a periodic nanoscaled wire array by an anisotropic plasma etching process after film deposition. The Fe film was grown on an $Al_2O_3(1\bar{1}02)$ substrate onto a 150 nm thick Nb buffer layer which has a (001)-orientation as can be derived from the three-dimensional epitaxial relationship between niobium and sapphire. Finally, an array of well separated Fe wires on top of a Nb buffer is obtained. The measurement confirms the regularity of the Fe nanoscaled wires having a width of 150 nm and a periodicity of 300 nm as well as that the wires are completely separated from each other. The stripes have a sinusoidal shape.

Due to the shape anisotropy the magnetization is expected to be oriented along the wire. The upper part of Fig. 13.2 shows the remanent Kerr signal θ_K^{rem} normalized to the Kerr signal at saturation θ_K^{sat} as a function of the angle of rotation χ about the surface normal of the Fe film which yields information about the squareness of the hysteresis loops. The signal θ_K represents the rotation of linearly polarized monochromatic light due to the reflection on a

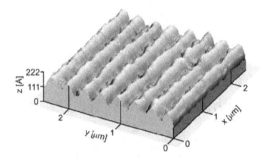

Fig. 13.1. Surface morphology of a periodic array of Fe nanowires on a Nb/sapphire substrate imaged with atomic force microscopy AFM. (Reprinted from [40] with permission of IOP)

ferromagnetic surface. This experimental technique is known as the magneto-optical Kerr effect (MOKE). According to Fig. 13.2 the remanent Kerr signal is significantly reduced at certain angles χ without reaching zero-values signifying the hard axis orientations (around 90° and 270°). For the corresponding angles χ along the easy axis orientations (0° and 180°) the ratio $\theta_K^{rem}/\theta_K^{sat}$

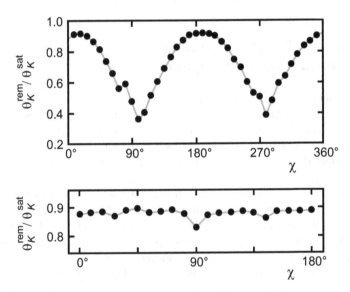

Fig. 13.2. The *upper panel* shows results from hysteresis loop measurements at different angles of rotation χ for the nanowire array as measured at remanence normalized to the Kerr rotation as measured at saturation. The *lower panel* depicts the results of MOKE hysteresis loop measurements as a function of the angle of rotation of the unpatterned sample where θ_K^{rem} as measured at remanence is normalized to θ_K^{sat} as measured at saturation and plotted as a function of the angle of rotation χ which is a measure of the magnetic anisotropy. (Adapted from [40] with permission of IOP)

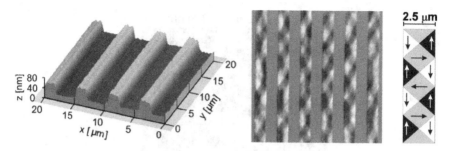

Fig. 13.3. Left: Atomic force microscopy image of a grating with $w_{Fe} = 2.1\,\mu m$. Middle: Kerr microscopic image in the demagnetized state of a corresponding grating with about the same stripe width being $w_{Fe} = 2.5\,\mu m$. The field direction during demagnetization was perpendicular to the stripes. **Right**: Orientation of the magnetization within the domains schematically depicted. (Reused with permission from [41]. Copyright 2002, American Institute of Physics)

measures almost unity. The twofold symmetry is clearly to be seen. In the lower part the behavior of an unpatterned polycrystalline Fe film is shown. The Kerr rotation is measured at remanence normalized to the Kerr rotation in saturation as a function of χ. As can be readily seen the remanent magnetization amounts to about 88% of the saturation magnetization and is almost independent of the angle of rotation. Thus, the overall in-plane magnetic anisotropy is negligible.

The part of the coercivity of magnetic nanoscaled wires due to their low dimensionality is reciprocally proportional to the width w:

$$H_c \propto \frac{M_S}{\pi} \cdot \frac{d}{w} \tag{13.1}$$

with d being the thickness. The width of the wire determines whether it is energetically favorable to stabilize a single domain particle or to introduce domain walls.

Let us discuss this behavior with a sample of a polycrystalline Fe film which was grown on $Al_2O_3(11\bar{2}0)$. The width amounts to about $2\,\mu m$, the height to $50\,nm$, and the periodicity to $5\,\mu m$. An AFM image is given in the left part of Fig. 13.3. The domain pattern observed for a grid with a width of $2.5\,\mu m$ in the demagnetized state is shown in the medium part of Fig. 13.3 using a Kerr microscope. This instrument enables to determine laterally resolved signals being obtained by means of MOKE. There is a very regular domain structure with closure domains at the stripe edges observed as depicted schematically in the right part of Fig. 13.3. In the remanent state one essentially observes similar domains with one magnetization direction in the interior of the stripes. Thus, for wires with a large width domains exhibiting the Landau state are found (cf. Fig. 12.1).

The influence of the aspect ratio is demonstrated in Fig. 13.4 for Co wires with a different width, constant height, and a uniaxial anisotropy along the

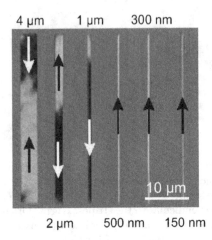

Fig. 13.4. MOKE images of Co wires with a different width. Wires exceeding a width of about 4 μm exhibit a complex domain pattern. Reducing the width results in vanishing closure domains, i.e. the wire only possesses two domains with opposite magnetization. Below a width of about 550 nm single domain wires occur. (From [42])

wire axis. Wires with a large width exhibit a complex domain formation which becomes easier with decreasing width w. For $w = 2\,\mu m$ the domain pattern consists of such which possess a magnetization along the wire and thus along the easy magnetization axis and are separated by 180° domain walls. At the ends of the wires closure domains are created. Exceeding a critical value of the aspect ratio, i.e. falling below a critical value of the width assuming a constant height, results in the occurrence of single domain wires.

Now we change the situation and discuss Co wires which exhibit an easy magnetization axis being perpendicular to the wire. Figure 13.5 shows the domain pattern of Co(1010) wires exhibiting a thickness of 60 nm and different widths w after in-plane and easy axis saturation. For $w = 800\,nm$ a pure transverse single-domain state is stabilized while for $w = 150\,nm$ the stripe-domain and transverse single-domain state coexist. For $w = 100\,nm$ a complete stripe structure is induced.

If such nanoscaled wires are not isolated interactions occur between neighbored wires. Such a behavior is presented in Fig. 13.6. The single domain stripes exhibit an alternating behavior of the magnetic domains in order to minimize the stray field.

An additional influence is caused by the shape which is exemplarily shown in Fig. 13.7a–c. A high remanence state is stabilized for interacting rectangular elements (see Fig. 13.7a) whereas the isolated particle prefers a demagnetized state in equilibrium. This state is favored by effective flux distribution configurations at the sample ends. Small closure domains are also found in elliptical particles in high-remanence states (see Fig. 13.7b) but they are much smaller

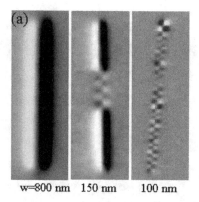

w=800 nm   150 nm   100 nm

Fig. 13.5. Magnetic force microscopy MFM images taken on Co(1010) wires after in-plane easy-axis saturation for wires being 60 nm thick with widths of 800, 150 and 100 nm. (Reprinted from [43] with permission of IOP)

and obviously less effective in reducing the stray field. This may explain why interacting elliptical particles prefer a demagnetized state in equilibrium. The pointed elements (see Fig. 13.7c) are designed with parabolic contours to mimic ellipsoidal particles in their cross sections along the axis. They prefer the saturated state and under the influence of interactions even a regularly alternating arrangement can be observed.

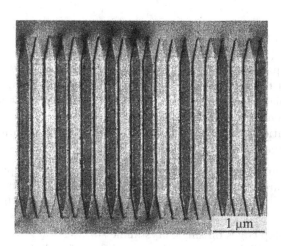

1 μm

Fig. 13.6. Switching in an interacting array of pointed elements with a thickness of 26 nm consisting a of soft magnetic NiFe alloy. About every second particle has switched in a field along the particle axis showing lighter contrast. (From [10] (used with permission))

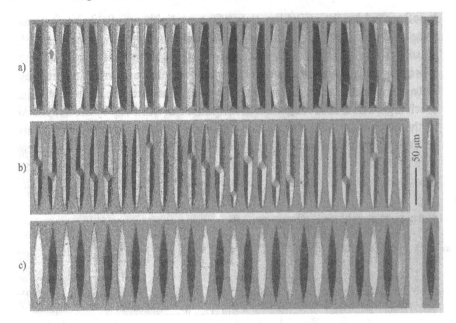

Fig. 13.7. Equilibrium state of interacting elements of (**a**) rectangular, (**b**) elliptical, and (**c**) pointed NiFe alloy elements of 240 nm thickness generated in an alternating field of decreasing amplitude along the particle axis. The equilibrium demagnetized state of the isolated particles is shown on the *right side*. (From [10] (used with permission))

13.2 Wires Consisting of Single Atoms

In the previous part the width of the wires was in the range of several nanometers up to about two microns. Now, we want to examine the magnetic properties of wires which possess a width of only one to a few atoms and will follow the results presented in [44, 45, 46].

The preparation of such wires is experimentally a difficult task. Two possibilities allow the realization of such systems:

- Atom by atom by means of a scanning tunnelling microscope
  This procedure is rather time-consuming because it represents a serial process. Additionally, low temperatures below approximately 10 K are necessary.
- Growth at step edges on single crystalline surfaces
  Due to its parallel character this procedure is fast but an ideal growth does not succeed for every preparation. A crucial parameter is given by the temperature which is exemplarily shown in Fig. 13.8 for Co/Pt(997) exhibiting an average terrace width of about 20 Å.

The nucleation and growth of metals on densely stepped substrates allows to create arrays of 1D nanoscaled wires with precise morphological

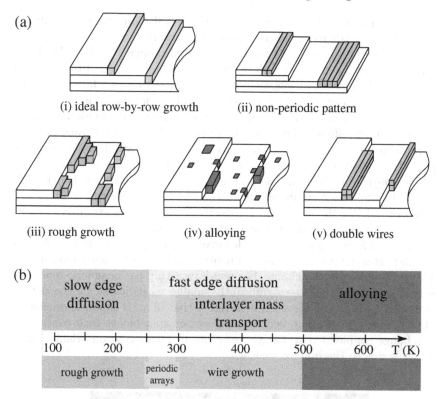

Fig. 13.8. (a) Different growth modes on a stepped substrate. (b) Co growth modes on Pt(997) as a function of the substrate temperature. Although the temperature scale refers to the Co/Pt(997) system this description applies to other metals such as Ag and Cu. (Reprinted from [44] with permission of IOP)

characteristics. Depending on the surface temperature adatoms on vicinal surfaces self-assemble into chain like structures by decorating the step edges. This is simply due to the increase of binding energy at the step sites. An advantage of this growth method is that by adjusting the adatom coverage and the average step spacing one can independently control the wire width and separation, respectively.

Growth proceeds either as a smooth step-wetting process or as nucleation of two-dimensional (2D) islands at the step edges provided that the adatom placement prior to nucleation is larger than the terrace width of the substrate. In Fig. 13.8 different scenarios of heteroepitaxy on a stepped substrate are shown. One can distinguish (i) the ideal case of row-by-row growth, (ii) wires of different widths due to interlayer crossing of the adatoms resulting in a non-periodic arrangement of the chains, (iii) formation of irregular 2D islands at the step edges, (iv) alloying, and (v) formation of double-layer wires. As a general trend wire formation is limited at low temperature by slow edge-diffusion processes and at high temperature by interlayer diffusion and, eventually, by alloying between the metal adspecies and the substrate.

Single Co atoms on Pt(111) terraces are mobile above T = 55 K. At higher temperature, as the terrace width of Pt(997) is small compared to the mean free path of Co adatoms, nucleation at the step sites occurs. The wire growth proceeds via incorporation of adatoms in 1D stable nuclei attached to the step edges. However, below 250 K the wire formation is kinetically hindered by slow edge- and corner-diffusion processes. Regular Co wires grow only above 250 K as shown in Fig. 13.9. A monatomic chain array is obtained as the coverage equals the inverse of the number of atomic rows in the substrate terraces, i.e. 0.13 monolayer for Pt(997). The average length of a continuous Co chain is estimated to be about 80 atoms from the average kink density per Pt step.

In Fig. 13.10a the magnetic response of a set of monatomic wires at $T = 45$ K is presented. The zero remanent magnetization reveals the absence of long range ferromagnetic order. However, the shape of the magnetization curve indicates the presence of short range order, i.e. of significant interatomic exchange coupling in the chains. For non-interacting paramagnetic moments the magnetization would be significantly smaller as indicated by the dotted line in Fig. 13.10a. The observed behavior is that of a 1D superparamagnetic system, i.e. a system composed by segments or spin blocks each containing N exchange-coupled Co atoms whose resultant magnetization orientation is not stable due to thermal fluctuations.

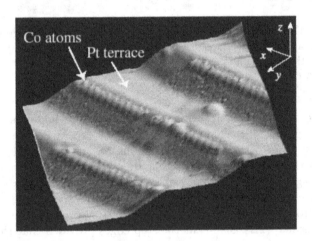

Fig. 13.9. Co monatomic chains decorate the Pt step edges following deposition of 0.07 ML Co at $T = 250$ K (the vertical scale has been enhanced for better rendering). The chains are linearly aligned and have a spacing equal to the terrace width. The protrusion on the terrace is attributed to Co atoms incorporated in the Pt layer. (Reprinted by permission from Macmillan Publishers Ltd: Nature (see [45]), copyright (2000))

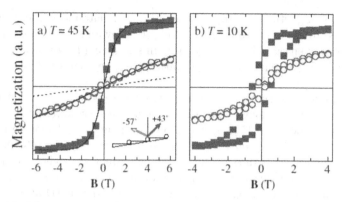

Fig. 13.10. Magnetization of a Co monatomic wire array as a function of the applied field B. (**a**) Magnetization at $T = 45$ K in the easy direction (*solid squares*, $+43°$) and $80°$ away from the easy direction (*empty circles*, $-57°$) in the plane perpendicular to the wire axis (see the inset). The (*solid curves*) are fits to the data. The *dashed curve* represents the magnetization expected for an isolated Co atom on Pt(997). (**b**) Magnetization at $T = 10$ K for the same geometry as in (a). Hysteretic behavior sets in due to long-range ferromagnetic order. The unsaturated zero-field magnetization is attributed to the inhomogeneous lengths of the chains (Reprinted by permission from Macmillan Publishers Ltd: Nature (see [45]), copyright (2000))

A noticeable dependence of the magnetization on the direction of the applied field is present as shown in Fig. 13.10. The strongest magnetic response is found in the $+43°$ direction with respect to the terrace normal. Clearly, the shape of the superparamagnetic curves depends on the magnetic anisotropy energy of each spin block $N E_{crys}$ as well as on N times the magnetic moment per Co atom. By fitting the curves in Fig. 13.10a assuming dominant uniaxial anisotropy and a classical model of the magnetization one obtains $N = 15 \pm 1$ and $E_{crys} = (2.0 \pm 0.2)$ meV/atom. Thus, on average, about 15 Co atoms are coupled in each spin block at $T = 45$ K. A simple argument shows that this result does not contradict the spin lattice models treating magnetic order in 1D.

Let us consider a chain consisting of N moments described by the Ising Hamiltonian given in (6.22). The ground state energy of the system is $E_0 = -J(N-1)$ and corresponds to the situation where all the moments are aligned. The lowest-lying excitations are those in which a single break occurs at any one of the N sites. There are $N - 1$ such excited states all with the same energy $E = E_0 + 2J$. At temperature T the change in free energy due to these excitations is $\Delta F = 2J - kT \ln(N - 1)$. For $N \to \infty$ we have $\Delta F < 0$ at any finite temperature and the ferromagnetic state becomes unstable against thermal fluctuations. For $(N - 1) < e^{2J/kT}$, however, ferromagnetic order is energetically stable. Assuming $2J = 15$ meV we get an upper limit of $N = 50$ exchange-coupled atoms at $T = 45$ K. Measurements of the magnetization in the Co monatomic chains agree with this limit.

The large magnetic anisotropy energy of the monatomic spin chains is directly related to the anisotropy of m_L along the easy and hard directions. The large anisotropy energy plays a major role in stabilizing long range ferromagnetic order in 1D in particular in inhibiting the approach to the thermodynamic limit described above.

As in bulk ferromagnetic systems anisotropy energy barriers can effectively pin the magnetization along a fixed direction in space. By lowering the sample temperature below $T_B = 15\,\mathrm{K}$ a transition to a long range ferromagnetically ordered state with finite remanence can be observed (see Fig. 13.10b). The threshold temperature is the so-called blocking temperature where the magnetization of each spin block aligns along the common easy axis direction and the whole system becomes ferromagnetic. Long range order in 1D atomic chains therefore enters as a metastable state thanks to slow magnetic relaxation.

As the wire width increases the average magneto anisotropy energy per Co atom E_{crys} changes in a non-monotonic way with n. Figure 13.11. presents the magnetization curves $M(\Phi_1)$ and $M(\Phi_2)$ where Φ_1 and Φ_2 represent two directions in the plane perpendicular to the wires close to the easy and hard axis, respectively. A fit of $M(\Phi_1)$ and $M(\Phi_2)$ in the superparamagnetic regime shows that $E_{\mathrm{crys}} = (2.0 \pm 0.2)$ meV/atom is largest for the monatomic wires. Since in small as well as large 2D clusters E_{crys} is a rapidly decreasing function of the local coordination number of the magnetic atoms it is not surprising that E_{crys} reduces abruptly to about 0.33 meV/atom in the two atom wide wires. The magneto crystalline anisotropy energy reduction is such that the low temperature hysteretic behavior almost vanishes going from the monatomic to the two atom wide wires (see Fig. 13.11b) despite the larger size of the superparamagnetic spin blocks in the two atom wide wires relative to the monatomic

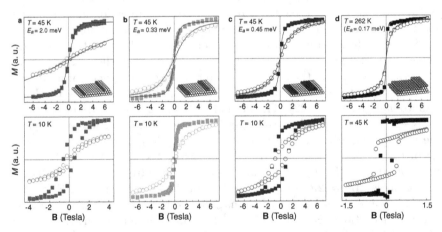

Fig. 13.11. Magnetization of (**a**) monatomic wires, $\Phi_1 = +43°$ (*solid squares*), $\Phi_2 = -57°$ (*open circles*); (**b**) two atom wide wires, $\Phi_1 = -67°$; $\Phi_2 = +23°$; (**c**) three atom wide wires, $\Phi_1 = -7°$; $\Phi_2 = +63°$; (**d**) 1.3 ML, $\Phi_1 = -7°$; $\Phi_2 = +63°$. *Solid lines* are fits to the data. (From [46] (used with permission))

wires. In the three atom wide wires, however, E_{crys} shows a significant and unexpected 35% increment up to 0.45 meV/atom. In concomitance with the increased size of the spin blocks in the three atom wide wires such increment favors again ferromagnetic order at $T = 10$ K (see Fig. 13.11c). This E_{crys} upturn is opposite to that expected for the increasing average coordination number of the Co atoms from two atom wide to three atom wide wires.

This example highlights that interaction on the atomic scale significantly influences magnetic properties of nanoscaled objects.

14

Magnetism in Reduced Dimensions – Single Thin Films

In the following chapter we will deal with the magnetic behavior of single thin metallic layers.

Our discussion will contain the influence of a capping layer. In this context we will distinguish between a non-magnetic and a magnetic capping material. Further, we will compare the ideal and real interface between the thin film and a substrate which the film is deposited on. Finally, the behavior of a specific situation, a ferromagnetic layer on an antiferromagnetic substrate, will be examined.

14.1 Single Thin Film on a Substrate

The magnetization across a thin film is characterized by constant magnetic moments within the film but deviations occur at both interfaces (see Fig. 14.1). At the interface between substrate and thin film the magnetic moments are reduced due to hybridization between the electronic states of the atoms in the ferromagnetic thin film and at the interface of the substrate. The surface exhibits an increased magnetization because of the reduced coordination number, i.e. a more atomic-like behavior is present. An increasing temperature results in a decreasing magnetization over the whole film. This reduction is more pronounced at the interface and surface compared to the inner layers of the thin film.

An important parameter represents the thickness of a thin film. Let us discuss its influence on the anisotropy, Curie temperature, (spin dependent) transport properties, and quantum well states.

Influence on the Anisotropy

The behavior of the anisotropy for different film thicknesses was already discussed in Chap. 7.6. The effective anisotropy constant:

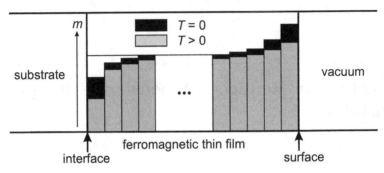

Fig. 14.1. Schematic illustration concerning the variation of the magnetic moment m per atomic layer near an interface to a substrate and near the surface at $T = 0$ and $T > 0$

$$K^{\text{eff}} = K^{\text{V}} + 2K^{\text{S}}/d \qquad (14.1)$$

with d being the thickness of the thin film opens up, e.g., the possibility of a spin reorientation transition from an in-plane to a perpendicular magnetization with increasing thickness.

Influence on the Curie Temperature

In thin film systems the Curie temperature T_C is often decreased compared to the bulk as exemplarily shown for Ni(111) films exhibiting various thicknesses d on a Re(0001) surface (see Fig. 14.2). This reduction is due to the

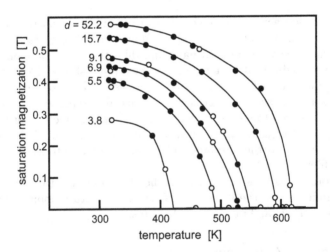

Fig. 14.2. Saturation magnetization of Ni(111) films on Re(0001) as a function of temperature with the number of atomic layers d as parameter. The data were taken with increasing (*closed symbols*) and decreasing temperature (*open symbols*). (Reprinted from [47]. Copyright 1984, with permission from Elsevier)

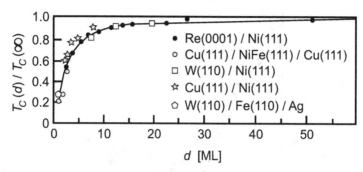

Fig. 14.3. Normalized Curie temperatures as a function of the atomic layers d for different densely packed cubic transition metal films (Reprinted from [48]. Copyright 1993, with permission from Elsevier)

absence of magnetic neighboring atoms. A quantitative description can be given using a finite-size scaling model which results in:

$$\frac{T_C(\infty) - T_C(d)}{T_C(\infty)} = \left(\frac{d}{d_0}\right)^{1/\nu} \tag{14.2}$$

The critical exponent $\nu = 0.7$ is nearly independent of the anisotropy. For cubic systems the normalized Curie temperature $T_C(d)/T_C(\infty)$ exhibits a uniform behavior as a function of the thickness (see Fig. 14.3). The full curve represents a fit to the Re(0001)/Ni(111) data by a power law. The critical thickness d_0 amounts to nearly two atomic layers. It is characterized by the minimum thickness which exhibits the loss of magnetic order already at $T = 0\,\mathrm{K}$.

But, (14.2) is not more valid for such thin films. Therefore, already the monolayer ($d = 1$) is often ferromagnetic.

Influence on the (Spin Dependent) Transport Properties

The decreasing intensity I of electrons moving through a ferromagnetic thin film depends on their kinetic energy. This function is nearly independent of the material and represents therefore a universal curve. The corresponding inelastic mean free path λ (see Fig. 14.4) is given by:

$$I(d) = I_0\, e^{-d/\lambda} \tag{14.3}$$

The intensity additionally depends on the magnetization of the magnetic thin film which is due to a spin dependence of the inelastic mean free path characterized by λ_+ and λ_- for spin up and spin down electrons, respectively. At a given kinetic energy the number of electrons leaving the interface of a non-magnetic substrate is the same for electrons with spins being parallel and antiparallel to the magnetization of the ferromagnetic thin film $I^{\uparrow\uparrow} = I^{\uparrow\downarrow}$.

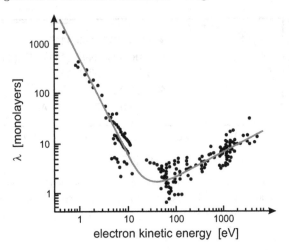

Fig. 14.4. Inelastic mean free path λ given in monolayers as a function of kinetic energy for electrons in solids of various elements (*black dots*) (From [49]. Copyright 1979. Copyright John Wiley & Sons Limited. Reproduced with permission.)

After passing the magnetic thin film both intensities are no more identical: $I^{\uparrow\uparrow} \neq I^{\uparrow\downarrow}$. Figure 14.5 presents this behavior for a thin ferromagnetic Co film on a W(110) substrate. Electrons with a binding energy of 3 eV (below $E_F = 0$) are only related to the substrate. After excitation with monochromatic radiation they pass several layers of ferromagnetic Co. We recognize that the intensity of the majority electrons is much less reduced compared to that of minority electrons. This spin filter effect is due to a spin dependence of the inelastic mean free path and is discussed on p. 227 in more detail.

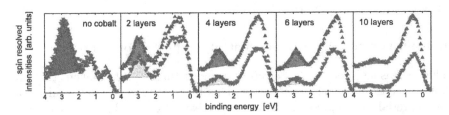

Fig. 14.5. Spin resolved photoelectron spectra for Co thin films on a W(110) substrate as a function of film thickness. The evaluated tungsten structures at a binding energy of 2.9 eV are shaded, the *upper curve* represents the majority and the *lower curve* the minority spin channel. The different intensities in the two channels are clearly visible at higher coverage. The peak areas which were obtained by integration assuming a linear background and indicated by light gray (for minority) and dark gray shaded areas (for majority electrons) allow to determine the inelastic mean free paths λ_+ and λ_- for spin up and spin down electrons, respectively

Influence on the Quantum Well States

The fundamental difference between the electronic structure of a thin film and of a three dimensional crystal is due to the boundary conditions at the interfaces.

In the ideal case of an isolated film composed of n layers the finite size of the system imposes a quantization of the electronic states resulting in n discrete levels. The energy level spectrum changes with film thickness as schematically indicated in the right part of Fig. 14.6 being in analogy with the simple case of a particle in a potential well. The number of allowed states increases with the film thickness while their electronic separation decreases. The electronic structure converges to that of a bulk material with increasing film thickness. For a thin film on a substrate or as a part of a multilayer the reflection at the interfaces (with the vacuum and with other materials) determines the degree of confinement of the electron wave functions in the film.

The electron reflection depends on the energy and wave vector of the corresponding electronic state and also on the spin character if an interface component is ferromagnetic. Strongly reflected electron waves remain confined within the film and form "quantum well states". Photoelectron spectroscopy allows to determine the electronic structure of the film as a function of binding energy and film thickness. The complexity (i.e. roughness, interdiffusion, clustering) of the film growth often makes the observation of the quantization effects difficult or even impossible. However, in a few cases which the film grows almost perfectly layer-by-layer in the formation of quantum well states can directly be observed in photoemission spectra.

As one example of such a situation the left part of Fig. 14.6 shows the photoemission spectra of thin Cu films epitaxially grown on fcc-Co(100). The spectra of the films exhibit several features derived from the Cu electronic states with binding energy varying with film thickness. The quantization effects on these levels are visible in the spectra up to a thickness of about 50 atomic layers.

Considering the magnetization of the substrate it can also be shown that the spin dependent reflection at the interface induces a magnetic character in these quantum well states. The polarization analysis of the emitted electrons by means of spin resolving photoelectron spectroscopy demonstrates that the quantum well states in the Cu film on Co(100) possess a predominantly minority spin character. Figure 14.7(a) shows the spectra of Cu films decomposed into the two spin components. The corresponding non-spin resolved spectra are also presented for comparison. The quantum well states appear as prominent structures in the minority spin channel whereas they are weak or absent in the majority spin spectra.

This behavior can easily be understood since the Co substrate acts as a spin dependent potential barrier for the confinement of the electronic wave function in the Cu film (see Fig. 14.7(b)). The electron waves are reflected at

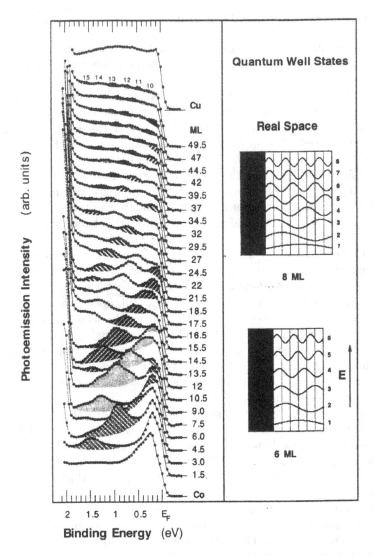

Fig. 14.6. Photoemission spectra of ultrathin Cu films on fcc-Co(100). The shadowed spectral structures with binding energies depending on the film thickness derive from the quantization of the energy levels due to electron confinement. They are observed up to 50 atomic layers. On the *right side* a simple representation of the electron states of a thin film is presented in analogy to quantum well states. The confinement of the wave functions is due to the reflection at the potential barriers at the interface with the substrate and the vacuum. (Reprinted from [50] with permission of the FZ Jülich)

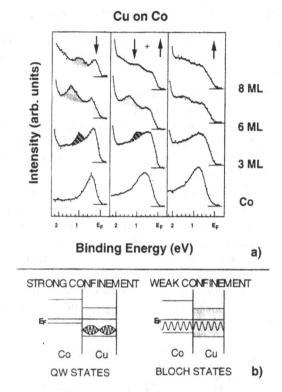

Fig. 14.7. (a) The photoemission spectrum (middle) of thin Cu films on fcc-Co(100) is decomposed into the two spin channels, minority ($\downarrow$) and majority ($\uparrow$), by the measurement of the photoelectron spin polarization. The quantized levels giving rise to the shadowed structures have predominantly minority spin character. (b) The ferromagnetic substrate acts as a spin dependent potential barrier and confines to a different degree the electronic wave function of opposite spin.(Reprinted from [50] with permission of the FZ Jülich)

the interface to a degree that depends on their spin character. The minority spin electrons which the potential is very different in the two metals for are strongly reflected at the interface and become effectively confined in the Cu film. Conversely, the majority spin electrons which possess a similar energy in the two metals are weakly reflected and become delocalized over the two materials. The electronic structure of the paramagnetic Cu film in contact with the ferromagnetic substrate acquires in this way a magnetic character, i.e. a dependence on the electronic spin orientation.

14.2 Influence of a (Non-Magnetic) Capping Layer

As already seen in Fig. 14.1 the free surface of a ferromagnetic thin film exhibits enhanced magnetic moments. Capping with a non-magnetic layer leads to the following changes:

- The increase of the coordination number of the surface atoms results in a reduction of the magnetic moments in the topmost layer.
- The overlap between the wave functions of atoms at the magnetic substrate and the non-magnetic capping layer results in a hybridization which leads to a decrease of the magnetic moments not only of the topmost but of several layers near the interface.

As one example of this behavior the reduction of the magnetic moment Δm is shown in Fig. 14.8 for the situation that a ferromagnetic Ni thin film is capped with a Cu film with various thickness. A decrease of the magnetic moment in units of that concerning one monolayer of Ni(111) occurs up to a thickness of one atomic Cu capping layer followed by a slight further increase in reduction to the saturation value of 0.79 units with further increasing thickness.

Additionally, this observation gives evidence that in this system no interdiffusion takes place which would result in a continuing reduction of the magnetic moment.

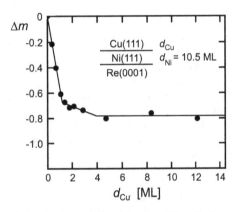

Fig. 14.8. Change of magnetic moment Δm in units of the magnetic moment of one monolayer Ni(111) caused by Cu(111) on Ni(111) thin films with a thickness of about 10 monolayers on Re(0001) as a function of the thickness d_{Cu} of the Cu coating layer. (Reprinted from [51]. Copyright 1985, with permission from Elsevier)

14.3 Influence of a Magnetic Capping Layer

The behavior of a magnetic capping layer on a ferromagnetic substrate which exhibit different Curie temperatures as well as different anisotropy contributions results in complex properties.

An example of this situation is given by a thin Fe film on a Gd(0001) substrate. Gadolinium possesses an in-plane magnetization and a Curie temperature of $T_C^{\text{Gd}} = 293\,\text{K}$. Because the Fe overlayer is exchange coupled to Gd the magnetization of the Fe film is held in-plane. The Fe/Gd interface, however, has a strong perpendicular magnetic anisotropy. Additionally, Fe films exhibit a thickness dependent Curie temperature being above T_C^{Gd}.

The determination of the spin polarization of secondary electrons permits to obtain independently the in-plane component of the magnetization as well as the out-of-plane component (see Fig. 14.9a).

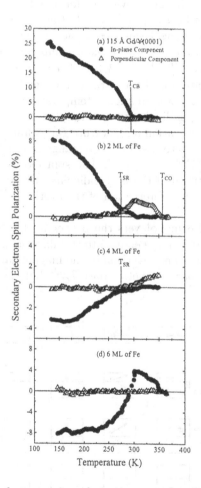

Fig. 14.9. Secondary electron spin polarization as a function of temperature for increasing thickness of Fe deposited on the Gd(0001) surface. The data is completely reversible on decreasing the temperature. The sample was remanently magnetized at each data point. (Reused with permission from [52]. Copyright 1998, AVS The Science & Technology Society)

The properties for increasing thickness of the Fe overlayer is shown in the lower panels of Fig. 14.9. A reduction of the in-plane component of the magnetization occurs at a thickness of 2 monolayers (note the different scales for the electron spin polarization) which proves an antiferromagnetic coupling between the Fe thin film and the Gd substrate at low temperatures. With increasing thickness a reversal of sign of the in-plane component takes place. This is due to the surface sensitivity for slow secondary electrons (cf. Fig. 14.4) which amounts to only a few layers.

The onset of an out-of-plane component at $T_{\mathrm{SR}} = 275\,\mathrm{K}$ for Gd being capped with a thin Fe film gives evidence for a spin reorientation transition in the Fe film for a thickness up to about 4 atomic layers. For this thickness regime the in-plane component vanishes at the Curie temperature of the bulk material $T_C^{\mathrm{bulk}} = 293\,\mathrm{K}$ and the out-of-plane component at the Curie temperature of the Fe overlayer $T_C^{\mathrm{overlayer}} \geq 320\,\mathrm{K}$. The first step at low temperature is a continuous reorientation of the surface moment from in-plane to canted out-of-plane. The second step at higher temperature is a rotation from this canted direction to perpendicular to the film plane.

The lower spin reorientation temperature T_{SR} compared to the bulk Curie temperature T_C^{bulk} points to a stronger exchange interaction at the interface than in the bulk: $E_{\mathrm{exch}}^{\mathrm{Fe-Gd}} > E_{\mathrm{exch}}^{\mathrm{Gd-Gd}}$. The spin reorientation transition results from the competition of the perpendicular surface anisotropy of the Fe overlayer and the exchange interaction of the overlayer with the in-plane Gd magnetization.

For a thickness of 6 monolayers the perpendicular component vanishes. Thus, the dipole energy is large enough to stabilize the magnetization within the plane of the Fe film. The inversion of the magnetization direction occurs at T_C^{bulk}. Gd becomes paramagnetic above T_C^{bulk} but the Fe film is magnetic

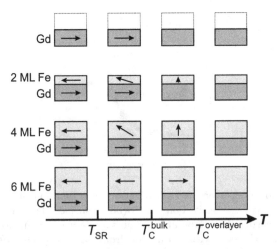

Fig. 14.10. Illustration of the magnetic properties of an Fe overlayer with various thickness on a Gd substrate as a function of temperature

up to $T_C^{\text{overlayer}}$. Therefore, the magnetic moments in the Fe film are no more influenced by the substrate in this temperature regime between T_C^{bulk} and $T_C^{\text{overlayer}}$.

An overview of this complex system for different temperatures with varying film thickness is given in Fig. 14.10.

14.4 Comparison Between an Ideal and Real Interface

The influence on the magnetic properties of a real, i.e. generally not ideal, interface is illustrated by a ferromagnetic thin film on an antiferromagnetic substrate (see Fig. 14.11). Assuming a ferromagnetic coupling at the ideal, i.e. smooth, interface six uncompensated spins occur with respect to the antiferromagnet in this example. But, a real interface is not smooth. Its roughness leads to the formation of terraces with opposite spin direction thus preventing a ferromagnetic coupling over the whole area. In order to reduce the number of uncompensated spins (here from six to two) frustration (marked by circles) or domain walls (white box) occur.

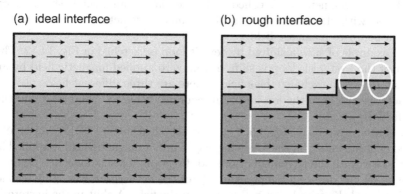

Fig. 14.11. (a) Ideal interface being smooth. Assuming a ferromagnetic coupling across the interface results in six uncompensated spins concerning the antiferromagnet. (b) In reality most of the interfaces are rough and exhibit steps which separate different terraces. Now, the ferromagnetic coupling across the interface cannot simultaneously be fulfilled. Thus, frustration (*circles*) and domain walls (*white box*) can occur which leads to a reduction of uncompensated spins, in this example from six to two

14.5 Exchange Bias

One of the most interesting interfaces for basic study and application is the interface between a ferromagnet and an antiferromagnet. A ferromagnet such as iron exhibits a large exchange parameter but a relatively small anisotropy. This makes ferromagnetic order stable at high temperatures but the orientation may not be, particularly if the dimensions are a few nanometers.

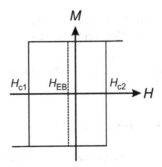

Fig. 14.12. Shifted hysteresis loop in an exchange biased ferromagnet. Characteristic fields are the exchange bias field, H_{EB}, and coercivities, H_{c1} and H_{c2}

Many antiferromagnets have large anisotropies and consequently very stable orientations. In heterostructures exchange coupling between the ferromagnet and antiferromagnet can, in principle, produce a ferromagnetic behavior with stable order combined with a high anisotropy. In such a structure the anisotropy may behave as unidirectional or uniaxial. This phenomenon is not found in ferromagnets and is called exchange bias because the hysteresis loop associated with the ferromagnet/antiferromagnet structure can be centered about a non-zero magnetic field.

An example of such a shifted hysteresis loop is sketched in Fig. 14.12. The center of the hysteresis loop is shifted from zero applied magnetic field by an amount H_{EB}, the exchange bias field. There are three different fields used to characterize the bias: the left and right coercive fields, H_{c1} and H_{c2}, and the bias field H_{EB}.

A shifted hysteresis loop, such as that sketched in Fig. 14.12, can be obtained experimentally in the following way. First, a magnetic field is applied in order to saturate the ferromagnet in field direction. This is done at a temperature above the ordering temperature T_N of the antiferromagnet. The second step is to cool the sample below T_N while in the field. A shift in the hysteresis loop can appear if measured after cooling.

This shift is due to a large anisotropy in the antiferromagnet and a weaker exchange energy coupling between the ferromagnet and antiferromagnet. A schematic diagram of the process is given in Fig. 14.13 for a ferromagnet with no anisotropy. In Fig. 14.13(a) the saturating magnetic field is applied for a temperature above T_N. This aligns the ferromagnet. After cooling the system while still in the field the magnetization remains pinned along the original direction for small negative fields (see Fig. 14.13(b)). A field large enough to overcome the interlayer exchange reverses the ferromagnet (see Fig. 14.13(c)). On the reverse path the ferromagnet rotates back into the original positive direction while the applied field is still negative. This gives a shifted magnetization curve as shown in Fig. 14.12.

The magnitude of the shift is equal to the effective field associated with the interlayer exchange. A hysteresis loop appears when anisotropy is included

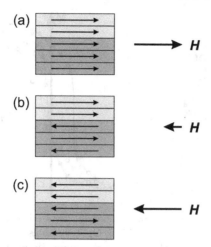

Fig. 14.13. (a) A saturating magnetic field is applied in order to align the ferromagnet above T_N. After cooling the system in field the magnetization remains pinned along the original direction when the field is reversed as shown in (b). A sufficiently large field reverses the ferromagnet as shown in (c)

in the ferromagnet. Bounds for the coercive fields H_{c1} and H_{c2} can be found by examining the stability of an energy per unit area of the form

$$E = -H M_S d_F \cos\theta - J \cos\theta + K \sin^2\theta \qquad (14.4)$$

In this model the applied magnetic field is H, M_S is the saturation magnetization of the ferromagnet, d_F the thickness of the ferromagnetic film, J the interlayer exchange constant between the ferromagnet and the antiferromagnet, and K a measure of the uniaxial anisotropy in the ferromagnet. The angle θ is taken between M and the uniaxial anisotropy easy axis. The field is aligned along the easy axis and the magnetization is assumed to remain uniform in this model. The most important restriction is that the antiferromagnet remains rigidly aligned along the direction of its easy axis assumed to lie also parallel to the ferromagnet easy axis.

Experimentally, exchange anisotropy has been discovered in fine Co particles exhibiting a CoO coating (see left part of Fig. 14.14). CoO is an antiferromagnet with a Néel temperature T_N of 293 K. When cooling the particles from 300 K (CoO in a paramagnetic state) to 77 K (CoO in an antiferromagnetic state) in a saturating magnetic field a unidirectional anisotropy was observed. The two hysteresis loops shown in the right part of Fig. 14.14 were taken at 77 K. The CoO coated Co particles were at first cooled from room temperature to 77 K in a zero external field. The corresponding hysteresis loop (drawn as a dashed line with open symbols) is symmetric about the vertical axis. When the specimen is cooled to 77 K in a strong magnetic field (so-called field cooling) the hysteresis loop (drawn as a full line with closed symbols) is displaced to the left along the applied field axis.

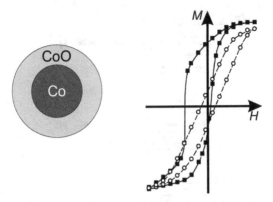

Fig. 14.14. Left: Co particles being coated by CoO. **Right**: Hysteresis loops of oxide coated particles of Co taken at 77 K. Open symbols show the hysteresis loop when the material is cooled in the absence of a magnetic field. The closed symbols show the hysteresis loop when the material is cooled in a saturating magnetic field. (Data taken from [53])

The energy given in (14.4) has extremum values corresponding to saturation in the directions of $\theta = 0$ and $\theta = \pi$. Stability of the $\theta = 0$ configuration is possible if $J + HM_S d_F + 2K > 0$ and stability of the $\theta = \pi$ configuration is possible if $2K - J - HM_S d_F > 0$. This corresponds to coercive fields

$$H_{c1} = -\frac{J + 2K}{M_S d_F} \tag{14.5}$$

and

$$H_{c2} = \frac{2K - J}{M_S d_F} \tag{14.6}$$

Because the coercive fields are not equal in magnitude the total hysteresis is biased. The bias field in this model can be defined as the midpoint of the hysteresis and is directly proportional to the exchange coupling:

$$H_{\mathrm{EB}} = \frac{H_{c1} + H_{c2}}{2} = -\frac{J}{M_S d_F} \tag{14.7}$$

The bias field is determined by competition between the Zeeman energy and the interlayer exchange energy and therefore depends on the thickness of the ferromagnet d_F. The coercive field H_c is given by:

$$H_c = \frac{|H_{c1} - H_{c2}|}{2} = \frac{K}{M_S d_F} \tag{14.8}$$

and thus proportional to the anisotropy constant K.

A typical temperature dependence of the hysteresis loop is presented in Fig. 14.15 for a system consisting of a layered structure of Co/CoO. The

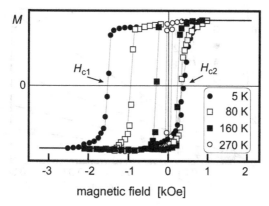

Fig. 14.15. Hysteresis loops for different temperatures of CoO(25 Å)/Co(120 Å). For each hysteresis the bilayer was cooled in a field of +2000 Oe from 320 K to the respective temperature. (Adapted from [54] (used with permission))

hysteresis loops show the following typical and general features: H_{c1} strongly increases with decreasing temperature while H_{c2} remains almost constant at a field value of about several hundreds Oe. The slope of the hysteresis loops at H_{c1} is steeper than at H_{c2} on the return path.

Figure 14.16 summarizes the analysis of the temperature dependence of H_{c1} and H_{c2} for a similar sample. Both coercive fields start to slightly increase just below the Néel temperature with the same rate. At the blocking temperature of about $T_B = 186$ K the slope increases drastically. Below T_B a bifurcation for the temperature dependence of H_{c1} and H_{c2} develops. While H_{c1} keeps rising with a rate of 11 Oe/K H_{c2} levels off and reaches saturation

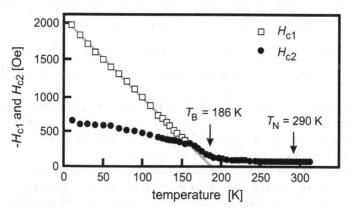

Fig. 14.16. Coercive fields H_{c1} and H_{c2} (cf. Fig. 14.15). The line is a fit to the linear region of H_{c1}. It intersects the abscissa at the blocking temperature $T_B = 186$ K. (Adapted from [54] (used with permission))

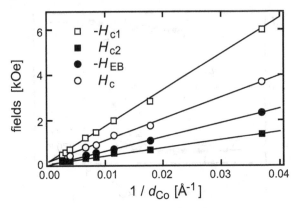

Fig. 14.17. $-H_{c1}$ (*open squares*), H_{c2} (*full squares*), $-H_{EB}$ (*full circles*), and H_c (*open circles*) are plotted as a function of the Co thickness. The samples were cooled down through the Néel temperature of CoO to 10 K in an applied magnetic field of +2000 Oe. The lines are linear fits to the data points. (Adapted from [54] (used with permission))

at the lowest temperature. Thus, there are three distinguishable temperature regimes.

First, from T_N to T_B the coercive fields are equal and increase slowly. Second, close to T_B the slopes increase drastically and H_{c1} is slightly smaller than H_{c2}. Third, below T_B both coercive fields develop linearly but with different slopes such that the absolute value of H_{c1} is larger than H_{c2}. Only in this last region a strong negative exchange bias is observed.

The dependence of the coercive fields as well as the exchange bias field on the thickness of the ferromagnetic film d_F is shown in Fig. 14.17. It is obvious that all fields exhibit a $1/d_F$ dependence which gives evidence that the exchange bias effect is due to the existence of an interface.

The most important characteristics concerning the exchange bias are:

- It only occurs for a ferromagnetic thin film on an antiferromagnetic substrate.
- It is necessary to cool the system in an external magnetic field below the blocking temperature $T_B < T_N$.
- A unidirectional anisotropy is induced due to the field cooling procedure.
- Generally, the exchange bias field is negative.
- The exchange bias field increases with decreasing temperature.
- An increase of exchange bias field leads to a rise of the coercive fields.
- The exchange bias scales with the reciprocal value of the film thickness.

Further discussions are given in Chap. 16.

Magnetism in Reduced
Dimensions – Multilayers

In this chapter we will discuss systems which consist of *a lot of* magnetic thin films (so-called *multilayers*).

The principle arrangement consists of a ferromagnetic bulk-like or thin film substrate covered by a non-magnetic thin film which itself is capped by a ferromagnetic layer. This stacking may be continued by additional non-magnetic and ferromagnetic thin films. The non-magnetic layer consists of a metal, an oxide, a semiconductor, or vacuum. The latter case stands for two ferromagnetic electrodes which are separated by several Å. An important feature is given by the coupling over the interface, the so-called interlayer exchange coupling (IEC).

Additionally, the electrical resistance or the electrical conductance of this layered system depends on the relative orientation of the magnetization of two neighbored ferromagnetic layers. The resistance of an antiparallel orientation R_{ap} is enhanced compared to a parallel alignment:

$$R_{\mathrm{ap}} > R_{\mathrm{p}} \tag{15.1}$$

This phenomenon is called magnetoresistivity and will be discussed in more detail in Chap. 16.

15.1 Interlayer Exchange Coupling (IEC)
Across a Non-Magnetic Spacer Layer

The coupling between two localized magnetic moments being separated by a non-magnetic material can be described by means of the RKKY exchange interaction (see Chap. 4.3):

$$J_{\mathrm{RKKY}}(R) \propto \frac{x\,\cos x - \sin x}{x^4} \xrightarrow{x \to \infty} \frac{1}{x^3} \tag{15.2}$$

with $x = 2k_F R$ and R the distance between the moments. This oscillatory behavior is exemplarily shown in Fig. 15.1 for Mn atoms embedded in a Ge

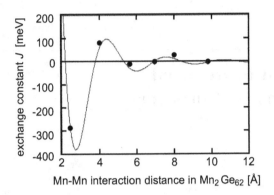

Fig. 15.1. Exchange interaction for Mn atoms as a function of their distance in Mn_2Ge_{62}. The *solid line* is the RKKY model fitted with $k_F = 1.02\,\text{Å}^{-1}$. (Adapted from [55] (used with permission))

matrix. This dependence is different in magnetic thin film systems being separated by a non-magnetic interlayer. Assuming an interlayer thickness of z the exchange coupling constant exhibits the dependence of:

$$J_{\mathrm{RKKY}}(z) \propto \frac{1}{z^2} \qquad (15.3)$$

which is schematically shown in Fig. 15.2. In layered systems the RKKY interaction is a pronounced effect and acts over long distances. A positive value of J means a ferromagnetic coupling between both ferromagnetic thin films whereas $J < 0$ results in an antiferromagnetic arrangement. We directly see that the thickness of the spacer layer determines the type of coupling.

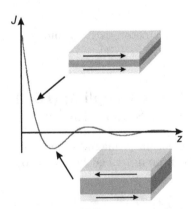

Fig. 15.2. Interlayer exchange coupling between two ferromagnetic thin films exhibiting an RKKY behavior. Depending on the thickness z of the spacer layer a ferromagnetic or antiferromagnetic alignment occurs for $J > 0$ or $J < 0$, respectively

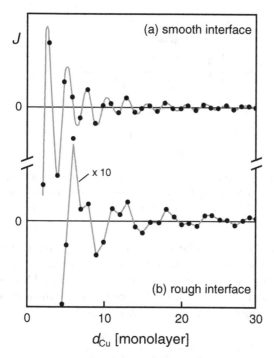

Fig. 15.3. Calculated interlayer coupling for a Cu(001) spacer as a function of the Cu thickness d_{Cu} at $T = 0$ K. The solid circles correspond to physically achievable thickness, i.e. d_{Cu} integer. **(a)** Zero roughness, **(b)** rough interface. (Adapted from [56] (used with permission))

A calculation of this RKKY interaction is exemplarily presented in Fig. 15.3 for a Co/Cu-system with smooth and rough interface. Figure 15.3(a) shows the calculated coupling exhibiting a strong short-period oscillation and a much weaker long-period oscillation. The layers were assumed to be atomically flat whereas real samples always possess some interfacial roughness. In order to consider the effect of the roughness the spacer layer of average thickness d is assumed to consist actually of large (compared to the spacer thickness) patches with local thickness equal to $d - 1$, d, and $d + 1$. The coupling behavior for this situation is shown in Fig. 15.3(b). The coupling strength is strongly reduced and the apparent period is increased. This is because the short period is almost suppressed by the roughness and only the weak long-period oscillation is seen. This simple example illustrates how important the influence of the roughness can be.

The determination of the interlayer exchange coupling can only be carried out at positions of the discrete lattice planes. The thickness of the interlayer amounts to:

$$z = (N + 1) \cdot d \tag{15.4}$$

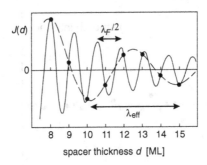

Fig. 15.4. *Full curve*: Coupling function for a fcc(100) metal calculated using the RKKY model. *Broken curve*: Actual coupling with the experimentally measured periodicity as a function of the spacer thickness. (Data taken from [57])

with d being the distance between lattice planes and N the number of planes. The Fermi surface of the non-magnetic interlayer determines the strength of the oscillation as well as the oscillation period $\lambda_F = \pi/k_F$. This oscillation period is often incommensurable with the lattice constant which results in an effective period λ_{eff} being larger than λ_F (see Fig. 15.4):

$$\lambda_{\text{eff}} = \frac{2\pi}{2k_F - 2\pi/d} = \frac{d \cdot \lambda_F}{|d - \lambda_F|} \tag{15.5}$$

Thus, the situation of interlayers consisting of different materials with identical density can result in a different relative orientation of the magnetization in both ferromagnetic films (see Fig. 15.5). The type of coupling, i.e. ferro- or antiferromagnetic, can be determined by the shape of the hysteresis loop (see Fig. 15.6).

An additional influence is given by magnetic anisotropy effects which can be realized by different values of the anisotropy constant K. This is exemplarily discussed for a system with a uniaxial anisotropy. The shape of the hysteresis loops and thus the magnetic properties are significantly different for $K = 0$, small value of K, and a large value of K. The system consists of two identical magnetic layers of thickness d which are antiferromagnetically coupled across the spacer layer with a strength J possessing an uniaxial

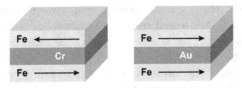

Fig. 15.5. Spacer layers of a given thickness can result in an antiferromagnetic and ferromagnetic coupling in dependence of the interlayer material

(a) antiferromagnetic coupling   (b) ferromagnetic coupling

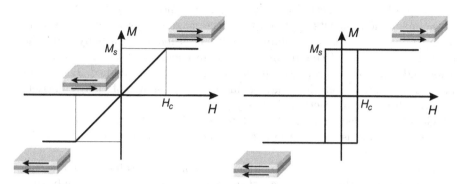

Fig. 15.6. The shape of the magnetization curve is significantly different for an (a) antiferromagnetic and (b) ferromagnetic coupling between two ferromagnetic thin films being separated by a spacer layer

anisotropy constant K and saturation magnetization M_S. The applied field H is directed along the easy axis which is in this case the film normal, i.e. $K > 0$.

Calculating the absolute minimum energy allows to determine the thermodynamically stable state and corresponding transitions which result in magnetization curves shown in Fig. 15.7. The shape of the magnetization curve depends on the ratio between the strength of the magnetic anisotropy and the magnitude of the antiferromagnetic coupling. For $K > -J/d$ and $K < -J/d$ the respective curves are shown in Fig. 15.7(a) and Fig. 15.7(b), respectively. The curve in Fig. 15.7(c) is a special case of the one shown in Fig. 15.7(b). The characteristic fields H_f, H_{sf}, and H_s occurring in Fig. 15.7 are given by $H_f = -2J/M_S$ for $K > -J/d$, $H_s = -2(K + 2J/d)/M_S$ for $K < -J/d$,

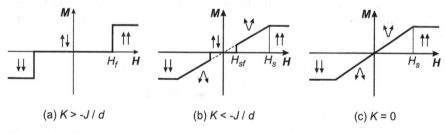

(a) $K > -J/d$ (b) $K < -J/d$ (c) $K = 0$

Fig. 15.7. Theoretical magnetization curves for two identical antiferromagnetically coupled ferromagnetic layers exhibiting a uniaxial magnetic anisotropy which is (a) larger than the coupling strength and (b) smaller than the coupling strength. The arrows schematically indicate the orientation of the magnetization directions relative to the vertical easy axis which the field is applied along. The situation of zero anisotropy (c) is a special case of (b). (Adapted from [58])

$H_{sf} = 2\sqrt{-K(K+2J/d)}/M_S$ for $K < -J/d$, and $H_s = -4J/dM_S$ for $K = 0$. These formulas are derived from absolute minimum energy calculations; no hysteresis occurred. However, if one allows for coherent rotation of the magnetic moments only and takes the magnetic layers always to be in a single domain state, i.e. excluding the mechanisms which usually drive the system to the state of the absolute minimum energy such as domain nucleation and domain wall propagation, the above results are modified. The curves necessarily show hysteretic behavior due to the existence of energy barriers resulting from the magnetic anisotropy.

Generally, measurements of the hysteresis loop allow to determine the oscillation period as well as the negative, i.e. antiferromagnetic, part of the coupling constant.

Experimental techniques which permit a lateral resolution enable the determination of the coupling constant and oscillation period in only **one** measurement if structures are investigated which are shaped as a wedge (see Fig. 15.8). This is exemplarily shown for the system of a Au wedge with an increasing thickness of 0 to 20 monolayers on a ferromagnetic Fe substrate. The Au layer itself is capped with an Fe film exhibiting a constant thickness of 12 monolayers. An image obtained by means of SEMPA (secondary electron microscopy with polarization analysis) is presented in Fig. 15.9 which shows the x-component of the magnetization M_x for the top surface of the Fe/Au/Fe wedge structure. White (black) indicates that M_x is directed in the $+x(-x)$-direction while gray indicates that $M_x = 0$. With no applied magnetic field the whisker is divided by a horizontal domain wall into two equivalent but oppositely magnetized domains. The bottom half of the image in Fig. 15.9 shows a pattern that oscillates between black and white with interspersed regions of gray. Since the substrate domain is oriented in the $+x$-, or white, direction, white regions correspond to ferromagnetic coupling, black to antiferromagnetic coupling, and gray to $90°$ coupling.

The corresponding coupling strength given by J and the oscillation behavior are shown in Fig. 15.10. Two different oscillations can be observed, one with a small period of $\lambda_1 \approx 2.5$ monolayers and one with a longer period of $\lambda_2 \approx 8.5$ monolayers.

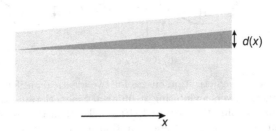

Fig. 15.8. Wedge-shaped structure. Due to the increasing thickness of the spacer layer $d(x)$ in dependence of the lateral position x the determination of thickness dependent properties can be achieved in only one measurement

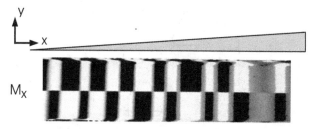

Fig. 15.9. (a) SEMPA measurements of the x-component of the magnetization M_x of the top Fe layer of the first 22 ML of an Fe/Au/Fe wedge structure (From [59] (used with permission))

The oscillation periods of the interlayer exchange coupling can be predicted by considering the Fermi surface of the spacer material. One finds that oscillatory coupling is related to a critical spanning vector Q in reciprocal space with the following properties:

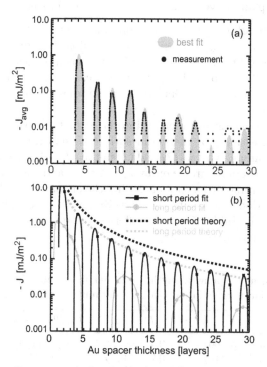

Fig. 15.10. Coupling strength for Fe/Au/Fe. (a) *Circles* indicate the measured values of J_{avg} and the (*gray areas*) show the best fit function on a semilog scale over a wide range of coupling strength and spacer layer thickness. (b) *Solid squares* (circles) show the variation in the short (long) period bilinear coupling strength determined by these experiments. The dark (light) (*dashed lines*) indicate the calculated maximum theoretical envelope for the short (long) period contribution to the bilinear coupling strength. (From [59] (used with permission))

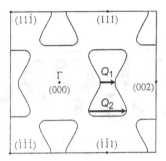

Fig. 15.11. Cross section of the Fermi surface of Au. Included are the critical spanning vectors Q_1 and Q_2 in the [100]-direction

- The critical spanning vector points perpendicular to the interface.
- The critical spanning vector connects two sheets of the Fermi surface which are coplanar to each other.
- The critical spanning vector is in the first Brillouin zone.

The last condition follows the Bloch theorem and reflects the atomic periodicity of the spacer material. The oscillation period is then given by $2\pi/Q$. For real materials several Q_i ($i = 1, 2, \ldots$) may exist each of them corresponding to a different oscillation period $2\pi/Q_i$. In this case the experimentally measured coupling as a function of the interlayer thickness is the superposition of all these oscillations.

As an example let us consider an interlayer material with fcc structure grown in [100]-direction. For the Fermi surface as shown in Fig. 15.11 there are two critical spanning vectors Q_1 and Q_2 in the [100]-direction. The periods of the oscillatory coupling are given by $\lambda_i = 2\pi/Q_i$ and thus are determined solely by the electronic properties of the interlayer material. The strength J and oscillation periods concerning the interlayer exchange coupling for various systems are given in Table 15.1. Two different periods seem to be related to a (100) interlayer interface.

Table 15.1. Bilinear coupling strength $-J$ in mJ/m^2 at an interlayer thickness z in nm and oscillation periods in monolayers (and nm) of various systems

System	$-J$	z	Oscillation periods
Co / Cu / Co(100)	0.4	1.2	2.6 (0.47), 8 (1.45)
Co / Cu / Co(110)	0.7	0.85	9.8 (1.25)
Co / Cu / Co(111)	1.1	0.85	5.5 (1.15)
Fe / Au / Fe(100)	0.85	0.82	2.5 (0.51), 8.6 (1.75)
Fe / Cr / Fe(100)	>1.5	1.3	2.1 (0.3); 12 (1.73)
Fe / Mn / Fe(100)	0.14	1.32	2 (0.33)
Co / Rh / Co(111)	34	0.48	2.7 (0.6)

15.2 Interlayer Exchange Coupling Across
an Antiferromagnetic Spacer Layer

The behavior of the exchange interlayer coupling drastically changes if the non-magnetic interlayer is replaced by a layered itinerant antiferromagnet like Cr. Chromium has a bcc structure. If Cr would have a commensurate antiferromagnetic structure the magnetic moment density at the corners would be opposite to the ones at the center of the bcc unit cell as sketched in the left part of Fig. 15.12 forming a commensurate spin density wave structure. Thus, the antiferromagnetic bcc structure consists of a sequence of ferromagnetic (001) planes with alternating spin direction. Pure Cr exhibits, in fact, a linearly polarized incommensurate spin density wave structure which consists of a sinusoidal modulation of the magnetic moments and is schematically shown in the right part of Fig. 15.12. It can be visualized as a spin lattice which is slightly expanded as compared to the crystal lattice yielding a beating effect between both.

Thus, one would expect that the IEC exhibits a period of 2 monolayers (see Fig. 15.13). For thin Cr films up to about 25 monolayers this expectation is in agreement with experiment (see Fig. 15.14) exhibiting many short-period oscillations with a period of two monolayers (in addition to a long-period oscillation).

A wedge-shaped structure Fe/Cr/Fe with a thickness of the Cr interlayer ranging from 0 to about 70 monolayers which is schematically shown in the left part of Fig. 15.15 exhibits an oscillation with a period of two monolayers (see right part of Fig. 15.15). In this SEMPA image white (black) corresponds to a magnetization to the right (left). The Fe substrate possesses a domain wall running along its length which provides a useful way to verify the zero of the magnetization. The short-period oscillation in the exchange coupling which causes the magnetization to change each layer is superposed on a long-period coupling. In the thinner parts of the wedge the short-period oscillations dominate only after the first antiferromagnetic transition at five layers and initially are not symmetric about zero which leads to the black and white

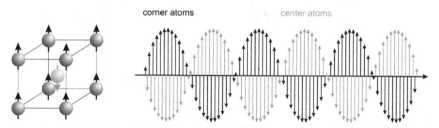

Fig. 15.12. Commensurate and incommensurate spin density wave structure of bcc Cr. In the right part a spin density wave is illustrated with the magnetic moments perpendicular to the wave vector

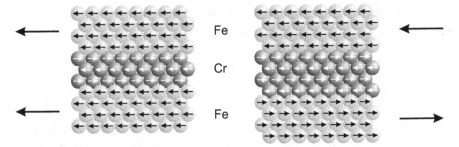

Fig. 15.13. If the spacer layer is realized by a topological antiferromagnet like Cr which additionally favors an antiferromagnetic coupling at the interface one would expect that the IEC exhibits a period of two monolayers depending on the number of spacer layers which can be odd (**left**) or even (**right**) leading to a ferro- or antiferromagnetic coupling, respectively, between both ferromagnetic thin films on both sides of the interlayer

stripes of slightly different width in the image. Coupling is observed through over 75 layers corresponding to a Cr thickness of over 10 nm.

There is a change in phase or "phase slip" apparent between layers 24 and 25, 44 and 45, and 64 and 65. Thus, just below the phase slip at 24 layers the Fe overlayer is coupled ferromagnetically to the substrate when the number of Cr layers is *even*. Just above the phase slip at 25 layers it is coupled fer-

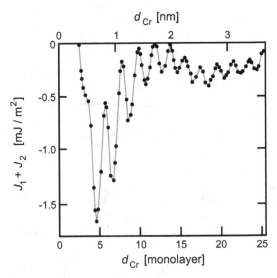

Fig. 15.14. Interlayer coupling in Fe/Cr/Fe as a function of the Cr thickness d_{Cr} measured at room temperature. The substrate temperature during preparation was 523 K. The thickness of each Fe film amounts to 5 nm. (Data taken from [60]. Copyright 1993, with permission from Elsevier)

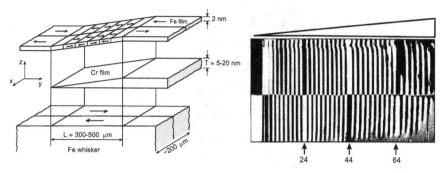

Fig. 15.15. Left: A schematic exploded view of the Fe/Cr/Fe(100) sample structure showing the Fe(100) single crystal substrate, the Cr wedge, and the Fe overlayer. The arrows show the magnetization direction in each domain. The z-scale is expanded approximately 5000 times. (From [61] (used with permission)) **Right**: SEMPA image of the magnetization in an Fe layer covering a varying thickness Cr film as a wedge grown on an Fe(100) single crystal substrate. The arrows mark the Cr interlayer thickness in atomic layers where phase slips in the magnetization oscillations occur due to the incommensurability of the spin density wave. The actual area imaged is approximately 0.1 mm high × 1 mm long. (From [62] (used with permission))

romagnetically when the number of Cr layers is *odd*. The phase slips are due to the fact the wave vector Q which governs the coupling is incommensurate with the lattice wave vector.

The Fe/Cr interface coupling is strong enough to place an antinode at the interface and induce proximity magnetism in the Cr layer. Whenever the Cr thickness is incremented by one monolayer the Fe in-plane magnetization switches direction. Thus, magnetic domains are not created in the Cr film but in the top Fe layer to overcome possible frustration effects following a step at the Fe/Cr interface. Stretching of the spin density wave seems not to take place since the distance between phase slips is constant. The highly regular pattern points to a rather rigid spin density wave period. The distance between the steps is larger than the average magnetic domain wall width in the Fe layer. Therefore, the magnetic domains in the top Fe layer can follow the directional change of the Cr spins at the Fe/Cr interface. We directly see that the surface of Cr is ferromagnetically ordered with a layered antiferromagnetic structure. Further, we can infer that in the Cr film the incommensurate spin density wave transverse with the wave vector being perpendicular to the plane and spins in the plane as sketched in Fig. 15.16.

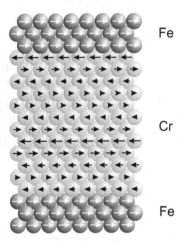

Fig. 15.16. Schematic representation of the ideal magnetic structure for a Cr spacer layer in-between two ferromagnetic Fe layers. The incommensurate spin density wave propagates perpendicular to the interface. The Cr spins are oriented within the plane which allows an antiferromagnetic exchange interaction at each Cr/Fe interface

15.3 Influence of the Interface Roughness on the IEC

The most important effects of the interlayer roughness on the interlayer exchange coupling are the reduction of the amplitude, a loss of fine structures of the oscillation period, the so-called orange peel effect, and frustration effects.

Reduction of the Amplitude

This behavior was already illustrated in Fig. 15.3.

Loss of Fine Structures of the Oscillation Period

This effect becomes obvious in a wash-out of the short-period structures. In Fig. 15.17 SEMPA images of the Fe top layer magnetization is shown for two different cases of the Cr wedge growth. The Cr wedges in the lower panel and upper panel were grown on Fe substrates at room temperature and elevated temperature, respectively. Measurements concerning the crystallographic order prove in both cases a perfect single crystal of the Fe substrate. The room temperature grown Cr exhibits indication of some disorder whereas Cr grown at higher temperatures is just as well ordered as the substrate. The difference in the crystallinity of the Cr layer drastically changes the coupling of the Fe layers as can be seen in Fig. 15.17. The lower panel shows the characteristic long-period oscillatory coupling of Fe through the Cr layer. In striking contrast the magnetization in well ordered Cr shown in the top panel changes to

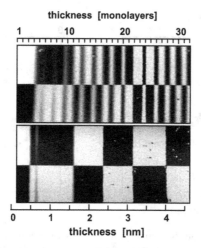

Fig. 15.17. The difference in the magnetic coupling of the Fe layers in the Fe/Cr/Fe sandwich for the Cr wedge grown on a substrate at room temperature (*lower panel*) and elevated temperature (*upper panel*) is directly obvious in these SEMPA images which characterize the M_y direction. (From [61] (used with permission))

the much shorter period of about two atomic layers. Thus, the magnetization changes with each atomic layer change in Cr thickness. The loss of fine structure (see left part of Fig. 15.18) is also obvious in graphical representations of the coupling strength (see Fig. 15.18).

Orange Peel Effect

Let us consider a correlated roughness at the interface exhibiting a topography which has been described as that of an orange peel. The flux closure due to stray field minimization within the system of the poles on one side by the poles on the other side of the interface as shown in Fig. 15.19 yields a parallel positive coupling between the magnetization of the two sides. This behavior gives rise to a ferromagnetic interlayer exchange coupling.

Frustration Effects

Frustration effects are induced by variation of the thickness of an antiferromagnetic interlayer (see Fig. 15.20). Let us discuss this behavior in more detail on the example of an Fe film on a Cr substrate.

Covering a Cr(001) substrate with a thin Fe layer one expects that the in-plane magnetization of the Fe layer matches the Cr spin structure such that at the interface the Fe and Cr spins lie antiparallel as schematically depicted in Fig. 15.21(a). This requires a reorientation of the spin density wave from a longitudinal out-of-plane propagation to a transverse out-of-plane propagation with spins in the plane. Surface roughness introduces steps of varying

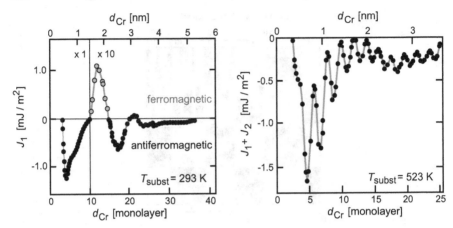

Fig. 15.18. Interlayer coupling in Fe/Cr/Fe as a function of the Cr thickness d_{Cr} measured at room temperature. The thickness of each Fe film is 5 nm. The substrate temperature during preparation was 293 K (**left**) and 523 K (**right**), respectively. A positive coupling strength J_1 corresponds to a ferromagnetic arrangement whereas a negative one to an antiferromagnetic coupling. (Data taken from [60] with permission from Elsevier)

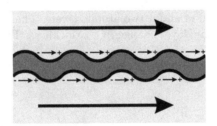

Fig. 15.19. Illustration of ferromagnetic coupling between two magnetic layers due to interface roughness. Stray fields emerging from protrusions are presented by "magnetic charges" − and +. The given parallel alignment of the two magnetic layers minimizes the stray field energy because charges with different sign from the upper and lower film oppose each other. This so-called "orange peel" coupling is of dipolar nature

heights. Any step height with an odd number of atomic layers introduces frustration to the interlayer exchange coupling at the Fe/Cr interface due to the antiferromagnetic order of the Cr (see Fig. 15.21(b)–(e)).

How the system overcomes this frustration depends on the relative magnitude of the intralayer exchange energies J_{Cr-Cr} and J_{Fe-Fe} and the interlayer exchange energy J_{Fe-Cr}. In case of a very small magnitude of J_{Fe-Cr} the frustration at the interface can be overcome by breaking the antiferromagnetic coupling at the Fe/Cr interface (see Fig. 15.21(b)) thus avoiding the formation of domain walls in Fe or Cr. On the other hand, in the case of a very large J_{Fe-Cr} a domain wall could form either in the Fe or the Cr layer

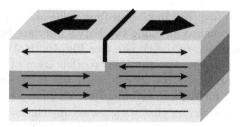

Fig. 15.20. If two ferromagnetic thin films are separated by an antiferromagnetic coupling spacer layer the variation of the thickness can result in a frustration within one of the ferromagnetic layers. This is exemplarily shown for a monatomic step in the spacer which leads to an additional layer. In order to maintain the antiferromagnetic coupling at the interface one ferromagnetic film exhibits two domains being separated by a domain wall where the step is located

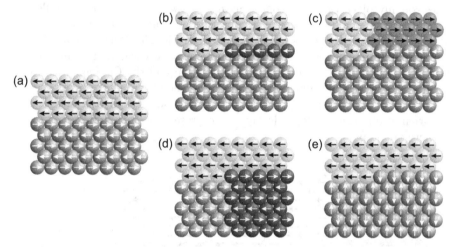

Fig. 15.21. Schematic and simplified representation of the interface between a thin ferromagnetic Fe layer and a thicker antiferromagnetic Cr film. The black arrows indicate the orientation of the magnetic moments in the Fe film, the white arrows the Cr magnetic moments. (**a**) represents an ideal and flat interface with antiferromagnetic coupling between the Fe and Cr moments. (**b**)–(**e**) show interfaces with monatomic high steps causing frustration of the interface exchange coupling. The frustration can be overcome by formation of a domain in the Fe layer (**c**), in the Cr film (**d**), or by a reorientation of the spin density wave (**e**). The latter case is experimentally observed in thick Cr films. (Adapted from [63])

(see Fig. 15.21(c) and (d)). For an intermediate value of J_{Fe-Cr} the system can react by forming a domain wall along the Fe/Cr interface by reorienting the Cr moments perpendicular to the Fe film (see Fig. 15.21(e)). Reorientation of the Cr moments requires less energy than reorientation of the Fe moments. In the latter case work has to be done against the shape anisotropy energy which is not required for the antiferromagnetic Cr film. The reorientation of the spin density wave indicates that with decreasing thickness more energy is gained by forming domain walls in the Cr (see Fig. 15.21(d)) than by forming a 90° wall along the Fe/Cr interface (see Fig. 15.21(e)). The case of domain walls in Fe (see Fig. 15.21(c)) which would lead to a vanishing magnetization of the Fe layers is experimentally not observed. Since the energy gained by domain wall formation in the Cr film scales with the Cr film thickness d_{Cr} while the energy gained by reorienting the Cr moments perpendicular to the Fe/Cr interface scales with the separation $\mathcal{L}$ of the steps and kinks at the interface one expects that the crossover from out-of-plane to in-plane spin orientation takes roughly place when the condition $J_{Cr-Cr} \cdot d_{Cr} = J_{Fe-Cr} \cdot \mathcal{L}$ is fulfilled.

To make this model more realistic interdiffusion at the Fe/Cr interface in addition to well defined steps can be included. In Fig. 15.22 the calculated corresponding ground state spin configuration is shown. Clearly, the frustration induces an effective 90° coupling between the Fe and the Cr magnetization. Together with the small uniaxial anisotropy of the Cr atoms this leads to an orientation of the Cr spins perpendicular to the surface. In this model the 90° orientation occurs independently of the presence of interdiffusion as long as there are steps.

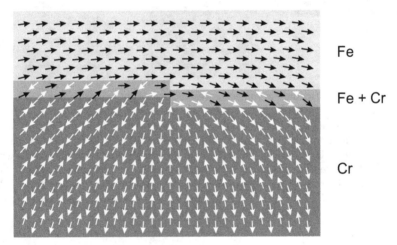

Fig. 15.22. Ground state spin structure near an Fe/Cr interface with monatomic steps from computer simulations assuming interdiffusion over two layers. Note that the Fe moments are also affected. (Adapted from [64])

Two length scales are important, the thickness of the Cr layer and the separation of the steps at the interfaces. When d_{Cr} is reduced below the distance between steps more energy can be gained by the exchange interaction at the interface than is lost by roughness induced domain wall formation within the Cr. Consequently, for thin Cr films the Cr moments are oriented in the film plane with domain walls in the Cr layer connecting the interfacial steps.

15.4 "Spin Engineering" by Interlayer Exchange Coupling

In a three layer or multilayer system the exchange behavior can be tuned by setting the different exchange coupling constants J_i to specific values.

If we assume a system consisting of three ferromagnetic layers which are separated by non-magnetic thin films as schematically shown in Fig. 15.23 and layer "FM 0" represents the dominating layer, e.g. due to its thickness, with its magnetization M_0 being aligned along a given direction determined by an external magnetic field then the coupling constants J_1 and J_2 are responsible for the shape of the hysteresis loop. The setting of $J_1 < 0$, i.e. antiferromagnetic coupling between the dominating and neighboring layer, $J_2 > 0$, i.e. ferromagnetic coupling between layers 1 and 2, and $|J_1| > |J_2|$ result in a complex behavior which is shown in Fig. 15.24. The variation of J_1 and J_2 allows to tune the shape of the magnetization curve.

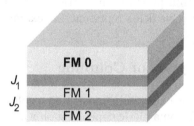

Fig. 15.23. Spin engineering samples consist of a thick ferromagnetic film "FM 0" which is coupled to the layered system to be studied, here the structure "FM 1"/interlayer/"FM 2" via an additional spacer layer

15.5 Spin Valves

Spin valve systems consist of ferromagnetic layers being separated by non-magnetic thin films and are therefore nearly identical to systems which show the giant magneto resistance effect (GMR) which will be discussed in more detail in Chap. 16.3. The only difference is given by the type of coupling.

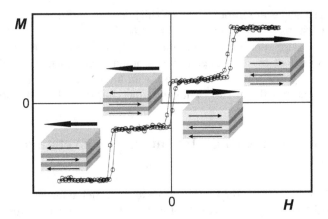

Fig. 15.24. Exemplary magnetization curve assuming a strong antiferromagnetic coupling between the topmost layer which follows the external magnetic field and the medium one. In moderate external fields layer "FM 2" (cf. Fig. 15.23) exhibits a parallel orientation with respect to layer "FM 1" assuming $J_2 > 0$ and an antiparallel orientation to the external field. Increasing the magnitude of the external field leads to a rotation of the magnetization of layer "FM 2" into field direction

Whereas in GMR systems an antiferromagnetic coupling between adjacent ferromagnetic layers is present the GMR effect also occurs in spin valve systems but the ferromagnetic thin films are magnetically decoupled. Ferromagnetic layers with different coercive fields represent one example.

15.6 Additional Types of Coupling

In the discussion above we have only considered a collinear type of coupling. It can be ferro- or antiferromagnetic depending on whether the angle between both directions of the magnetization M_1 and M_2 is $0°$ or $180°$, respectively. The energy density of the interlayer exchange coupling assuming only this bilinear coupling can be given by:

$$E_{\mathrm{IEC}} = -J_{\mathrm{BL}} \frac{M_1 \cdot M_2}{|M_1| \cdot |M_2|} \tag{15.6}$$

$$= -J_{\mathrm{BL}} \cos(\phi_1 - \phi_2) \tag{15.7}$$

with ϕ_i the angle of the magnetization of one magnetic film with respect to a given direction and thus $(\phi_1 - \phi_2)$ the angle between the magnetization of the films on both sides of the interlayer.

Taking into account an additional quadratic term the energy density amounts to:

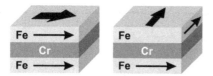

Fig. 15.25. Left: Bilinear coupling between two ferromagnetic Fe layers being separated by a Cr spacer layer results in an (anti)ferromagnetic alignment. **Right**: Biquadratic coupling may lead to an angle of 90° between both ferromagnetic layers

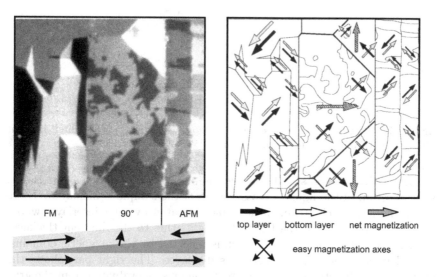

Fig. 15.26. Magnetic domains of an Fe/Cr/Fe sample. A cross section is shown in the lower panel of the left part. The thickness of the Cr spacer layer increases from 0.44 nm at the left to 0.57 nm at the right. The directions of the magnetization concerning both ferromagnetic Fe layers which exhibit a thickness of 10 nm each are given by *black arrows* for the top layer and white arrows for the bottom layer. A small thickness of the Cr interlayer results in a ferromagnetic coupling whereas a large thickness leads to an antiferromagnetic coupling. The net magnetization in the middle region where a 90° coupling is present is indicated by the *gray arrow*. The easy magnetization axes are shown in the lower right part. (Reprinted from [65]. Copyright 1995, with permission from Elsevier)

$$E_{\mathrm{IEC}} = -J_{\mathrm{BL}} \frac{\boldsymbol{M}_1 \cdot \boldsymbol{M}_2}{|\boldsymbol{M}_1| \cdot |\boldsymbol{M}_2|} - J_{\mathrm{BQ}} \left(\frac{\boldsymbol{M}_1 \cdot \boldsymbol{M}_2}{|\boldsymbol{M}_1| \cdot |\boldsymbol{M}_2|} \right)^2 \qquad (15.8)$$

$$= -J_{\mathrm{BL}} \cos(\phi_1 - \phi_2) - J_{\mathrm{BQ}} \cos^2(\phi_1 - \phi_2) \qquad (15.9)$$

with J_{BQ} being the coupling constant of this so-called biquadratic term. Now, it is possible to realize a 90° coupling between both ferromagnetic layers (see Fig. 15.25). The type of coupling depends on the thickness of the interlayer. For the determination of the type two different experimental techniques can be used: the observation of magnetic domains and magnetization curves.

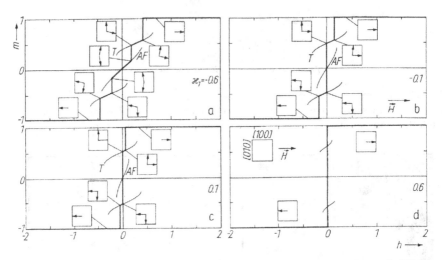

Fig. 15.27. Calculated easy axis magnetization curves for $\kappa_2 = 0.1$ and different values of κ_1. (From [66] with permission from Wiley)

Figure 15.26 presents an Fe/Cr/Fe sample with a Cr wedge as interlayer. At the bottom of the figure a cross section of the sample is displayed. There are two Fe films each 10 nm thick being separated by a Cr interlayer whose thickness increases from the left to the right. Due to the different thickness various types of coupling occur which is reflected by the magnetic domains. In the left part the original picture is displayed. The right part shows the magnetization directions in the different domains as evaluated from the gray tones of the original data. The relative orientation of the magnetization of the two films reveals the associated types of coupling. It is ferromagnetic on the left hand side, of 90° type in the middle, and antiferromagnetic on the right hand side.

The detailed shape of a magnetization curve also depends on the values of the different coupling constants. Let us exemplarily assume K as the anisotropy constant and set $\kappa_1 = J_{\mathrm{BL}}/K$ and $\kappa_2 = J_{\mathrm{BQ}}/K$. In this situation different hysteresis loops can be calculated for fixed $\kappa_2 = -0.1$ and different values of κ_1 which is shown in Fig. 15.27. Generally, plateaus at $M \approx 0$ are related to antiferromagnetic coupling whereas those at $M \approx \pm M_S/2$ are connected with biquadratic coupling. Taking into account a fourfold in-plane anisotropy the plateaus at $M \approx \pm M_S/2$ can be explained by assuming that one of the magnetization is aligned along an easy axis parallel to the external magnetic field while the other is aligned along the easy axis perpendicular to the field. Figure 15.27(b) shows the behavior of a relatively strong 90° coupling which overrides the plateau due to antiferromagnetic coupling at $M \approx 0$. In Fig. 15.27(a) the antiferromagnetic coupling is again strong enough to develop its own plateau at $M \approx 0$. Figure 15.27(d) represents a pure ferromagnetic coupling.

16

Magnetoresistivity

Magnetoresistivity deals with the influence of the static magnetic properties of solid states on the dynamic electronic behavior, i.e. the influence of the spin arrangement in a magnetic layer on the motion of electrons.

In an external magnetic field the spin arrangement changes and subsequently the mobility of the electrons. Thus, a variation of the resistance occurs which is referred to as magnetoresistivity. The magnitude of this effect is defined as:

$$\text{MR} = \frac{R(H) - R(H = 0)}{R(H = 0)} = \frac{\Delta R}{R} \tag{16.1}$$

with $R(H)$ being the resistance in an external field H. This value is often expressed as a percentage. The sensitivity concerning magnetoresistive effects is defined by:

$$S = \frac{\Delta R}{R} / H_s \tag{16.2}$$

with the unit of S given in %/Oe and H_s being the minimum external field which leads to a saturation of the magnetoresistive effect.

The most important magnetoresistive effects are:

- Normal magnetoresistance
- AMR: Anisotropic magnetoresistance
  It is caused by anisotropic scattering in the bulk. The effect amounts to several %. The sensitivity is about 1. H_s is rather small.
- GMR: Giant magnetoresistance
  It is due to a spin dependent scattering of electrons at interfaces. The magnitude is about 5% up to about several 100% at room temperature. The sensitivity is between $S = 0.01 \ldots 3$. H_s exhibits medium values.
- CMR: Colossal magnetoresistance
  This effect is caused by a spin dependent scattering of electrons in the bulk of specific materials. The effect is several 100% at room temperature and

reaches values up to $10^8\%$ at low temperatures. The sensitivity is rather small with about $S = 0.001$. H_s is large.

- TMR: Tunnelling magnetoresistance
  It is due to spin dependent tunnelling processes. The effect can reach values up to about 50% at room temperature with a sensitivity up to 1.5. The values of H_s are small.

A general differentiation must be carried out concerning:

- Transversal magnetoresistance
  The external magnetic field is perpendicular to the electric current.
- Longitudinal magnetoresistance
  The magnetic field is oriented along the direction of the electric current.

16.1 Normal Magnetoresistance

The general situation is characterized by an increase of the resistance when applying an external magnetic field. Thus, it is called positive or normal magnetoresistance. It occurs in non-magnetic metals as a consequence of the Lorentz force. The external field forces the electrons on spiral trajectories. Thus, the effective mean free path between two collisions is reduced which leads to an increase of the resistance.

Below the Curie temperature ferromagnetic transition metals exhibit a reduced resistance compared to non-magnetic transition metals like Pd (see Fig. 16.1). This occurrence is therefore called negative magnetoresistance. For transitions metals the electric current is mainly due to s electrons exhibiting a small effective mass. The resistance can be explained by scattering of the

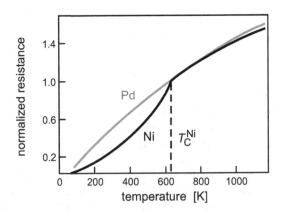

Fig. 16.1. Reduced resistance as a function of temperature for Ni and Pd. The curves are normalized with respect of the Curie temperature of Ni. (Data taken from [67]. Used with permission from Taylor & Francis Ltd. (http://www.informaworld. com))

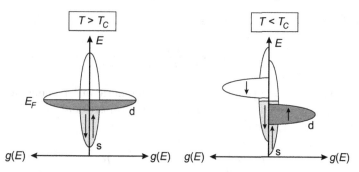

Fig. 16.2. Schematic illustration of the spin resolved density of states for a non-magnetic or a ferromagnetic material above T_C (**left**) and for a ferromagnet below the Curie temperature (**right**). The latter one exhibits a shift between majority and minority electrons

s electrons into empty states of the d band near the Fermi level E_F. Transition metals are characterized by a large density of states (DOS) of the d band at E_F (see Fig. 3.8) which causes a large scattering cross section and thus a high resistance. Contrarily, nobel metals exhibit a low DOS at E_F (see Fig. 3.10) which results in a low resistance.

In the following we will distinguish between majority and minority electrons. For a non-magnetic material or a ferromagnet above T_C the spin resolved DOS is identical for both types of electrons (see left part of Fig. 16.2) But, an exchange splitting of the bands is present for a magnetic material below the Curie temperature as schematically depicted in the right part of Fig. 16.2. As a consequence majority s electrons can no longer be scattered into d states which leads to an increased mobility and a reduced resistance. Applying an external magnetic field increases the degree of spin order which results in a reduction of the resistance. Therefore, we observe a negative magnetoresistance.

A descriptive explanation of this situation was carried out by Mott using a two spin channel model. An illustration is given by the following circuit diagram (see Fig. 16.3). The total resistance R_{total} is determined by the parallel connection of the resistance for majority electrons $R_\uparrow$ and minority

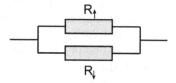

Fig. 16.3. Circuit diagram concerning the two spin channel model by Mott with $R_\uparrow$ being the resistance for majority electrons and $R_\downarrow$ the resistance for minority electrons allowing the determination of the total resistance R_{total}

electrons $R_\downarrow$ and therefore amounts to:

$$R_{\text{total}} = \frac{R_\uparrow \cdot R_\downarrow}{R_\uparrow + R_\downarrow} \tag{16.3}$$

Using the approximation that a variation of magnetic order being characterized by the parameter a induces an (inverse) proportional behavior of the resistances $R_\uparrow$ and $R_\downarrow$, i.e.:

$$R_\uparrow = \frac{1}{a} R \quad \text{and} \quad R_\downarrow = aR \tag{16.4}$$

the total resistance can be expressed as:

$$R_{\text{total}} = \frac{\frac{1}{a}R \cdot aR}{\frac{1}{a}R + aR} = R \cdot \frac{a}{1 + a^2} \tag{16.5}$$

Without magnetic order we get $a = 1$ and thus $R_{\text{total}} = 1/2 \cdot R$. The maximum resistance as a function of magnetic order can be calculated by:

$$0 = \frac{\partial R_{\text{total}}}{\partial a} = R \cdot \frac{1 - a^2}{(1 + a^2)^2} \tag{16.6}$$

Thus, the maximum is reached if $a = 1$ which corresponds to a vanishing magnetic order. Consequently, the occurrence of magnetic order reduces the resistance.

This negative magnetoresistance behaves isotropically, i.e. it does not depend on the direction of the electric current relative to the magnetization direction, to the direction of the external magnetic field, and to the orientation of the crystalline axes.

16.2 AMR – Anisotropic Magnetoresistance

The anisotropic magnetoresistance effect is current induced and exists in ferromagnetic metals such as Ni, Co or Fe upon application of an external field H. The physical origin of the AMR effect is spin orbit coupling on the $3d$ orbitals caused by an applied magnetic field. Without spin orbit interaction, i.e. $\boldsymbol{L} \cdot \boldsymbol{S} = 0$, an s-d scattering of majority electrons cannot occur. For a non-vanishing spin orbit interaction spin-flip scattering is allowed, i.e. majority s electrons can be scattered into empty minority d states which results in an increase of the resistance. Additionally, the momentum of the scattered electrons must be conserved. Thus, the scattering cross section is different due to the orbital anisotropy of the empty d states for a parallel and a perpendicular orientation between the magnetization direction and the direction of the electric current which is schematically depicted in Fig. 16.4.

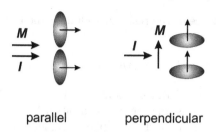

parallel   perpendicular

Fig. 16.4. Origin of the anisotropic magnetoresistance

The magnitude of the AMR effect depends on the angle between the electric current I being θ_I and the angle of the magnetization M being θ_M in the ferromagnetic metal with respect to a given direction. The angle θ_M is usually the same as the angle of the applied field θ_H but can be influenced by other factors such as shape or magneto crystalline anisotropy in the ferromagnetic thin film.

The resistance of a ferromagnetic thin film is at a maximum when the current is parallel to the magnetization direction $R_\parallel$ and is at a minimum when the current is perpendicular to the magnetization direction $R_\perp$. The magnetic field dependence of the AMR effect for a ferromagnetic thin film is shown Fig. 16.5. There is no difference in resistance at zero field between the two current field orientations. Upon application of a magnetic field the difference in resistance between the $R_\parallel$ and $R_\perp$ becomes immediately apparent. The resistance with the current set perpendicular to magnetic field direction decreases while the resistance increases with the current set parallel to the field direction. With increasing magnitude of the magnetic field the difference in resistance between the two orientations rapidly reaches a maximum at H_s.

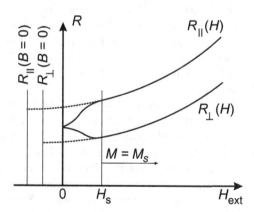

Fig. 16.5. Resistance as a function of an external magnetic field for a parallel $R_\parallel$ and perpendicular orientation $R_\perp$ between the direction of the electric current and the magnetic field

The angle dependence of the resistance change due to the AMR effect can be expressed as follows:

$$R(\theta) = R_{\min} + \Delta R_{\mathrm{AMR}} \cdot \cos^2(\theta_M - \theta_I) \qquad (16.7)$$

with ΔR_{AMR} being the resistance change due to the AMR effect. According to this relationship the resistance change exhibits a $180°$ periodicity when the ferromagnet is rotated in a constant applied magnetic field. Equation (16.7) can be expanded to include the terms $R_{\parallel}$ and $R_{\perp}$:

$$\Delta R_{\mathrm{AMR}} = R_{\parallel} - R_{\perp} \qquad (16.8)$$
$$R_{\min} = R_{\perp} \qquad (16.9)$$
$$R(\theta) = R_{\perp} + (R_{\parallel} - R_{\perp}) \cdot \cos^2(\theta_M - \theta_I) \qquad (16.10)$$

The determination of $R_{\parallel}$ and $R_{\perp}$ can be carried out by extrapolation to a vanishing slope. This procedure is necessary in order to eliminate the influence of the external field and the inner field due to the magnetization being caused by the asymmetric charge distribution due to spin orbit interaction.

If the external magnetic field is larger than H_s the positive or negative magnetoresistance occurs, respectively, as discussed in Chap. 16.1.

16.3 GMR – Giant Magnetoresistance

Whereas the normal, the anisotropic, and the colossal magnetoresistance occur in bulk material the giant magnetoresistance is restricted to systems which consist of magnetic thin layers. The giant magnetoresistance is present for ferromagnetic layers which are separated by a non-magnetic or an antiferromagnetic metallic thin film. Thus, the giant magnetoresistance is closely related to the interlayer exchange coupling which deals with the observation that the relative orientation of the magnetization in ferromagnetic multilayer systems depends on the thickness of the spacer layer and is discussed in Chap. 15.

The pioneering work was carried out by P. Grünberg [68] and A. Fert [69] on Fe/Cr/Fe multilayers.

The magnitude of the electric current and thus the resistance depend on the relative orientations of the magnetization in the thin film system. The resistance is low (large) for a parallel (antiparallel) alignment of the magnetization of neighbored magnetic layers (see Fig. 16.6). Thus, in layered systems with an antiferromagnetic coupling the parallel orientation forced by an external magnetic field significantly reduces the resistance. The variation of the resistance $\Delta R/R(B = 0)$ can be extremely large (about 80% in the example) which gives rise for this "giant" magnetoresistance GMR.

The GMR effect cannot be based on the AMR because it is significantly larger and exhibits no dependence on the angle between the magnetization and the direction of the electric current.

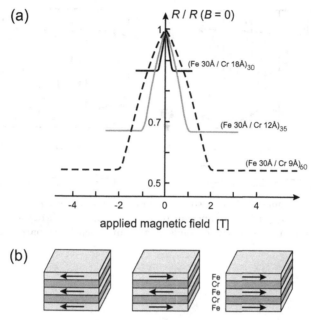

Fig. 16.6. (a) Magnetoresistance of three Fe/Cr superlattices at 4.2 K. The resistance is referenced to that without magnetic field. The current and the applied field are along the same axis in the plane of the layers. (Adapted from [69] (used with permission)) (b) The corresponding relative magnetization orientation for zero and applied field for a part of the Fe and Cr layers within the stack. Whereas a high external magnetic field results in a parallel orientation of all Fe layers which corresponds to a low resistance an antiparallel arrangement occurs without applying an external field which significantly increases the resistance

The magnitude of the GMR effect is defined by:

$$\text{GMR} = \frac{\rho_{\text{ap}} - \rho_{\text{p}}}{\rho_{\text{p}}} = \frac{\sigma_{\text{p}}}{\sigma_{\text{ap}}} - 1 \tag{16.11}$$

with ρ_{p} (ρ_{ap}) being the resistivity for a parallel (antiparallel) alignment of the magnetization in neighbored layers and σ_{p} (σ_{ap}) the corresponding specific conductance.

The effect can be explained using the Mott model of two spin channels. The scattering cross section and thus the resistivity is different for a parallel ρ^+ and an antiparallel orientation ρ^- of the spins and thus the magnetization in neighbored layers. The parallel connection for majority and minority electrons in multilayer systems is shown in Fig. 16.7. Neglecting the resistivity ρ of a non-magnetic interlayer and assuming a parallel alignment of the magnetization ρ_{p} is given concerning two magnetic layers by:

$$\rho_{\text{p}} = \left(\frac{1}{2\rho^+} + \frac{1}{2\rho^-} \right)^{-1} \tag{16.12}$$

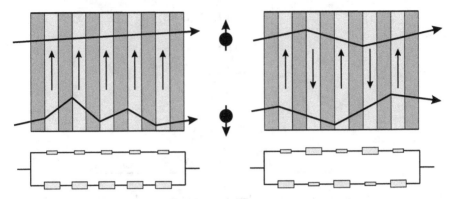

Fig. 16.7. Mott's model of the two spin channel for illustrating the resistivity in the situation of a parallel (**left**) and antiparallel orientation (**right**) in a multilayer system. Each upper channel corresponds to spin up, each lower part to spin down electrons

$$= \frac{2\rho^+\rho^-}{\rho^+ + \rho^-} \tag{16.13}$$

and assuming an antiparallel alignment ρ_{ap} by:

$$\rho_{ap} = \left(\frac{1}{\rho^+ + \rho^-} + \frac{1}{\rho^- + \rho^+} \right)^{-1} \tag{16.14}$$

$$= \frac{\rho^+ + \rho^-}{2} \tag{16.15}$$

The resistivity of the antiparallel configuration is larger than for the parallel alignment:

$$\rho_{ap} - \rho_p = \frac{\rho^+ + \rho^-}{2} - \frac{2\rho^+\rho^-}{\rho^+ + \rho^-} = \frac{(\rho^+ - \rho^-)^2}{2(\rho^+ + \rho^-)} \geq 0 \tag{16.16}$$

Thus, the magnitude of the GMR effect amounts to:

$$\text{GMR} = \frac{\rho_{ap} - \rho_p}{\rho_p} = \frac{(\rho^+ - \rho^-)^2}{4\rho^+\rho^-} \tag{16.17}$$

and is large if the scattering cross sections are significantly different for a parallel and antiparallel alignment.

Let us discuss this behavior of ferromagnetic $3d$ metals using the spin resolved density of states as depicted in Fig. 16.8 which is known as the Stoner model. As already mentioned the electric transport is carried out by the $4s$ electrons due to their reduced effective mass compared to $3d$ electrons. A large DOS of the $3d$ electrons at the Fermi energy results in an enlarged scattering of $4s$ electrons into empty $3d$ states followed by a high resistivity. Due to the exchange splitting of the DOS for $3d$ electrons the scattering

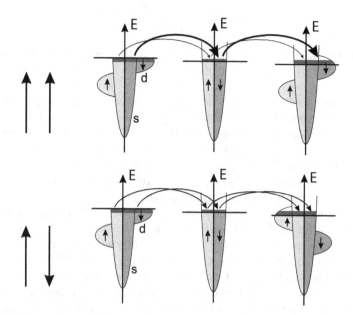

Fig. 16.8. Dependence of the resistivity on the relative orientation of the magnetization. For a parallel alignment one spin channel exhibits a low resistivity which gives rise to a low total resistivity. For an antiparallel alignment an alternating change in the resistivity occurs for each spin channel which leads to large total resistivity

probability for an antiparallel alignment in both ferromagnetic thin films is significantly higher than for the parallel alignment. This behavior gives directly evidence for the equivalent circuit diagrams shown in Fig. 16.7.

Now we take into account the resistivity ρ of the non-magnetic interlayer with a cross-sectional area A and thickness d assuming a thickness of the magnetic layer of d_m. Introducing the parameters:

$$\gamma = \frac{d_m}{d} \tag{16.18}$$

$$a = \frac{\rho^+}{\rho} \tag{16.19}$$

$$b = \frac{\rho^-}{\rho} \tag{16.20}$$

we obtain:

$$R_{\mathrm{p}} = \frac{(1 + a\gamma)(1 + b\gamma)}{2 + a\gamma + b\gamma} \cdot \frac{2\rho d}{A} \tag{16.21}$$

$$R_{\mathrm{ap}} = (2 + a\gamma + b\gamma) \cdot \frac{\rho d}{2A} \tag{16.22}$$

Thus, the magnitude of the GMR effect amounts to:

$$\text{GMR} = \frac{(a-b)^2}{4(a+1/\gamma)(b+1/\gamma)} \qquad (16.23)$$

Using the example given in Fig. 16.6 with a thickness of the Fe layer of $d_m = 3\,\text{nm}$ and of the Cr layer of $d = 0.9\,\text{nm}$ which led to a GMR effect of 83% we get $\gamma = 3/0.9 = 3.3$. Assuming $\rho^+ = \rho$ we obtain $a = 1$. This allows to determine b to $b = 6.5$. Thus, the resistivity of an antiparallel configuration is 6.5 times larger than for the parallel orientation.

In the discussion above we assumed that the direction of the electric current is perpendicular to the layered system. But, in most of the experiments this direction is parallel along the thin film system. Therefore, two different configurations must be distinguished: CIP (current in plane) and CPP (current perpendicular to plane) configuration (see Fig. 16.9). The CIP configuration allows easier measurements whereas the CPP one requires lithographic procedures or shadow masks for the production of the layered stack. Additionally, the resistivity for CPP is significantly smaller compared to CIP. Concerning the magnitude of the GMR effect each layer in CIP geometry exhibits its own conductivity and can shunt the effect. This is not possible for the CPP configuration where each spin current must go through the whole stack. Therefore, the GMR effect is larger for the CPP than for the CIP configuration.

The spin dependent scattering occurs at the interfaces. Thus, the magnitude of the GMR effect depends on the temperature and the number of interfaces. Figure 16.10 shows the magnitude of the GMR effect as a function of temperature for $(\text{Fe/Cr})_n/\text{Fe}$ samples with $n = 1, 2, 4$. It is obvious that the effect increases if the temperature is reduced and scales with the number of interfaces.

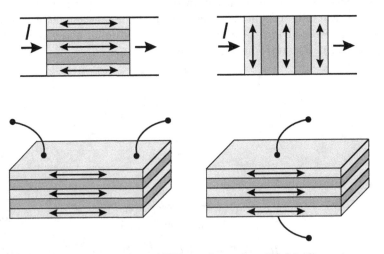

Fig. 16.9. Left: Current in plane (CIP) configuration. **Right:** Current perpendicular to plane (CPP) configuration

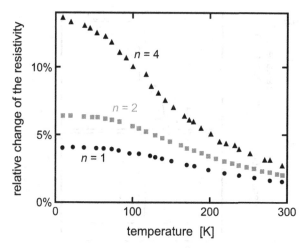

Fig. 16.10. Temperature dependence of the GMR effect, i.e. of the relative change of the resistivity $(\rho_{ap} - \rho_p)/\rho_p$ in epitaxial $(Fe/Cr)_n/Fe$ structures with $n = 1, 2, 4$ (Data taken from [70])

The GMR effect consists of a decrease in the resistivity when the magnetization of the ferromagnetic layers being separated by a non-magnetic or antiferromagnetic layer rotate from the antiparallel to the parallel alignment. In the Fe/Cr layered structures described in Fig. 16.10 the antiparallel alignment was induced by an antiferromagnetic exchange coupling between neighboring Fe layers across the Cr interlayers. The origin of the antiparallel alignment, however, is not important for the description of the effect. One may thus expect similar effects in double layers with no antiferromagnetic coupling between the ferromagnetic layers but with the antiparallel alignment obtained by other means. In this situation we have spin valve systems.

One of these possibilities is given by different coercivities H_c of both ferromagnetic films. The realization of this situation can be carried out by Co/Au/Co double layers with the Au interlayer thick enough that there is no exchange coupling between the Co films. The first Co layer is evaporated on a GaAs substrate whereas the second one is grown on the Au spacer layer. Owing to this fact both Co films possess different coercive fields. In Fig. 16.11(a) the hysteresis loop obtained via the magneto-optical Kerr effect at room temperature is shown. As one can see there is a range of magnetic fields where the directions of magnetization of both Co layers are aligned antiparallel being indicated by arrows. In Fig. 16.11(b) the resistance is shown scanning through the hysteresis loop given above.

The experimental configuration is the same as in the case of the Fe/Cr structures being discussed above with the current being directed parallel to the external magnetic field. At sufficiently high magnetic field the directions of the magnetization of both ferromagnetic films are parallel. We see that the

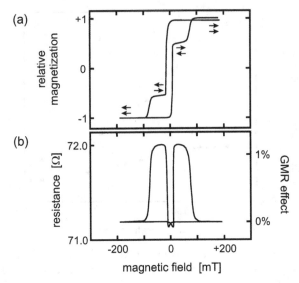

Fig. 16.11. (a) Relative magnetization (determined by MOKE technique) and (b) resistance of a Co/Au/Co structure. The corresponding magnitude of the GMR effect is given at the right scale. The measurements were carried out at $T = 294$ K. (Data taken from [70])

resistance increases each time the antiparallel alignment is achieved during the scan through the hysteresis loop. At the right scale of Fig. 16.11(b) the corresponding relative change of resistance is given. We therefore observe that the GMR effect amounts to about 1.1%.

16.4 CMR – Colossal Magnetoresistance

The colossal magnetoresistance was found in mixed valence Manganese oxides which exhibit Mn ions in different oxidation states and localized magnetic moments. It is related to a negative magnetoresistance effect due to spin disorder.

Let us discuss this effect using the compound $LaMnO_3$ which occurs as $La^{3+}Mn^{3+}O_3^{2-}$, i.e. only Mn^{3+} ions are present which exhibit 4 electrons in the $3d$ shell. Due to Hund's rules (see Chap. 2.4) a parallel alignment of the four spins is energetically favored. The compound represents an isolator and behaves as an antiferromagnet below 140 K.

Replacing a part of the La atoms by Sr atoms results in the compound $La_{1-x}Sr_xMnO_3$ exhibiting the oxidation states $La_{1-x}^{3+}Sr_x^{2+}Mn_{1-x}^{3+}Mn_x^{4+}O_3^{2-}$, i.e. mixed valences of the Mn ions are present: Mn^{3+} and Mn^{4+}. In the regime $0.15 < x < 0.5$ a transition from paramagnetic and semiconducting to ferromagnetic and metallic behavior occurs. The Curie temperature depends on the stoichiometry.

The transport of electrons in this type of compound can be understood as follows. The remaining $3d$ electron of the Mn^{3+} ion is mobile whereas the other ones as well as the three of the Mn^{4+} ions are localized. This free electron of the Mn^{3+} ion can hop via the O^{2-} ions to neighbored Mn^{4+} ions, i.e. the oxidation state is reversed and the hopping can continue. Thus, the electron possesses a high mobility.

Now, the spin orientation is additionally taken into consideration. Due to Hund's coupling the spin of the mobile electron must be parallel to the spins of the localized electrons. Therefore, hopping can only occur if the spins of the localized electrons at neighbored sites are parallel. We see that a high mobility is correlated with a ferromagnetic arrangement of neighbored Mn atoms. The transition from para- to ferromagnetism denotes an isolator-metal transition. The disorder of spin moments in the paramagnetic regime inhibits the transport of electrons and gives rise to a high resistivity.

The temperature dependence of this resistivity is shown in Fig. 16.12 for a different type of a Manganese oxide $Nd_{1-x}Pb_xMnO_3$ with $x = 0.5$. This compound exhibits a Curie temperature of 184 K. We see that the colossal magnetoresistance mainly occurs near the Curie temperature in an external magnetic field which leads to a reduction of the disorder. Thus, the resistivity is decreased. Well below T_C the spin system is already ordered and the influence of the external field vanishes.

The magnitude of the colossal magnetoresistance amounts up to 100000% and even more!

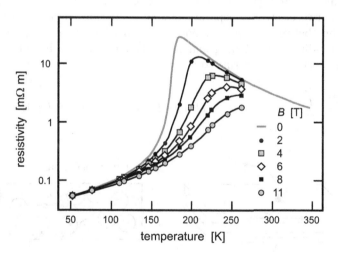

Fig. 16.12. Resistivity of $Nd_{1-x}Pb_xMnO_3$ with $x = 0.5$ as a function of temperature and applied magnetic field. (Data taken from [71]. Copyright 1989, with permission from Elsevier)

16.5 TMR – Tunnelling Magnetoresistance

For the occurrence of the GMR effect it was essential that the two ferromagnetically or antiferromagnetically coupling layers were separated by a metallic and non-magnetic thin film. Now, we change this situation by replacing the metallic spacer with an insulator between both magnetic films (see left part of Fig. 16.13). The discussion below can also be extended to ferromagnetic nanoparticles being located in an insulating matrix (see right part of Fig. 16.13).

The electric current can no longer be due to diffusive transport of electrons but only through the insulating barrier. As a prerequisite the thickness of the barrier must be small enough to allow quantum mechanical tunnelling. It is essential for our discussion that we assume that this process conserves the spin orientation.

The dependence of the tunnelling current on the relative magnetization is shown in Fig. 16.14 for two ferromagnetic thin layers. The total resistivity for the parallel alignment is less than for the antiparallel orientation as already found for the GMR effect. But, the TMR represents a band structure effect which relies on the spin resolved DOS at the Fermi level whereas the GMR is caused by a spin dependent scattering at the interfaces.

In the following we want to discuss different situations in dependence of the properties concerning both electrodes. We will use the abbreviations "N" for normal metallic conductors, "SC" for superconductors, "FM" for ferromagnetic metals, and "I" for insulators. Especially, the tunnelling current between the two electrodes separated by an insulating barrier as a function of the voltage between both electrodes represents an important property for the understanding of tunnelling magnetoresistance and can be calculated on the basis of Fermi's golden rule to be:

$$I = 2e\frac{2\pi}{\hbar} \int\limits_{-\infty}^{\infty} |M(E)|^2 \, n_1(E - eU) \, n_2(E) \, [f(E - eU) - f(E)]\mathrm{d}E \quad (16.24)$$

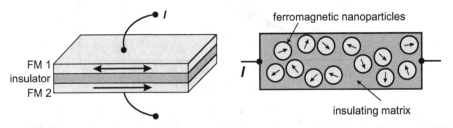

Fig. 16.13. The tunnelling magnetoresistance can occur for ferromagnetic thin films being separated by an insulating layer (**left**) and for ferromagnetic nanoparticles being embedded in an insulating matrix (**right**)

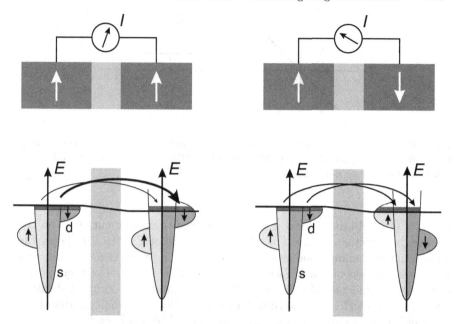

Fig. 16.14. Dependence of the tunnelling current on the relative magnetization of two ferromagnetic layers. For a parallel orientation a large amount of spin down electrons at the Fermi energy can tunnel into empty down states which results in a high tunnelling current whereas for an antiparallel orientation the amount of empty down states is significantly lower leading to a reduced tunnelling current

with E being the energy with respect to the Fermi energy E_F, U the applied bias voltage, M the tunnelling matrix element, $n_{1,2}$ the density of states of the first and second electrode, respectively, and $f(E)$ the Fermi function:

$$f(E) = \frac{1}{\exp(E/kT) + 1} \tag{16.25}$$

with E being referenced to the Fermi energy (cf. (3.13)).

Planar N-I-N Contacts

Both electrodes are normal metallic conductors. Assuming that near E_F the density of states does not depend on the energy and that the tunnelling matrix element is independent on E which is given if eU is negligible compared to the barrier height the tunnelling current amounts to:

$$I = 2e\frac{2\pi}{\hbar} n_{N1}\, n_{N2}\, |M|^2 \int\limits_{-\infty}^{\infty} (f(E - eU) - f(E))\, \mathrm{d}E \tag{16.26}$$

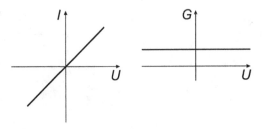

Fig. 16.15. Left: Linear dependence of the tunnelling current I on the applied bias voltage U. **Right**: Thus, the differential tunnelling conductance $G = dI/dU$ is constant, i.e. it does not depend on U

For $T = 0$ we obtain a linear dependence of the tunnelling current I on the applied bias voltage U, i.e. $I \propto U$ (see left part of Fig. 16.15). Thus, the differential tunnelling conductance $G = dI/dU$ is proportional to $n_{N1} \, n_{N2}$ and does not depend on U (see right part of Fig. 16.15). These properties can easily be understood using Fig. 16.16. The number of occupied states in one electrode which are responsible for the tunnelling into empty states of the other electrode linearly increases with the applied bias voltage U.

The typical tunnelling current as a function of the applied bias voltage between real non-magnetic metals is shown in the left part of Fig. 16.17. The right part of Fig. 16.17 represents the corresponding conductance which possesses an approximately parabolic shape. This behavior is due to the bias dependence of the tunnelling matrix element which typically leads to an order-of-magnitude increase in the tunnelling current for each volt increase in magnitude of the applied bias voltage.

Nevertheless, the linear correlation between I and U is still given for a small bias voltage.

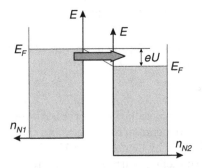

Fig. 16.16. Schematic illustration of the tunnelling processes between two electrodes which are energetically shifted by eU

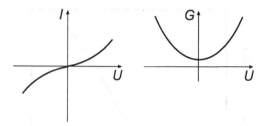

Fig. 16.17. Left: Typical tunnelling current I on the applied bias voltage U between real non-magnetic metals. **Right**: The corresponding conductance exhibits an approximately parabolic shape

Planar N-I-SC Contacts

Superconductors exhibit a gap in the density of states at the Fermi energy. Let us assume the width of the gap to be 2Δ. In this energy region no allowed electronic states of the quasiparticles are present. Thus, if we replace one of the electrodes of an idealized N-I-N contact with a superconductor the tunnelling current I as a function of the applied bias voltage U is zero at the corresponding values of U within the gap (see left part of Fig. 16.18). The linear dependence remains at larger values of U. Consequently, the conductance G also vanishes within the gap and is constant for $e|U| > \Delta$ (see right part of Fig. 16.18 and cf. Fig. 16.15).

More realistically the density of states of the quasiparticles n_{SC} can be described in the BCS theory (Bardeen, Cooper, Schrieffer) by:

$$n_{SC}(E) = \begin{cases} \dfrac{|E|}{\sqrt{E^2 - \Delta^2}} & \text{if} \quad |E| \geq \Delta \\ 0 & \text{if} \quad |E| < \Delta \end{cases} \tag{16.27}$$

Applying a low voltage U, i.e. $eU < \Delta$, no tunnelling current occurs for $T = 0$ (curve 2 in Fig. 16.19) or only a small one for $T > 0$ (curve 3). In the

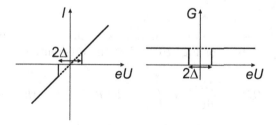

Fig. 16.18. Left: Linear dependence of the tunnelling current I on the applied bias voltage U between a superconductor and a normal metal for larger values of U. The tunnelling current vanishes in a gap around $U = 0$ exhibiting a width of 2Δ. **Right**: The conductance also vanishes within the gap and is constant with a positive value outside the gap

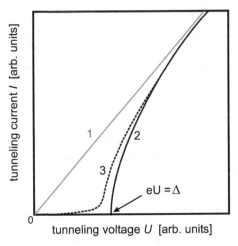

Fig. 16.19. Characteristic tunnelling current for a junction between (1) normal metal and normal metal, (2) normal metal and superconductor at $T = 0$, and (3) normal metal and superconductor at $0 < T \ll T_C$. (Adapted from [72] with permission from Wiley)

latter case only a few thermally filled states in one electrode face a similar number of empty states in the other one. When eU exceeds the gap energy the tunnelling current rapidly increases. In this situation electrons from the filled states at E_F of the normal conductor can tunnel into a large number of empty states in the superconductor. A further increase of eU results in the linear dependence of I as a function of U comparable to the N-I-N contact (curve 1).

Assuming that the density of states of the normal conductor n_N is independent on E it can be removed from the integral and the tunnelling current is given by:

$$I = 2e\frac{2\pi}{\hbar}\, n_N \,|M|^2 \int\limits_{-\infty}^{\infty} n_{SC}(E)\,(f(E - eU) - f(E))\,\mathrm{d}E \qquad (16.28)$$

The density of states of the quasiparticles n_{SC}, characterized by (16.27), is shown in the upper part of Fig. 16.20. For the differential conductance we obtain:

$$G(U) = \frac{\mathrm{d}I(U)}{\mathrm{d}U} \propto \int\limits_{-\infty}^{\infty} n_{SC}(E)\,K(E - eU)\mathrm{d}E \qquad (16.29)$$

which is shown in the lower part of Fig. 16.20. In this Figure $G(U)$ is normalized to the density of states of a normal metal n_N and thus reaches the corresponding values outside the gap (see dashed line in Fig. 16.20(c)). It results from the convolution of n_{SC} and the derivative of the Fermi function $f(E - eU)$ with respect to U:

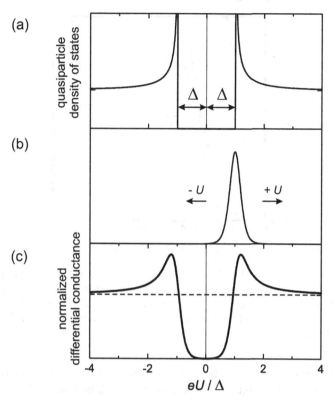

Fig. 16.20. Superconductor–normal metal tunnelling. (a) BCS density of states of a superconductor exhibiting a gap of 2Δ as a function of voltage. (b) Temperature dependent function K. (c) Theoretical normalized differential conductance $(\mathrm{d}I/\mathrm{d}U)_{\mathrm{SC}}/(\mathrm{d}I/\mathrm{d}U)_{\mathrm{N}}$. The voltage is given with respect to the Fermi energy of the superconductor. The dashed line represents the differential conductance of a normal metal assumed to be independent on the energy

$$K = \frac{E}{kT} \cdot \frac{\exp((E - eU)/kT)}{(1 + \exp((E - eU)/kT))^2} \tag{16.30}$$

This function K is exemplarily shown in the middle part of Fig. 16.20.

Planar N-I-SC Contacts in an External Magnetic Field

Before we start our discussion concerning FM-I-SC contacts we deal with the behavior of a superconductor in a magnetic field. Pioneering experiments and related discussions are given in [73, 74].

It was experimentally observed that for the planar $Al/Al_2O_3/Ag$ system far below the critical temperature of 2.4 K which Al becomes superconducting at the quasiparticle density of states exhibits a splitting if a magnetic field is applied (see Fig. 16.21). At $B = 5.2$ T the field strength is high enough to

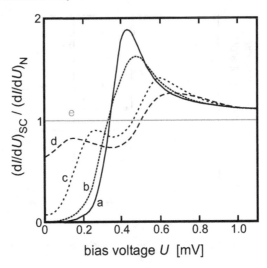

Fig. 16.21. Experimental plots for several values of magnetic field of the superconductor conductance $(dI/dU)_{SC}$ divided by the normal state conductance $(dI/dU)_N$ for planar Al/Al$_2$O$_3$/Ag contacts. *Curve a*: without external magnetic field; *curve b*: $B = 1.1$ T; *curve c*: $B = 3.0$ T; *curve d*: $B = 4.5$ T; *curve e*: $B = 5.2$ T. (Data taken from [75])

suppress the superconducting behavior (curve e). Thus, the superconductor conductance $(dI/dU)_{SC}$ equals the normal state conductance $(dI/dU)_N$.

Plotting the magnitude of the bias voltage U which the differential conductance possesses the maximum at shows a linear dependence with respect to the applied magnetic field (see Fig. 16.22). This Zeeman splitting of the density of states for spin up and spin down electrons is in good agreement to the theoretical value of $U = (\Delta \pm \mu\mu_0 H)/e = (\Delta \pm \mu B)/e$ with $\mu = g\mu_B S/\hbar \approx \mu_B$ being the magnetic moment of the electron.

Analogously to the interpretation carried out concerning Fig. 16.20 for N-I-SC contacts Fig. 16.23 allows to understand the differential conductance for N-I-SC contacts with an additional external magnetic field. We see in the upper part of Fig. 16.23 that the density of states for spin up electrons is shifted by μB to the left (gray line) whereas that for spin down electrons is shifted by the same value to the right (black dashed line). Thus, due to the Zeeman splitting spin up electrons are primarily responsible for the tunnelling current at the energy $\Delta - \mu B$ whereas the tunnelling current at the energy $\Delta + \mu B$ is mainly due to spin down electrons.

The density of states $n_{SC}(E)$ can be divided into two parts concerning the spin up $n^\uparrow(E)$ and spin down electrons $n^\downarrow(E)$:

$$n_{SC}(E) = n^\uparrow(E) + n^\downarrow(E) = \frac{1}{2}\left(n_{SC}(E - \mu B) + n_{SC}(E + \mu B)\right) \quad (16.31)$$

Thus, the differential conductance amounts to:

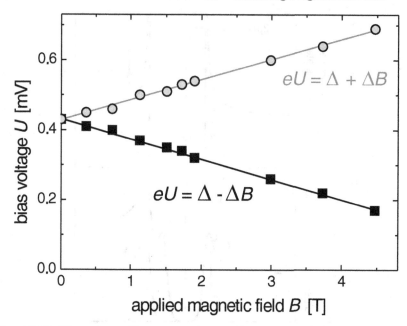

Fig. 16.22. Bias voltage U corresponding to the maxima of the spin up and spin down density of states curves determined from Fig. 16.21 as a function of the applied magnetic field B. The lines represent the theoretically expected results of $eU = (\Delta \pm \mu B)$ (Data taken from [75])

$$G(U) \propto \int_{-\infty}^{\infty} n_{\mathrm{SC}}(E + \mu B)\, K(E - eU)\mathrm{d}E + \int_{-\infty}^{\infty} n_{\mathrm{SC}}(E - \mu B)\, K(E - eU)\mathrm{d}E$$

$$(16.32)$$

Both contributions are shown in the lower part of Fig. 16.23 represented by the solid gray and black dashed line. The black solid line corresponds to the total differential conductance.

This straightforward analysis is based on some assumptions. Neglecting spin orbit or spin flip scattering in the superconductor allows to assume that the quasiparticle density of states exhibits the same functionality for each spin direction in an applied magnetic field and is merely shifted by $\pm\mu B$. A further requirement is given by the spin conservation in the tunnelling process, i.e. a spin flip during tunnelling does not occur.

Planar FM-I-SC Contacts

Now, we replace the normal metal electrode by a ferromagnetic one. Concerning the ferromagnetic electrode it is important that the density of states at the Fermi energy is different for majority electrons $n^\uparrow(E_F)$ and minority electrons $n^\downarrow(E_F)$. Majority or spin up electrons exhibit magnetic moments being

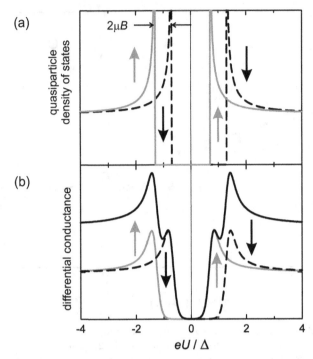

Fig. 16.23. (a) Magnetic field splitting of the density of quasiparticle states into spin up (*gray line*) and spin down intensities (*black dashed line*). (b) Spin up conductance (**gray line**), spin down conductance (*black dashed line*), and total conductance (*black solid line*) for N-I-SC contacts in an external magnetic field

parallel to the applied magnetic field. Thus, minority or spin down electrons possess magnetic moments which are antiparallel to the external field. The fraction of majority electrons is given by $a = n^\uparrow(E_F)/(n^\uparrow(E_F) + n^\downarrow(E_F))$ and that of minority electrons by $1 - a = n^\downarrow(E_F)/(n^\uparrow(E_F) + n^\downarrow(E_F))$. This allows to define the spin polarization by:

$$P = \frac{n^\uparrow(E_F) - n^\downarrow(E_F)}{n^\uparrow(E_F) + n^\downarrow(E_F)} = a - (1 - a) = 2a - 1 \qquad (16.33)$$

Therefore, if we assume that in the system $Al/Al_2O_3/Ag$ being discussed above the non-magnetic Ag electrode is replaced by a ferromagnetic material the Zeeman splitting results in an asymmetric behavior of the dI/dU characteristic which is shown for $Al/Al_2O_3/Ni$ in Fig. 16.24 and obvious near $U = 0$. In this situation the differential tunnelling conductance is given by an expression being similar to that in (16.32) and amounts to:

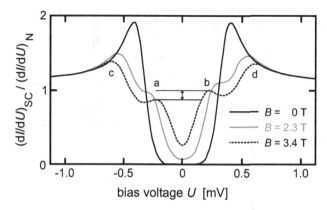

Fig. 16.24. Normalized conductance, i.e. superconductor conductance $(dI/dU)_{SC}$ divided by the normal state conductance $(dI/dU)_N$, of an $Al/Al_2O_3/Ni$ junction measured as a function of the bias voltage applied to the Al film for three values of the applied magnetic field. The asymmetry of the conductance peaks a and b as well as c and d for $B = 3.4\,T$ result from the polarization of the Ni carriers. (Data taken from [76])

$$G(U) \propto \int_{-\infty}^{\infty} a \cdot n_{SC}(E+\mu B)K(E-eU)dE + \int_{-\infty}^{\infty} (1-a) \cdot n_{SC}(E-\mu B)K(E-eU)dE$$

$$(16.34)$$

if we assume that n_{SC} is only shifted by $\pm\mu B$ and no spin-flip process is present. Thus, the interpretation of (16.34) using Fig. 16.25 is similar to that of (16.32) by means of Fig. 16.23. The important difference concerns the spin polarization of the ferromagnetic electrode being included in both integrals in (16.34).

The qualitative behavior of the differential conductance becomes obvious if the density of states of the majority electrons is dominant, i.e. a high degree of a positive spin polarization is present. For a quantitative analysis we compare the maxima of the differential conductance shown in the lower part of Fig. 16.25.

Let us define $g(U)$ as the differential conductance as a function of the applied bias voltage U *without* Zeeman splitting. The contribution of spin up electrons to the conductance with an energy shift by the Zeeman term of $h = \mu B/e$ amounts to $a \cdot g(U - h)$. The contribution of spin down electrons is thus given by $(1 - a) \cdot g(U + h)$. The total differential conductance $G(U)$ amounts to the sum of both contributions and is shown in the lower part of Fig. 16.25. For any value of U we obtain four equations for the total differential conductance at the four maxima G_i with $i = 1, 2, 3, 4$ at the energies $-U - h$, $-U + h$, $U - h$, and $U + h$, respectively, which can be expressed by the function g:

$$G_1 = G(-U - h) = a \cdot g(-U) + (1 - a) \cdot g(-U - 2h) \qquad (16.35)$$

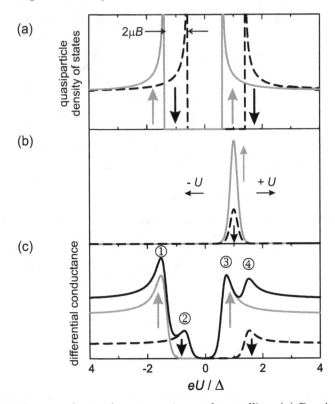

Fig. 16.25. Superconductor–ferromagnetic metal tunnelling. **(a)** Density of states of a superconductor as a function of voltage in a magnetic field. **(b)** Temperature dependent function K for each spin direction assuming $a = 0.75$ which corresponds to a spin polarization of 50%. **(c)** Normalized conductance for each spin direction (*gray solid* and *black dashed lines*, respectively) and the total conductance (*black solid line*). The four maxima 1–4 correspond to the four values G_i

$$G_2 = G(-U + h) = a \cdot g(-U + 2h) + (1 - a) \cdot g(-U) \qquad (16.36)$$
$$G_3 = G(U - h) \;\; = a \cdot g(U) + (1 - a) \cdot g(U - 2h) \qquad (16.37)$$
$$G_4 = G(U + h) \;\; = a \cdot g(U + 2h) + (1 - a) \cdot g(U) \qquad (16.38)$$

Due to $g(U) = g(-U)$ we obtain:

$$G_4 - G_2 = a \cdot (g(U + 2h) - g(U - 2h)) \qquad (16.39)$$
$$G_1 - G_3 = (1 - a) \cdot (g(U + 2h) - g(U - 2h)) \qquad (16.40)$$

which allows to define the tunnelling spin polarization P_t as:

$$P_t = \frac{(G_4 - G_2) - (G_1 - G_3)}{(G_4 - G_2) + (G_1 - G_3)} = 2a - 1 \qquad (16.41)$$

The second equals sign, i.e. $P_t = P$, is only valid if the tunnelling matrix element M does not differ for majority and minority electrons.

Analyzing experimental values [73] using the scheme shown in Fig. 16.25 results in a spin polarization for Ni of $P = 11\%$. These early measurements relied on oxidizing the Al films in laboratory air saturated with water vapor to form the tunnelling barrier and resulted in low values of P for Ni. Later, barriers were formed in-situ with a glow discharge in pure oxygen leading to values of the polarization for Ni from 17% to 25%. It was conjectured that in the older method, OH^--ions were present in the Al_2O_3 leading to a contamination of the Ni surface. New measurements with improved quality of the tunnel junctions yield larger values of 33% for Ni, 42% for Co, and 44% for Fe.

It is astonishing that the values of the spin polarization are positive which is in disagreement to the negative spin polarization for ferromagnetic bulk materials. This discrepancy is related to the electronic structure of the interface between the ferromagnetic electrode on the one hand and the insulating barrier material on the other hand because it is comparable neither to the pure bulk nor to the clean surface.

Planar FM-I-FM Contacts

For the discussion of the behavior of two ferromagnetic electrodes separated by an insulating barrier we use the model of Jullière [77] which exhibits the assumptions that the tunnelling process is spin conserving and the tunnelling current is proportional to the density of states of the corresponding spin orientation in each electrode. Thus, the tunnelling current for a parallel magnetization is given by:

$$I^{\uparrow\uparrow} \propto n_1^\uparrow n_2^\uparrow + n_1^\downarrow n_2^\downarrow \tag{16.42}$$

with n_i being the electron density of electrode i at the Fermi level E_F. For the antiparallel orientation the tunnelling current amounts to:

$$I^{\uparrow\downarrow} \propto n_1^\uparrow n_2^\downarrow + n_1^\downarrow n_2^\uparrow \tag{16.43}$$

With $a_i = n_i^\uparrow/(n_i^\uparrow + n_i^\downarrow)$ being the part of majority electrons of electrode i and $1 - a_i = n_i^\downarrow/(n_i^\uparrow + n_i^\downarrow)$ that of the minority electrons the spin polarization P_i of electrode i is analogously to (16.33) given by:

$$P_i = \frac{n_i^\uparrow - n_i^\downarrow}{n_i^\uparrow + n_i^\downarrow} = 2a_i - 1 \tag{16.44}$$

This allows to express the differential conductance for a parallel orientation as:

$$G^{\uparrow\uparrow} = G_p \propto a_1 a_2 + (1 - a_1)(1 - a_2) = \frac{1}{2}(1 + P_1 P_2) \tag{16.45}$$

and for an antiparallel orientation as:

$$G^{\uparrow\downarrow} = G_{ap} \propto a_1(1 - a_2) + (1 - a_1)a_2 = \frac{1}{2}(1 - P_1 P_2) \qquad (16.46)$$

The magnitude of the TMR effect is given by:

$$\text{TMR} = \frac{G^{\uparrow\uparrow} - G^{\uparrow\downarrow}}{G^{\uparrow\downarrow}} = \frac{G_p - G_{ap}}{G_{ap}} = \frac{R_{ap} - R_p}{R_p} \qquad (16.47)$$

Using the spin polarization the TMR effect can be written as:

$$\text{TMR} = \frac{\Delta R}{R_p} = \frac{2P_1 P_2}{1 - P_1 P_2} \qquad (16.48)$$

The difference of the differential conductance can also be normalized to the differential conductance of the parallel alignment. This situation defines the junction magnetoresistance JMR which exhibits the magnitude of:

$$\text{JMR} = \frac{G^{\uparrow\uparrow} - G^{\uparrow\downarrow}}{G^{\uparrow\uparrow}} = \frac{G_p - G_{ap}}{G_p} = \frac{R_{ap} - R_p}{R_{ap}} = \frac{\Delta R}{R_{ap}} = \frac{2P_1 P_2}{1 + P_1 P_2} \qquad (16.49)$$

Both effects are often given as a percentage. We directly see that for a total spin polarization of both electrodes ($P_1 = P_2 = 1$) the JMR reaches its maximum value with 100% whereas the TMR becomes infinite. FM-I-FM contacts with identical material of the electrodes allow the determination of the tunnelling spin polarization ($P_1 = P_2 = P$) if the conductance for parallel and antiparallel orientation of the magnetization is known:

$$P = \sqrt{\frac{G_p - G_{ap}}{G_p + G_{ap}}} \qquad (16.50)$$

Additionally, it is possible to quantify the spin polarization of one electrode if that of the other one is known provided that G_p and G_{ap} can experimentally be determined.

The TMR effect significantly depends on the temperature, bias voltage, and barrier properties like width and purity.

As one example for the influence of temperature on the TMR effect Fig. 16.26 shows the resistance curves measured at $T = 77\,\text{K}$ and $T = 295\ \text{K}$ using a Co-Al$_2$O$_3$-NiFe junction. Due to the strongly different coercivities of both ferromagnetic electrodes an antiparallel alignment is achieved in an intermediate field range whereas strong magnetic fields induce a parallel orientation of the magnetization. At $T = 77\,\text{K}$ we obtain a magnitude of the JMR effect of 27% whereas it is reduced to about 20% at room temperature. For the understanding of this behavior we introduce a variation of the electrode polarizations P_i with temperature. It is well established that in the case of alloys P scales approximately with the magnetic moment of the alloy as its composition is varied. A logical extension of this proportionality is to adopt a polarization P that varies with T as does the magnetization.

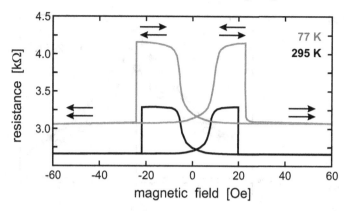

Fig. 16.26. Resistance as a function of an applied magnetic field for a Co/Al$_2$O$_3$/Ni$_{80}$Fe$_{20}$ junction at room temperature (*black curve*) and 77 K (*gray curve*) showing JMR values of 20.2% and 27.1%, respectively. Arrows indicate the magnetization configuration of the two ferromagnets; the upper arrow corresponds to Co whereas the lower one to Ni$_{80}$Fe$_{20}$. (Data taken from [78] (used with permission))

The temperature dependence of the magnetization was already discussed in Chap. 6.5 and is fairly well described by thermal excitation of spin waves for T far below the Curie temperature. This results in a term proportional to $T^{3/2}$ in the magnetization (see (6.102)). For the polarization we thus can write:

$$P(T) = P_0 \cdot (1 - \alpha T^{3/2}) \tag{16.51}$$

which leads to reduced spin polarization in the ferromagnetic electrodes with increasing temperature and thus to a decreased magnitude of the TMR and JMR effect.

Additionally, the TMR and JMR effect becomes reduced with increasing bias voltage. Figure 16.27(a) shows the magnitude of the JMR as a function of the bias voltage U for a Co-Al$_2$O$_3$-NiFe junction. The JMR monotonically decreases with increasing $|U|$. The data being normalized to the magnitude of the JMR effect at zero bias voltage which are shown in Fig. 16.27(b) prove a negligible dependence of the reduction on temperature. The magnitude of the reduction additionally depends on the quality of the insulating barrier and of the interfaces.

Figure 16.28 shows the influence of the Al barrier thickness on the JMR for two different junctions Co-Al$_2$O$_3$-CoFe and Co-Al$_2$O$_3$-NiFe. Both types exhibit a broad maximum centered around 1 nm and 1.5 nm. Beyond 1.8 nm the JMR magnitude rapidly decreases. If the barrier thickness is too low the effect also becomes reduced.

In order to understand this behavior we have to deal with the making of the Al$_2$O$_3$ spacer layer. It is produced by deposition of a metallic thin Al film on the base electrode with a subsequent oxidation. Thus, if the layer is

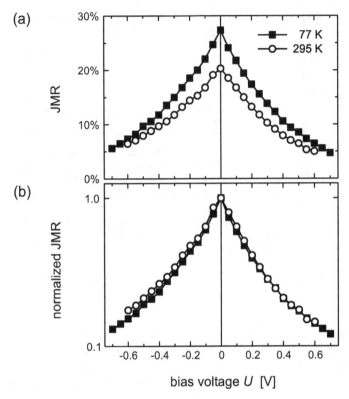

Fig. 16.27. JMR as a function of the applied bias at 77 K (*closed symbols*) and 295 K (*open symbols*) for the same junction as in Fig. 16.26. Data shown are (**a**) the actual percentages and (**b**) normalized at zero bias voltage. (Data taken from [78])

too thin an oxidation layer also occurs on the ferromagnetic electrode which results in a reduced magnetization at the interface and thus in a reduced polarization of electrons being responsible for the tunnelling current. If the layer is too thick the insulating Al_2O_3 spacer additionally exhibits a metallic Al layer within which leads to spin-flip processes of the tunnelling electrons. Thus, the magnitude of the TMR effect also becomes reduced.

For the investigation of the influence of barrier impurities on the magnetoresistance Co-Al_2O_3-NiFe junctions were prepared with submonolayer amounts of metals incorporated into the middle of the insulating oxide layer. Figure 16.29 shows the magnetoresistance curves for tunnel junctions containing different amounts of Ni. For comparison the values of the resistance are multiplied by constant factor for each curve to have an equal value in the high magnetic field state. A significant reduction of the JMR is obvious with increasing Ni content. The shape of the JMR curves practically remains unchanged. This reduction does also occur for non-magnetic impurities.

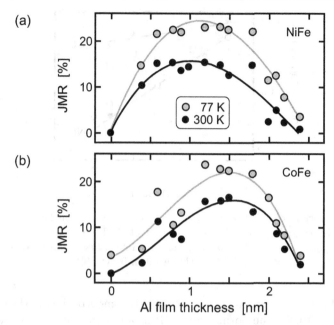

Fig. 16.28. Junction magnetoresistance plotted as a function of the thicknesses of the Al metal overlayer used to form the Al_2O_3 barrier in $Co/Al_2O_3/Ni_{80}Fe_{20}$ and $Co/Al_2O_3/Co_{50}Fe_{50}$ tunnelling junctions in (**a**) and (**b**), respectively. JMR percentage is the change in junction resistance normalized to its highest value in an applied magnetic field. The lines are only a guide to the eye. Uncertainties in the determination of Al film thicknesses and the JMR values are smaller than the size of the data points. (Data taken from [79])

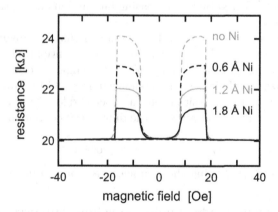

Fig. 16.29. Magnetoresistance curves for tunnelling junctions containing 0.6, 1.2, and 1.8 Å of Ni, respectively, together with the corresponding control junction without Ni. Data are obtained at 77 K and the resistances are multiplied by a constant factor for each curve to have an equal value of 20 kΩ at high fields. (Adapted from [80] with permission. Copyright 1998, American Institute of Physics)

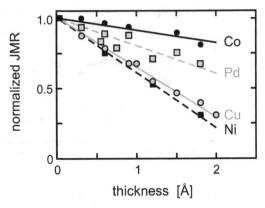

Fig. 16.30. Normalized JMR as a function of the thickness of the layer of impurities present in the tunnel barrier. Data, measured at 77 K, are shown for Co (*filled circles*), Pd (*open squares*), Cu (*open circles*), and Ni (*filled squares*), together with corresponding linear fits (*solid lines*). (Data taken from [80])

The influence of their amount given as the thickness of the interlayer is presented in Fig. 16.30. The normalized JMR as a function of thickness exhibits a linear decrease and can be understood assuming that, for the submonolayer thicknesses being used, the covered junction area increases linearly with the thickness. This is a reasonable assumption considering the fact that the impurity atoms were deposited with the substrate cooled with liquid nitrogen thereby minimizing surface diffusion and clustering effects. The net result is that the fraction of tunnelling electrons that experience a spin-flip linearly scales with the thickness provided that the spin scattering properties of the impurities are not affected by the increasing interaction among the impurities at higher coverage.

In our considerations above we have assumed that the magnetization in both ferromagnetic electrodes are oriented parallel or antiparallel. But, the differential conductance additionally depends on the angle Θ between both directions of magnetization. This behavior is shown in Fig. 16.31 for an Fe-Al$_2$O$_3$-Fe junction. Thus, up to now the situation was discussed for $\Theta = 0$ and $\Theta = 180°$, respectively. For an arbitrary angle Θ the differential conductance can be expressed as:

$$G = G_0 \cdot (1 + P_1 P_2 \cos \Theta) \tag{16.52}$$

with $G_0 = (G_\mathrm{p} + G_\mathrm{ap})/2$ being the spin averaged conductance.

FM-I-FM Contacts in a Scanning Tunnelling Microscope

The substitution of one of the ferromagnetic electrodes by a ferromagnetic probe tip represents the situation of spin polarized scanning tunnelling microscopy SPSTM. The insulating barrier is realized by the vacuum between

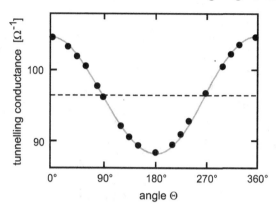

Fig. 16.31. Dependence of the tunnelling conductance (inverse resistance) of a planar Fe–Al$_2$O$_3$–Fe junction on the angle Θ between the magnetization vectors of both electrodes (Data taken from [27])

sample and tip which are separated by several Å. This allows the laterally resolved determination of magnetic properties.

In pioneering experiments the Cr(001) surface was imaged by using non-magnetic W tips as well tips made from the half-metallic ferromagnet CrO$_2$ which is highly spin polarized (nearly 100%) at about 2 eV below the Fermi level. The Cr(001) surface which the topological step structure is directly linked for to the magnetic structure represents a topological antiferromagnet, i.e. each terrace exhibits a ferromagnetic alignment of the magnetic moments but between two adjacent terraces the magnetization possesses an antiparallel orientation. Some line sections across several steps of the Cr(001) surface measured with either of both tip materials are shown in Fig. 16.32. While a uniform step height of 0.14 nm was measured using a non-magnetic tungsten tip (see Fig. 16.32(a)) two different apparent step heights of 0.12 and 0.16 nm, alternately, can be recognized in the line sections measured with the magnetic CrO$_2$ tip (see Fig. 16.32(b)).

Due to the spin polarization of the probe tip the tunnelling current is enhanced for a parallel alignment of the magnetization of sample and tip:

$$I_{\uparrow\uparrow} = I_0 \left(1 + P\right) \qquad (16.53)$$

with I_0 being the tunnelling current without the contribution due to spin polarized tunnelling and P the effective spin polarization of the junction. Contrarily, the tunnelling current for an antiparallel orientation is reduced:

$$I_{\uparrow\downarrow} = I_0 \left(1 - P\right) \qquad (16.54)$$

The line scans shown in Fig. 16.32 were taken in the constant current mode, i.e. the magnitude of the tunnelling current was held constant by varying the distance s between sample and probe tip. Thus, the parallel alignment results

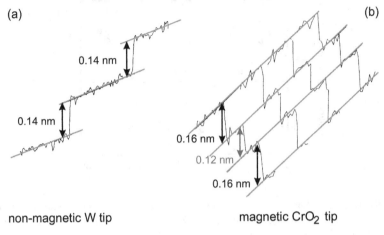

(a)   (b)

0.14 nm

0.14 nm

0.16 nm

0.12 nm

0.16 nm

non-magnetic W tip   magnetic CrO$_2$ tip

Fig. 16.32. (a) In the left part a single line scan over two monatomic steps taken with a non-magnetic tungsten tip is shown. In this case the measured step height value is constant with 0.14 nm and corresponds to the topographic monatomic step height. (b) Single line scans over the same three monatomic steps being obtained with a magnetic CrO$_2$ tip. The same alternation from the step height values (0.16, 0.12, and again 0.16 nm) in all single line scans is evident. (Adapted from [81] (used with permission))

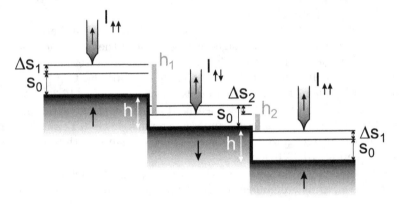

Fig. 16.33. Schematic drawing of a ferromagnetic tip scanning over alternately magnetized terraces separated by monatomic steps of height h. An additional contribution from spin polarized tunnelling leads to alternating step heights $h_1 = h + \Delta s_1 + \Delta s_2$ and $h_2 = h - \Delta s_1 - \Delta s_2$ which is therefore based on the relative orientation between tip and sample magnetization being either parallel or antiparallel. In the first case the effective tunnelling current is given by $I_{\uparrow\uparrow}$, in the latter one by $I_{\uparrow\downarrow}$. (Adapted from [81] (used with permission))

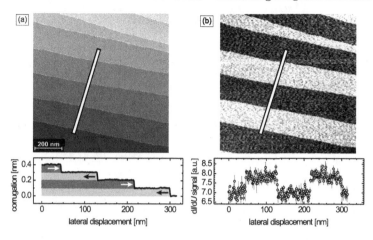

Fig. 16.34. (a) Topography and (b) spin resolved map of the dI/dU signal of a clean and defect free Cr(001) surface as measured with a ferromagnetic Fe-coated tip. The bottom panels show averaged sections drawn along the line. Adjacent terraces are separated by steps of monatomic height. (From [82]) (used with permission)

in an increase of s by Δs_1 with respect to the distance s_0 using a non-magnetic tip whereas the antiparallel configuration leads to a reduction of the distance by Δs_2 (see Fig. 16.33).

If the magnetization changes at a step exhibiting a height of h the measured step height amounts to $h_{\mathrm{p \to ap}} = h + \Delta s_1 + \Delta s_2$ for the transition of a parallel to an antiparallel orientation of the magnetization. At the next step which thus exhibits the transition of an antiparallel to a parallel orientation the apparent step height is given by $h_{\mathrm{ap \to p}} = h - \Delta s_1 - \Delta s_2$.

Using the scanning possibilities the antiferromagnetic coupling between neighbored terraces of a Cr(001) surface can directly be imaged. The topography (see Fig. 16.34(a)) presents a regular step structure with terrace widths of about 100 nm. The line section in the bottom of Fig. 16.34(a) reveals that all step edges in the field of view are of single atomic height, i.e. 1.4 Å. This topography should lead to a surface magnetization that periodically alternates between adjacent terraces. Indeed, this is experimentally observed (see Fig. 16.34(b)). The line section of the differential conductance drawn along the same path as in (a) indicates two discrete levels with sharp transitions at the positions of the step edges.

17

Applications

Industrial products which are based on magnetoresistive effects are already established at the market. In this chapter we will discuss a few typical examples which are non-volatile data storage elements, read-heads in hard disks, and various sensors.

17.1 MRAM – Magnetic Random Access Memory

Magnetic random access memories are data storage elements which are based on magnetoresistive effects. They exhibit the following important advantages:

- They are non-volatile in contrast to common dynamic random access memories DRAM.
- Thus, MRAM elements can simultaneously be used as main memory and mass storage device. This allows a simplification of the chip architecture which leads to reduced dimensions of the chip. Therefore, the equipment can be produced much cheaper.
- MRAM elements can be used in stationary systems like computers as well as mobile systems like laptops or mobile phones.
- No charging time is necessary. Thus, computers are in the same state after switching on they had at the moment they were switched off. The raising time can be neglected. Similarly, mobile phones are ready to be used directly after switching on.
- No refresh cycles are necessary like for DRAM elements. This allows longer operating times of batteries and accumulators.

The principle setup of an MRAM cell consists of a magnetic tunnel junction TMJ as shown in Fig. 17.1.

Types of MRAM Cells

Figure 17.2 illustrates structures used to engineer the response of magnetic tunnel junctions in ways beneficial for memory applications. Figure 17.2(a)

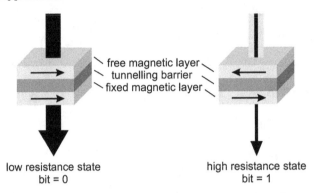

low resistance state
bit = 0

high resistance state
bit = 1

Fig. 17.1. The heart of a magnetic tunnel junction consists of two layers of magnetic material such as nickel iron and cobalt iron that sandwich a very thin insulator typically a layer of aluminum oxide only a few atoms thick. Current flows down through the layers and it meets less resistance when the two magnetic layers are magnetized in the same direction and more resistance when they are magnetized in opposite directions. The two states serve to encode a "0" or a "1". (Adapted from [83]. Reprinted with permission from AAAS.)

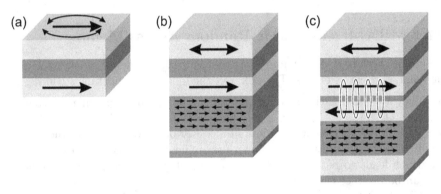

Fig. 17.2. Tunnel junctions engineered for MRAM applications. (**a**) Basic magnetic tunnel junction structure consisting of two ferromagnetic metals separated by a thin insulating layer. With the same anisotropy direction for both magnetic film layers, the junction has a hysteretic TMR response characteristic like that shown in Fig. 16.11. (**b**) By exchange coupling one of the magnetic layers to an antiferromagnetic layer, i.e. by "pinning" the layer, the TMR response reflects the hysteresis of the other so-called "free" layer and has a response curve more suitable for memory. (**c**) The magnetic offset caused by fields emanating from the pinned layer can be avoided by replacing a simple pinned layer with a synthetic antiferromagnetic pinned layer which consists of a pair of ferromagnetic layers antiferromagnetically coupled through a ruthenium (Ru) spacer layer. The lower layer in this artificial antiferromagnet is pinned via exchange bias as shown in (b). This flux closure increases the magnetic stability of the pinned layer and reduces coupling to the free layer. (Adapted from [84]. Image reproduced by permission of IBM Research. Unauthorized use not permitted.)

shows the basic magnetic tunnel junction structure with specific characteristics which were already illustrated in Fig. 16.11. While in principle this
structure could be made to work for a memory if the coercivity of one of the
layers, a "reference" layer, is much higher than that of the other one difficulties
would arise with this approach.

First, field excursion would have to be restricted to being lower than a
maximum value so that the high-coercivity layer would never be disturbed.
Even so it is possible that repeated low-field excursions could reverse small
domains in the higher-coercivity reference layer that have no way of returning
to their original state. The possibility of upsetting the reference layer could be
avoided by pinning one of the magnetic electrodes via exchange coupling to an
adjacent antiferromagnet as illustrated in Fig. 17.2(b). Only the other "free
layer" electrode responds to the field. The low-field electrical response of such
a structure would very directly reflect the memory function of the magnetic
hysteresis of the free layer. In subsequent structures such as the one illustrated
in Fig. 17.2(c) the antiferromagnetic material is replaced by a synthetic antiferromagnet (SAF) sandwich comprising, for example, CoFe/Ru/CoFe, with
the Ru thickness being about 7 Å. In this thickness range the Ru exchange-
couples the moments of the two ferromagnetic layers in opposite directions.
Thus, a "fixed" reference layer, the SAF, could be produced with no net magnetic bias on the other, "free", magnetic layers which is important if a suitable
response is to be obtained for a magnetic memory in the absence of a magnetic
field bias.

MRAM Array Architecture

Using these elements an array of magnetic storage cells can be built up which
are based on the tunnelling magnetoresistance. There exist two basic architec-
tures for constructing an MRAM array, the cross-point ("XPT") architecture
and the one-transistor one-MTJ ("1T1MTJ") architecture.

In the XPT architecture (see Fig. 17.3) the MTJs lie at the intersection
of the word lines and the bit lines which connect directly to the fixed and
free layers (or vice versa). This arrangement allows for a considerable packing
density. Since no contact is made to the silicon within the cell it is possible
to stack such arrays thus further increasing MRAM density. In addition, it
is possible to place peripheral circuits under the array which increases the
density even further.

In the 1T1MTJ array architecture (see Fig. 17.4) each MTJ is connected
in series with an n-type field effect transistor (n-FET). The n-FET the gate of
which is the read word line is used to select the cell for the read operation. The
write word line runs directly below but does not actually contact the MTJ.
The read and write word lines run parallel to each other and perpendicular
to the bit line which contacts the free layer of the MTJ. The source of the
n-FET is grounded whereas the drain connects to the fixed layer of the MTJ
via a thin local interconnect layer. This layer and the dielectric below it are

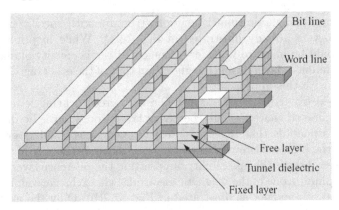

Fig. 17.3. Cross-point architecture for an MRAM array (From [85]. Image reproduced by permission of IBM Research. Unauthorized use not permitted.)

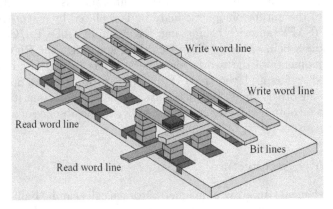

Fig. 17.4. MRAM architecture with an additional field effect transistor as a switch (From [84]. Image reproduced by permission of IBM Research. Unauthorized use not permitted.)

relatively thin in order to ensure good magnetic coupling from the write word line to the MTJ.

Read Operation

The MTJ device is read by measuring the effective resistance of the structure which is a function of the magnetic state of the MTJ free layer. This can be achieved by applying a voltage and sensing the current (current sensing) or by applying a current and sensing the voltage (voltage sensing). In either case the sensed parameter (assumed to be current in the following) is compared to a reference value to determine the state of the device.

The fractional value change in effective resistance or MR is not constant but rather decreases with increasing read voltage. Therefore, the relative signal or fractional difference between the data and reference currents decreases with increasing voltage. However, the absolute signal or absolute difference between the data and reference currents vanishes as the voltage approaches zero. Both relative and absolute signals are critical for a robust and high-performance design. Therefore there exists an optimum value of the read voltage which appears to be approximately 200 to 300 mV.

The reference value must be designed to compensate for process-related variations in MTJ parameters (R_0 and MR) and for environmental variations such as voltage and temperature. Three general methods are known for generating the reference value: the twin cell, reference cell, and self-referenced methods.

Because of its attractive combination of high density, high performance, low power, and high degree of symmetry the current-sensing two-reference-cell design is the most popular approach. With this method the raw signal must be sufficiently large to compensate for parameter mismatch between the data and reference cells as well as offsets within the sense amplifier. This requirement places strict requirements on the MR, MTJ parameter matching, and design of the sense amplifier.

Write Operation

The MRAM write operation is illustrated in Fig. 17.5. The selected MTJ is situated between the selected word line and the selected bit line which are orthogonal to each other. During the write process currents (depicted by arrows) are forced along the selected word line and the selected bit line creating magnetic fields in the vicinity of these wires. The vector sum of the

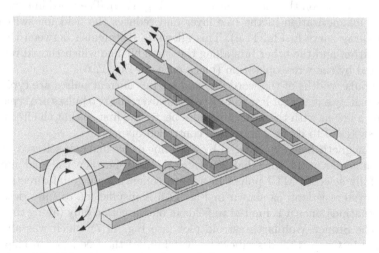

Fig. 17.5. MRAM write operation (From [85]. Image reproduced by permission of IBM Research. Unauthorized use not permitted.)

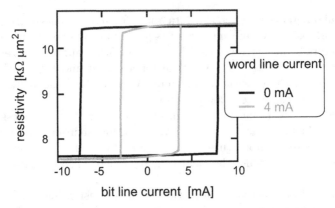

Fig. 17.6. Resistivity as a function of current of an TMR device. When a particular word line current is applied the hysteresis loop collapses so less bit line current is required to change the resistance state of a bit. A bit line current of ±5 mA will write the selected bit (*gray curve*) but not the other bits along the same bit line (*black curve*) (Adapted from [86] with permission from Wiley)

fields at the selected MTJ must be sufficient to switch its state. However, the field generated by the word line or bit line alone must be small enough that it never switches the state of the so-called half-selected MTJs that lie along the selected word line and bit line.

The process is designed in such a way that the word lines and bit lines are as close as possible to the MTJs for good magnetic coupling to the MTJs. Nonetheless, currents of the order of 5 mA are typically required to switch the state of an MTJ. This behavior can directly be seen in the corresponding resistivity response on the bit line current (see Fig. 17.6). The word line current tilts the magnetization in the free layer and reduces the field for switching the bit (gray curve in Fig. 17.6). This allows to distinguish between the bit to be written and the other bits along the same bit line which should remain unaffected by the write operation (black curve in Fig. 17.6).

The pulse widths of the word line and bit line current pulses are typically approximately equal to or less than 10 ns. However, the two pulses are typically offset by a few ns with the word line pulse beginning first so that the free layer can be switched to its new state in a controlled manner.

The magnetic field experienced by word line or bit line half-selected MTJs is perpendicular to the wire that generates the field. Further, the field applied to the fully selected MTJ points in a third somewhat diagonal direction. A typical hysteresis loop as shown in Fig. 17.6 is insufficient to fully describe these situations since it is limited to fields in one direction only (along the long axis). The Stoner–Wohlfarth astroid plot (see Fig. 17.7) which was already discussed in Chap. 12.5 describes the switching of the free layer in response to both field strength and direction. The x- and y-axes represent the x- and y-components of the magnetic field applied to the MTJ. In Fig. 17.7 the long

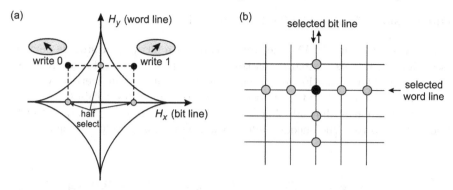

Fig. 17.7. Ideal astroid plot describing the switching of the free layer of a magnetic tunnel junction (**a**) and the corresponding MRAM array (**b**) (Adapted from [85]. Image reproduced by permission of IBM Research. Unauthorized use not permitted.)

axis of the MTJ and the word line are assumed to be horizontal and the bit line to be vertical. Since the word line field applied to the MTJ is perpendicular to and proportional to the word line current the y-component of the field is proportional to the word line current. Similarly, the x-component of the field is proportional to the bit line current.

The astroid plot is interpreted in the following manner. If the applied field begins at the origin (no applied field), moves to a point to the right of the y-axis and the diamond-shaped region or astroid, and returns to the origin the free layer will point to the right (data state "1"). Similarly, if the applied field begins at the origin, moves to a point to the left of the y-axis and the astroid, and returns to the origin the free layer will point to the left (data state "0"). If the applied field remains inside the astroid the state of the MTJ remains unchanged.

The fully selected MTJ experiences both x- and y-field components placing it in the first or second quadrant of the figure depending on the data state to be written. Since the polarity of the x-field component or bit line current determines the written data state the bit line write circuitry must support a bidirectional current. The y-field component is independent of the data state to be written simplifying the design of the word line write circuitry because bidirectional currents are not required. In order to write successfully the fully selected field points must always lie outside the astroid.

As indicated in Fig. 17.7 word line half-selected MTJs experience a y-field component only whereas bit line half-selected MTJs experience an x-field component only. The polarity of the field experienced by a bit line half-selected MTJ depends on the state being written to the fully selected MTJ. To avoid half-select disturbs (data loss between an MTJ being written and read) the half-select field points must always lie inside the astroid.

Old Version of MRAM-Like Storage Cells

An old version of an MRAM-like storage cell was a ferrite core storage device (see Fig. 17.8). This system needs three wires. Two are used for addressing and writing the information and are arranged horizontally and vertically. The "bits" are represented by the opposite magnetization states within the ferrite rings as exemplarily shown by black and white arrows (see Fig. 17.9). A third line utilizes to read the encoded information by means of inductive signals.

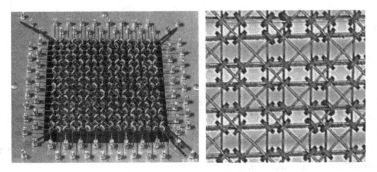

Fig. 17.8. Photographs of ferrite cores acting as non-volatile memories

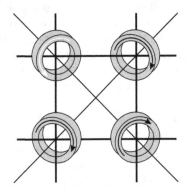

Fig. 17.9. Schematics of a magnetic core memeory which can be switched by currents through the horizontally and vertically arranged wires. Reading the stored information is carried out using a third line by means of inductive signals

Comparison between Different Storage Technologies

A comparison of different storage technologies is given in Table 17.1.

Table 17.1. Comparison of typical parameter values for different storage technologies

	DRAM	EEPROM	Flash	FeRAM	MRAM
Read endurance	$> 10^{15}$	$> 10^{15}$	$> 10^{15}$	$10^{12} - 10^{15}$	$> 10^{15}$
Write endurance	$> 10^{15}$	$10^4 - 10^6$	$10^5 - 10^6$	$10^{10} - 10^{15}$	$> 10^{15}$
Operating voltage	$2.5 - 5$ V	$12 - 18$ V	$10 - 18$ V	$0.8 - 5$ V	–
Programming time	10 ns	1 ms	$1\,\mu s$	10 ns	10 ns
Read time	10 ns	40 ns	40 ns	50 ns	10 ns
Cell size	$8f^2$	$40f^2$	$8 - 12f^2$	$9 - 13f^2$	$4f^2$
Cell area (f=0.25μm)	$0.5\ \mu m^2$	$2.5\ \mu m^2$	$0.5 - 0.7\ \mu m^2$	$0.6 - 0.8\ \mu m^2$	$0.25\ \mu m^2$
Retention time	volatile	>10 years	>10 years	>10 years	>10 years

17.2 Read Heads in Hard Disks

In the recording process information is written and stored as magnetization patterns along concentric tracks in the magnetic recording medium being on a spinning disk. This is done by scanning the write head over the medium and energizing the write head which is basically an electromagnet with appropriate current waveforms. Next, in the read-back process the stored information is retrieved by scanning a read head over the recording medium. The read head intercepts magnetic flux from the magnetization patterns on the recording medium and converts it into electrical signals which are then detected and decoded.

A very important performance criterion for a disk recording system is the amount of information it can store per unit area. Since information is typically stored as abrupt magnetization changes designated as transitions and characterized by a transition parameter along a track on the disk the areal density is the product of the linear bit density and the track density. The former is the density with which magnetic transitions can be packed along a track; the latter is the density with which these tracks can in turn be packed together. A high track density therefore implies recording with narrow tracks. The areal density performance of disk recording systems has increased consistently and dramatically for the last thirty years culminating in an improvement by more than five orders (see Fig. 17.10). This continuous and tremendous increase of storage density of hard disks is possible by improvement of:

- storage media
  They consist of thin film layers. Modern developments additionally include granular systems and ferromagnetic nanoparticles .
- write heads
- read heads

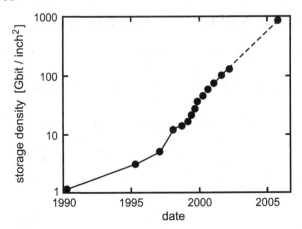

Fig. 17.10. Growth in the typical storage density of laboratory demos for magnetic disk drives since 1990

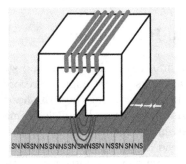

Fig. 17.11. Schematic view of an inductive read/write element

Traditionally, the recording head is a single inductive element energized as an electromagnet for writing and used according to the Faraday effect for reading (see Fig. 17.11). Early inductive heads were made primarily from individually machined polycrystalline ferrites wound with fine wires as write and read coils.

In the early 1990s dual-element heads with inductive write elements and magnetoresistive read elements were introduced. In these dual-element heads writing is performed by producing a writing field in the gap between the magnetic poles P_1 and P_2 as before. But, the stray fields arising at the transitions between areas of opposite magnetization direction is read back by a magnetoresistive sensor (see Fig. 17.12). This leads to a modulation of the sensor resistance through the magnetoresistance effect which is in turn converted into voltage signals by passing a sense current through the sensor. In comparison with the single inductive recording heads the dual-element heads have the advantages of separate optimization of read and write performance as well as a signal sensitivity which is several times larger because of the use of magnetoresistive sensors.

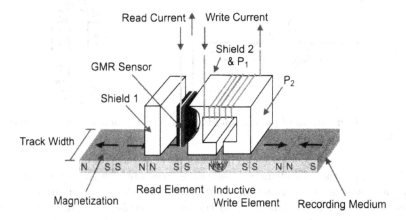

Fig. 17.12. Schematic illustration of longitudinal magnetic recording. Bit patterns are recorded along tracks using an inductive write head. A shielded GMR read head senses the stray fields from transitions between bit cells of opposite magnetization directions. (From [87] with permission from Wiley)

The stray field which is used to read the information of a bit is shielded between two soft magnetic layers to avoid interferences with stray fields from neighboring transitions. The magnetic pole P_1 may also serve as one of the sensor shields. The geometry of a shielded read head which is commonly used is shown in Fig. 17.13. The magnetoresistive element is present in between the two soft magnetic shields that confine the length of the region along a

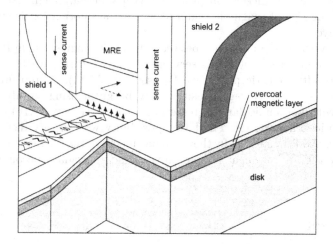

Fig. 17.13. Schematic view of the read principle for a shielded head in a hard disk recording system (Reprinted from [88]. Copyright 2003, with permission from Elsevier)

(a)   (b)

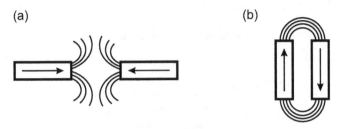

Fig. 17.14. Comparison of (a) longitudinal and (b) perpendicular arrangement of bit cells

track to which the MR element is sensitive to essentially the bit length. For that purpose the optimal distance between the shields is approximately two times the bit length. In addition to the magnetoresistive material the read gap contains two dielectric gap layers that electrically isolate the MR material from the metallic shields. The head is mounted on a slider which flies at a very small fixed distance above the disk (typically 10 nm) due to the formation of an air film that functions as an air bearing in between the head and the slider.

In contrast to longitudinal recording (see Fig. 17.12) where the magnetization lies in the plane of the recording layer information can also be stored using a perpendicular or vertical recording scheme where the medium is magnetized perpendicular to the film plane. In longitudinal recording the demagnetizing fields between adjacent magnetic bits tend to separate the bits making the transition parameter large. Perpendicular recording bits do not face each other and can therefore be written with higher density (see Fig. 17.14). In comparison to a longitudinal medium a perpendicular medium supporting the same areal density can be thicker and problems with superparamagnetism due to the decreasing size of magnetic material which the information is stored in can be delayed.

A narrow write gap is formed between the right magnetic pole P_2 and a soft magnetic underlayer which is serving as a flux return path towards the much wider left magnetic pole piece P_1 (see Fig. 17.15). A step further to achieve ultimate high storage densities may be given using suitable media materials with perpendicular magnetic anisotropy (see Fig. 17.16) or using ferromagnetic nanoparticles. The latter possibility was already discussed in Chap. 12.9. Additionally, a further enhancement of the storage density is possible by the improvement of spin valves which nowadays become rather complex.

17.3 Sensors

Sensors which are used to transform magnetic or magnetically coded information into electric signals already play an important role in nowadays applications due to their robust design, their non-contact and thus wear free

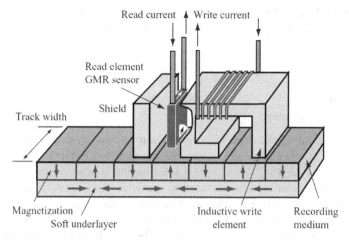

Fig. 17.15. Schematic illustration of perpendicular magnetic recording using a probe head and a soft magnetic underlayer in the medium (From [90]. Image reproduced by permission of IBM Research. Unauthorized use not permitted.)

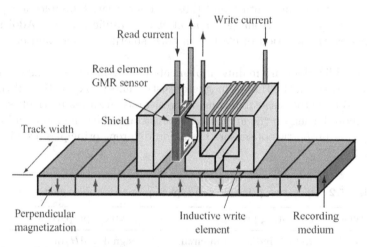

Fig. 17.16. Schematic illustration of perpendicular magnetic recording using a ring head and no soft magnetic underlayer (From [90]. Image reproduced by permission of IBM Research. Unauthorized use not permitted.)

operation, and their low manufacturing costs, e.g. solid state sensors which can be fabricated using batch processes. Many different parameters can be recorded like

- lateral position
- velocity
- angularity
- rotational frequency

One main field of application is given for automotive systems, in particular the sensing of speed and position, e.g. the wheel speed detection for anti-lock braking systems (ABS). Wheel speed information is also needed in modern vehicle dynamics control and navigation systems. Additionally to the wheel speed both require the steering angle as an input value which is also often provided by magnetic sensors. A classic field of application is the power train. The magnetic sensors deliver information about the cam and crank shaft positions as well as the transmission speed.

Besides these applications more and more position sensors are replaced by magnetic sensors in safety-relevant and drive- or fly-by-wire systems because they work without any contact and thus without wear. Examples are pedal position sensors for electronic gas and electrohydraulic braking. Additional quantities are the position of electric motors, steering torque, and electrical current.

These applications are mainly implemented using inductive sensors or using sensors based on the Hall or magnetoresistive effects like AMR and GMR. Sensors basing on the anisotropic magnetoresistance are already widely spread whereas those basing on the giant magnetoresistance become more and more used. Typical properties of sensors basing on different principles are given in Table 17.2.

Table 17.2. Operating principle and typical properties of magnetic sensors

Sensor type	Operating principle	Typical properties
Inductive	voltage induction in circuit loops due to a change of the magnetic field	signal $\propto \mathrm{d}H/\mathrm{d}t$
Hall	cross voltage of a current in semiconductor devices due to a magnetic field	signal $\propto H$ signal $\propto \cos\theta$
AMR	resistance change of the magnetic material due to an external magnetic field	signal $\propto H$ up to saturation signal $\propto \cos 2\theta$ $\Delta R/R \approx 2\%$
GMR	resistance change of a magnetic multilayer system due to an external magnetic field	signal $\propto H$ up to saturation signal $\propto \cos\theta$ $\Delta R/R \approx 10\%$

Whereas read heads must only distinguish between two opposite magnetization states encoding "0" and "1" the requirements on sensors are significantly higher. For the determination of mechanical parameters the readout is often carried out for continuously varying values with sufficient resolution. Thus, the sensors must satisfy requirements on:

- temperature stability
- shape of characteristics
- long-term stability

In the case of automotive sensors the specific application and the place in the car where the sensor is installed defines these requirements. Sensors in the interior for example have to withstand temperatures only up to about 80°C whereas near the engine block or near the brakes temperatures can reach 150°C. One very specific requirement for magnetic sensors is electromagnetic compatibility. Due to the fact that more and more electronic systems and electric actuators are being installed one has to deal with higher magnetic stray fields which may affect the sensor properties and falsify their output signal. Therefore, tomorrow's magnetic sensors need not only to have a better performance like enhanced accuracy, higher sensitivity, and longer lifetime but must be even more robust. Additionally, the sensor costs decide whether the developed solution will be successful in the market. An overview of relevant properties concerning different kinds of sensors is given in Table 17.3.

Table 17.3. Properties of GMR and AMR based sensors compared to Hall and inductive magnetic field sensors (+: fair, ++: good, + + +: very good)

Property	GMR	AMR	Hall	inductive
Temperature stability	++	++	+	++
Output signal	+ + +	++	+	size dependent
Sensitivity	+ + +	+ + +	++	++
Power consumption	+ + +	+	++	size dependent
Size	+ + +	+	+ + +	+
dc operation	yes	yes	yes	no
Costs	+ + +	+	+ + +	++

AMR Sensors

AMR sensors are based on the anisotropic magnetoresistive effect which occurs in ferromagnetic transition metals and was already discussed in Chap. 16.2. The electrical resistance is a function of the angle θ between the electric current and the direction of magnetization:

$$R(\theta) = R_0 + \Delta R_0 \cdot \cos^2 \theta \qquad (17.1)$$

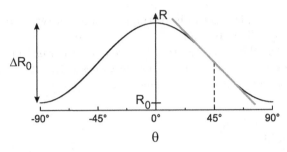

Fig. 17.17. Electrical resistance R as a function of the angle θ between the electric current and the direction of magnetization. A linear dependence is given around an angle of $\theta = 45°$ (see *gray line*)

This dependence is shown in Fig. 17.17. It is obvious from the concerning quadratic expression that the resistance behaves non-linear and in addition not unambiguous, i.e. one specific value of the resistance corresponds to more than one value of the angle θ. A linear dependence is given around an angle of $\theta = 45°$ (see gray line in Fig. 17.17).

An external magnetic field can change the direction of the magnetization, thus the angle θ and therefore the resistance allowing the AMR to be used as the transducer effect in magnetic field sensors.

We see that the AMR sensors utilize the effect that the internal magnetization vector can by rotated by an external magnetic field. In the simplified model of Stoner–Wohlfarth a magnetoresistive layer is always completely magnetized up to its saturation magnetization. Therefore, an external field can only change the direction of the magnetization and not its amplitude. In the ideal case the magnetization vector would directly follow the external field and is independent of its strength.

In this situation we are able to substitute the angle θ by the magnitude of the external field H_y in perpendicular direction to the electric current by:

$$\sin^2 \theta = \begin{cases} \dfrac{H_y^2}{H_0^2} & \text{if } H_y \leq H_0 \\ 1 & \text{if } H_y > H_0 \end{cases} \tag{17.2}$$

with H_0 being material dependent. Inserting this relationship into (17.1) results in:

$$R = R_0 + \Delta R_0 \cdot \left(1 - \frac{H_y^2}{H_0^2}\right) \tag{17.3}$$

for $H_y \leq H_0$. For $H_y > H_0$ we obtain $R = R_0$. This dependence of the resistance R on the y component of the external field is shown in Fig. 17.18.

But, in order to get a usable magnetic field sensor with a preferably linear characteristic a more sophisticated design is necessary.

The magnetoresistive effect can be linearized by depositing aluminum stripes (called barber poles) on top of the permalloy strip at an angle of 45° to

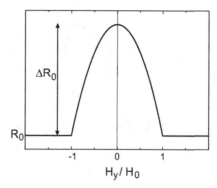

Fig. 17.18. Electrical resistance R as a function of the magnitude of the external field H_y in perpendicular direction to the electric current. This characteristic is non-linear. Additionally, each value of the resistance does not necessarily correspond to a particular magnetic field strength

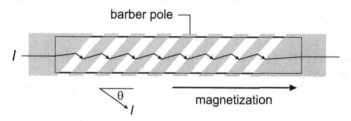

Fig. 17.19. Principle arrangement of a magnetoresistive element with additional barber poles which often consist of aluminum and are rotated by 45° with respect to the magnetization direction. Due to their much higher conductance compared to the magnetoresistive material like permalloy the electric current flows along the shortest way in the high-resistivity gaps thus forming an angle of 45° with regard to the direction of the magnetization (see *arrows*)

the strip axis. This choice is due to the linear dependence of the resistance as a function of the angle θ (cf. Fig. 17.17). The principle arrangement is shown in the right part of Fig. 17.19. As aluminum has a much higher conductance than permalloy the effect of the barber pole is to rotate the current direction by 45° thus effectively changing the angle between the magnetization and the electrical current from θ to $(\theta + 45°)$. Now, the resistance amounts to:

$$R = R_0 + \frac{\Delta R_0}{2} + \Delta R_0 \cdot \frac{H_y}{H_0} \cdot \sqrt{1 - \frac{H_y^2}{H_0^2}} \qquad (17.4)$$

This relationship is shown as curve 2 in Fig. 17.20. We directly see that we have significantly changed the situation. The change of resistance exhibits a *linear* dependence on the external field for small values of H_y (see gray line). For comparison the dashed curve 1 represents the case for $\theta = 0°$.

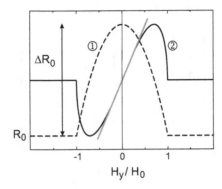

Fig. 17.20. Electrical resistance R as a function of the magnitude of the external field H_y in perpendicular direction to the electric current of a magnetoresistive element without (*dashed line 1*) and with barber poles being arranged at an angle of 45° with respect to the magnetization direction (*solid line 2*). For the latter situation, the resistance exhibits a *linear* dependence on the external field for small values of H_y (see *gray line*)

By this means the transfer curve is symmetric around zero field with maximum sensitivity and linear for small external fields.

For sensors using barber poles arranged at an angle of 45° the resistance is given by:

$$R = R_0 + \frac{\Delta R_0}{2} - \Delta R_0 \cdot \frac{H_y}{H_0} \cdot \sqrt{1 - \frac{H_y^2}{H_0^2}} \qquad (17.5)$$

which represents the mirror image of curve 2 in Fig. 17.20.

Therefore, a Wheatstone bridge arrangement consisting of four magnetoresistive elements is used to build up a complete sensor element. In this arrangement diagonal elements have barber poles of the same orientation. This means that one diagonal pair has barber poles orientated +45° to the strip axis while the other pair has an orientation of −45° (see Fig. 17.21). This ensures a doubling of the output signal while still having an almost linear output signal. Moreover, the inherent temperature coefficients of the four bridge resistances are mutually compensated.

AMR sensors often made of permalloy possess a strong temperature coefficient. To minimize drifts due to temperature changes four AMR sensors are usually arranged in a Wheatstone bridge configuration which also leads to an output signal twice as large as for a single resistor when constant voltage is applied. AMR resistors are meander-shaped to increase their base resistance which is typically in the kΩ range. The sensitivity can be tuned over a large range depending on the application. AMR sensors are used to detect magnetic fields in the range from a few μT, e.g. in compass applications, up to several 10 mT in position and angle sensors.

The AMR sensor elements fit into standard mold packages with iron-free leadframes. Often, the application necessitates a specific package design as for

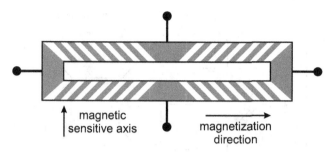

Fig. 17.21. Principle arrangement of a Wheatstone bridge configuration. One pair of diagonally opposed magnetoresistive elements exhibits barber poles at +45° with respect to the magnetization direction whereas in the other pair they are arranged at −45°. The magnitude H_y is determined along the specified magnetic sensitive axis

speed sensors which the reading configuration does not allow much space in. Today, the assembly and packaging technology is the limiting factor for the AMR temperature range. Although an AMR sensor by itself can operate at temperatures well above 200°C the standard sensor packages are specified for temperatures below 150°C.

AMR Angle Sensors

AMR angle sensors are operated in the saturating mode at magnetic fields typically above 50 mT. As discussed above the AMR effect turns into a pure angle dependency for large fields. Variations in the applied field do not affect the output signal. This measuring principle is fairly independent from assembly and magnet tolerances as well as from aging and temperature induced changes in the magnetic field strength. One example is given by the non-contact observation of a faultless spray arm rotation in dish washers.

The principle setup and layout of a conventional AMR sensor is shown in Fig. 17.22 which consists of 8 AMR resistors that are combined into two Wheatstone bridges. One of these bridges is turned by 45°. When an in-plane sensing field is rotated by, for example, simply rotating a permanent magnet above the sensor the first bridge delivers a sine output signal whereas the other bridge a cosine signal according to

$$U_1 = A(T) \cdot \sin(2\theta) \quad \text{and} \quad U_2 = A(T) \cdot \cos(2\theta) \qquad (17.6)$$

Using the arc tangent function the angle of the external magnetic field can be calculated as

$$\theta = \frac{1}{2} \cdot \arctan\left(\frac{U_1}{U_2}\right) \qquad (17.7)$$

This calculation is usually done in an evaluation circuit. The amplitudes of both bridges are almost the same because the bridges are placed into

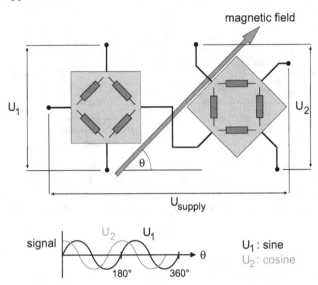

Fig. 17.22. Principle setup of an AMR angle sensor for a measuring range of 180°

one another and thus have experienced the same manufacturing process and will be operated at the same temperature. The different sensitive elements of both Wheatstone bridges are marked by 1a to 1d and 2a to 2d (see Fig. 17.23). Therefore, in this operation the temperature dependent signal amplitudes $A(T)$ are removed resulting in delivery of pure angle information largely without any temperature drifts. Of course, small asymmetries in the

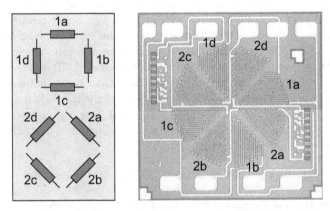

Fig. 17.23. Schematic arrangement (**left**) and layout (**right**) of a sensor which exhibits two Wheatstone bridges being placed into one another. 1a to 1d and 2a to 2d indicate the sensitive elements of Wheatstone bridge 1 and 2, respectively. (Layout reproduced with permission from NXP Semiconductors Germany GmbH)

bridges always remain leading to offset voltages with their own temperature dependencies.

As obvious from (17.6) and (17.7) the AMR output signal depends on twice the external magnetic field angle θ restricting the absolute measuring range to 180°. Many applications, e.g. in steering systems, need absolute angle information over a full rotation which cannot be provided using this AMR sensor. This becomes possible by replacing the AMR sensor by a type which is based on the GMR effect.

GMR Sensors

GMR-based sensors are often used in CIP (current in plane) configuration because the resistance of the element in CPP (current perpendicular to plane) is too low. The market requirement on sensors are a wide temperature window of about $-40°C$ to $+150°C$ and low field strengths for save and optimum working conditions.

Different types of GMR-based sensors were already shown in Fig. 17.2. The working principle of that in Fig. 17.2(a) is characterized by a thickness of the non-magnetic interlayer which enables an antiferromagnetic coupling of the ferromagnetic thin layers without an external magnetic field. Applying such a field leads to a parallel alignment which corresponds to a reduction of the resistance. In type (b) the antiferromagnetic substrate forces the neighboring ferromagnetic thin film to a given direction. This reference layer is therefore magnetically hard. The thickness of the non-magnetic interlayer is so large that the coupling between both ferromagnetic thin films can be neglected. Thus, the soft magnetic top layer acts as the testing layer. The sensors shown in Fig. 17.2(b) and (c) are very similar. The latter one differs in a synthetic antiferromagnet which exhibits an enhanced temperature stability.

GMR Angle Sensors

This type of sensor is used for the determination of a rotation angle over the whole range of 360° (see Fig. 17.24). In GMR angle sensors only spin valves can be used as the sensing material because the GMR effect in coupled multilayers and granular systems is isotropic.

Analogously to AMR sensors GMR resistors are arranged in two Wheatstone bridges. The elements are characterized by positive and negative signals at the same external magnetic field. This situation can be realized by different orientations of the synthetic antiferromagnetic layers. All these reference layers can be sensitive to a parallel alignment. In this situation, given e.g. in the sensor GMR B6 from Siemens/Infineon (see left part of Fig. 17.25), only the orientation of the external field can be determined due to the cosine-like characteristic curve (see Fig. 17.26(a)). In sensor C6 the reference layers exhibit a perpendicular orientation with respect to each other (see right part of Fig. 17.25). In a rotating external field one of these bridges generates a sine

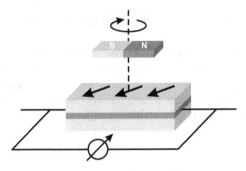

Fig. 17.24. Typical application of a GMR angle sensor

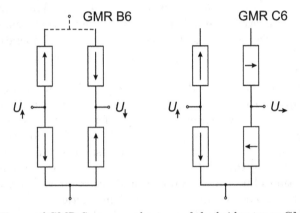

Fig. 17.25. Types of GMR Sensors and set-up of the bridge types GMR B6, GMR C6; the direction of internal magnetization is indicated by the arrows. (Used with permission from Siemens Application Note "Giant Magneto Resistors")

signal whereas the other one a cosine signal (see Fig. 17.25(b)). The angle of the external field is derived by applying the arc tangent function to both signals.

The main advantage of GMR is that it represents a uniaxial effect, i.e. the resistance is proportional to the cosine of the angle θ of the external field with respect to the pinning or reference direction:

$$R = R_0 - \Delta R_0 \cdot \cos\theta \qquad (17.8)$$

Therefore, GMR angle sensors exhibit a "natural" 360° measuring range being in contrast to the 180° range covered by AMR sensors. The angle signal from the arc tangent does not repeat after 180° as occurs with AMR sensors.

To create a sine- and a cosine-like signal producing GMR bridge the single resistors need at least two pinning directions with a 90° phase difference and additionally two directions to double the output signal. After deposition the pinning in spin valves containing antiferromagnets usually has a preferred direction. To change it the material has to be heated above the Néel tem-

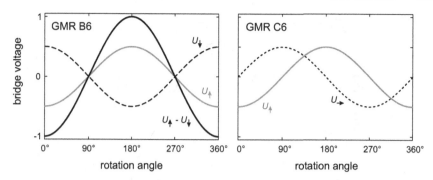

Fig. 17.26. Bridge voltages of full bridges of sensor B6 ($U_\uparrow - U_\downarrow$) and half bridges of sensor B6($U_\uparrow$, $U_\downarrow$) and sensor C6 ($U_\uparrow$, $U_\rightarrow$), respectively, while the external field is rotated by 360° (Adapted from Siemens Application Note "Giant Magneto Resistors" (used with permission))

perature of the antiferromagnet while applying an external magnetic field in the desired pinning direction. During cooling the pinning direction is fixed. The challenge is to do this imprinting of different pinning directions on the micrometer scale. Two different methods were developed for industrially produced devices. The first method consists of current flowing through the GMR resistor with an additional circuit which provides local heating. The second method utilizes short laser pulses which are focussed on the selected sensor region.

Figure 17.27 shows an example layout of a GMR angle sensor. The full bridges are split into 16 single elements that are arranged in a circle. The rotational symmetry further minimizes angle errors. An accuracy better than 1.5° over the complete temperature range from −40°C to +150°C is achieved. Since the spin valve material is extremely robust another feature of the sensor is its immunity against interfering fields exceeding 100 mT at 200°C.

Fig. 17.27. Example layout of a GMR angle sensor (Image used with permission from Robert Bosch GmbH, Germany)

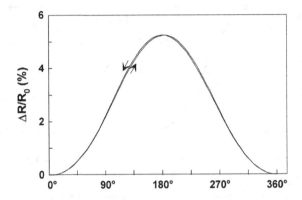

Fig. 17.28. Relative change of the resistance giving rise to small hysteretic effects (Used with permission from Siemens Application Note "Giant Magneto Resistors")

The angular resolution is given by hysteretic effects which is shown in Fig. 17.28. Using the entire range for the determination of the angle, i.e. the regime between 0° and 360°, the resolution amounts to about 2°. If only the first quadrant is important (angles between 0° and 90°) the angular resolution is significantly better with about 0.1°.

An essential advantage of magnetoresistive sensors for the determination of rotation angles is the large distance between sensor and rotating permanent magnet. Thus, a non-contact readout is enabled. The operating range concerning the distance is given by the "magnetic" window:

$$H_{\min} < H_{\text{function}} < H_{\max} \tag{17.9}$$

If the external magnetic field is too large ($H > H_{\max}$) the synthetic antiferromagnet becomes influenced, i.e. the reference layer changes its properties. If the external field is too small ($H < H_{\min}$) the soft magnetic sensing layer is not saturated. Within this range a constant amplitude of the signal is given. Thus, a second window occurs for the working distance (see Fig. 17.29). A different but constant amplitude exists in the range of several mm. This allows a large air gap as well as large tolerances concerning lateral deviations. The sensitivity is only given for the rotation angle because the amplitude of the signal is nearly independent on the field strength of the external magnetic field.

Rotational Speed Sensor

Rotational speed measurement using magnetoresistive sensors is achieved by counting ferromagnetic marks such as teeth of a passive gear wheel or the number of magnetic elements of a magnetized ring. Beside magnetoresistive sensors also the inductive sensors and Hall Effect sensors can be used for this task. However, the magnetoresistive effect offers some essential advantages.

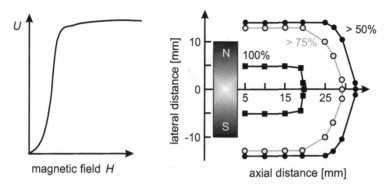

Fig. 17.29. Left: Magnitude of the GMR sensor signal as a function of the magnetic field strength. **Right**: Relative strength of the GMR sensor signal as a function of the relative position of magnet and sensor. (Adapted and used with permission from Siemens Application Note "Giant Magneto Resistors")

First, the output signal level of a MR sensor does not vary with rotation speed as it is the case in inductive sensor systems. Inductive sensors show a direct relation between the rotational speed and the output amplitude and therefore require sophisticated electronics to evaluate the large signal voltage range, especially in applications requiring low jitters. Magnetoresistive sensors, in contrast, are characterized by the fact that the sensor is static and the output signal is generated by the bending of magnetic field lines according to the position of the target wheel. This principle is shown in the lower part of Fig. 17.30. As bending of the magnetic field lines also occurs when the target is not moving magnetoresistive sensors can measure very slow rotations even down to 0 Hz. The necessary magnet is already attached to the sensor element so that the sensor modules are ready for use. Costs are further reduced as ferrite magnets can be used rather than the expensive samarium cobalt magnets required for Hall effect sensors. All these advantages recommend magnetoresistive sensor modules for rotational speed measurements in a wide range of both automotive and industrial applications

The magnetoresistive sensor cannot directly measure rotational speed but is sensitive to the motion of toothed wheels made from ferrous material (passive targets) or rotating wheels having alternating magnetic poles (active targets).

The general arrangement for a passive target wheel is shown in Fig. 17.31. The sensor is fitted with a permanent magnet. Without a ferromagnetic target or a symmetric position of the toothed wheel no component of the magnetic field would be in the sensitive direction and therefore the sensor output would be zero. For non-symmetric positions, for example if the passive target rotates in front of the sensor, the magnetic field is bent according to the actual wheel position and an alternating field component in the sensitive direction arises. This alternating field component is used to generate an output signal that

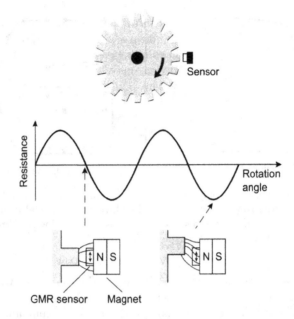

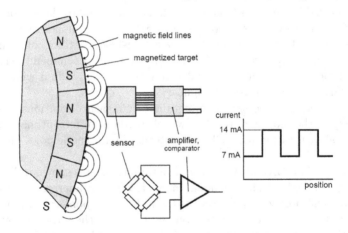

Fig. 17.30. Measurement of the rotation of an axis using a magnetoresistive sensor. The sensor is positioned between a soft magnetic gear wheel and a permanent magnet (top part) with its sensitivity direction in the tangential direction. A varying position of the teeth of the wheel gives rise to a varying tangential component of the field at the sensor position (lower part). The number of peaks per unit time of the approximately sinusoidal output signal is a measure of the rotation speed. (Reprinted from [88]. Copyright 2003, with permission from Elsevier)

Fig. 17.31. Rotational speed measurement using an active target wheel (From Application Note AN98087 from Philips Semiconductors; used with permission)

varies according to the wheel position. The amplitude of the sensor output voltage depends on the magnetic field strength of the biasing magnet, the distance between sensor and target and, obviously, on the structure of the target. Large solid targets will give stronger signals at larger distances from the sensor than small targets.

In contrast to passive targets that are not magnetized active targets show alternating magnetic poles as depicted in Fig. 17.31. Here the target provides the "working" field and no magnet is required for operation. However, a small stabilization magnet is still applied to the sensor. The structure of an active target can be expressed similarly to that for passive targets. In this case a north-south magnetic pole pair represents a tooth-valley pair.

Non-Contacting Rotary Selector

Another example is given by non-contacting rotary selectors. It represents the special case of an absolute determination of the rotation angle being supplemented by a fixed number of switching points. Typical applications are the use in control panels of washing machines, dish washers, tumble driers, electric kitchen stoves, and equipment of the consumer electronics. Different switch settings represent, for example, different radio channels, different washing programs, or adjustments of hot plates. The rotary switch can be used to detect the movement and position of objects located on the other side of a glass or plastic screen. Even a distance up to 3 cm away from the sensor presents no problems for the evaluation (see Fig. 17.32). The advantages are huge lifetimes due to neglecting wear, low expenses for assembling, and use

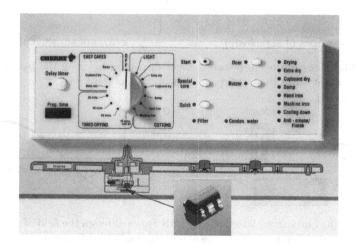

Fig. 17.32. Thanks to the large mounting tolerances, GMR sensors are well suited for insertion in the panels of household appliances (Used with permission from Application Note "Sensors" of Siemens/Infineon)

of the same kind of switch for different types of equipment which allows the realization of different functions by variation of the corresponding software.

Linear Sensor

Due to their high sensitivity GMR sensors can effectively provide positional information of actuating components in machinery, proximity detectors, and linear position transducers. Figure 17.33 illustrates two simple sensor/permanent magnet configurations used to measure linear displacement. In the left diagram displacement along the y-axis varies the field magnitude detected by the sensor that has its sensitive plane lying along the x-axis. The right diagram has the direction of displacement and the sensitive plane along the x-axis.

Using the sensor within the operating range the sensitivity is only given on the direction of the magnetic field and not on the amplitude. Thus, the signal amplitude varies for a linear motion of the sensor caused by the curved lines of magnetic flux (see Fig. 17.34).

Linear sensors can be used to determine a linear position, for measurements of lengths, and liquid level indicators. A subsequent comparator circuit for

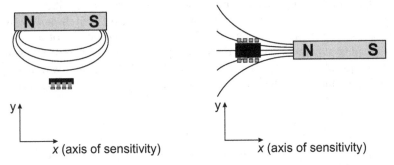

Fig. 17.33. Configuration to measure linear displacement along the y-axis (**left**) and x-axis (**right**) with the latter one being the axis of sensitivity

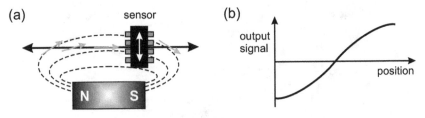

Fig. 17.34. (a) Linear translation of the GMR sensor through the field of a magnet. The white arrow indicates the sensitive axis of the sensor. (b) Corresponding output signal. Directly above the magnet the signal vanishes due to the perpendicular orientation between the magnetic flux lines and the direction of the sensor's sensitive axis

readout allows the realization of, e.g., proximity switches or, more generally, non-contact switches.

Position Sensor

Nowadays inkjet printers exhibit a high resolution. As a prerequisite the position of the print head must very accurately be determined. Such a position sensor can be realized using GMR-based sensors directly at the print head in combination with a guide bar which exhibits a continuous magnetization consisting of alternating magnet poles (see Fig. 17.35). This type of position sensor is characterized by its durability and reliability which is caused by the non-contact measurement principle and insensitiveness against contaminations from ink.

Instead of using one sensor combined with a lot of magnets one can also realize the opposite situation (see Fig. 17.36). Now, the varying magnetic field of only one magnet is measured with several magnetoresistive sensors. A subsequent read unit enables the accurate determination of the position of, e.g., a valve actuator stem.

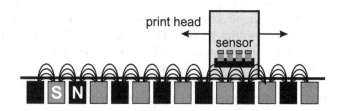

Fig. 17.35. Positioning of a print head

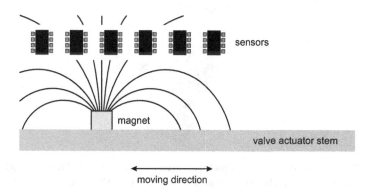

Fig. 17.36. Magnetoresistive sensors monitoring the axial position of a valve actuator stem

Sensors in String Instruments

Conventional pickup elements for guitars consist of a permanent magnet and a coil of isolated wire. The amount of flux through the coil is modulated by the vibration of the nearby soft-magnetic string and therefore induces an alternating current in the coil. Usually, one magnet or pole piece is used per string while the coil surrounds all six magnets together. The sound of the guitar depends in a subtle way on many parameters like the magnet material, the number of windings in the coil, etc.

A new type of pickup element uses magnetoresistance instead of inductance as the sensing principle. The sensors were mounted on small ferrite bias magnets and positioned close to the guitar strings. The vibration of the string modulates the magnetic field at the location of the MR sensor and thus modulates its resistance resulting in an oscillating electronic output signal. Of course, this signal may subsequently be amplified and/or processed electronically.

The inductive pickup of the electric guitar was replaced by six of these packaged GMR sensors which were glued on a ferrite bias (see Fig. 17.37). In order to warrant a symmetric and identical magnetic environment for each of the strings, two additional bias magnets without sensors were added at the sides. A small printed circuit board with some electronics was included in order to sum the sensor output signals. The required small amount of power for the electronics was taken from the amplifier but could also be supplied by a small battery in the guitar as is also quite common, e.g., for piezo guitar pickups. The output can directly be connected to the standard guitar amplifier. One of the biggest advantages of the magnetoresistive pickup elements over the conventional inductive elements is the dramatically lower sensitivity to disturbing external electromagnetic fields.

Fig. 17.37. Photograph of the magnetoresistive pickup elements (Reused with permission from [91]. Copyright 2002, American Institute of Physics)

Vehicle Detection

Vehicle detection technology has evolved quite a bit in the last couple decades. From air hoses to inductive loops embedded in roadways most legacy detection methods were concentrated on getting vehicle presence information to a decision making set of control systems. Today, much more information is required such as speed and direction of traffic, the quantity of vehicles per time on a stretch of pavement, or just very reliable presence or absence of a class of vehicles which will be discussed in the following.

Due to the fact that almost all road vehicles have significant amounts of ferrous metals in their chassis magnetic sensors are a good candidate for detecting vehicles. Nowadays, they are fairly miniature in size and thanks to solid state technology both the size and the electrical interfacing have improved to make integration easier.

But not all vehicles emit magnetic fields that magnetic sensors could use in detection. This fact eliminates most "high field" magnetic field sensing devices like Hall effect sensors. Thus, "low field" magnetic sensors are used to determine this field and also the field disturbances that nearby vehicles create. The upper part of Fig. 17.38 shows a graphical example of the lines of flux from the earth between the magnetic poles and the bending they receive as they penetrate a typical vehicle with ferrous metals.

As the lines of magnetic flux group together (concentrate) or spread out (deconcentrate) a magnetic sensor placed nearby will be under the same magnetic influence the vehicle creates to the earth's field. However because the sensor is not intimate to the surface or interior of the vehicle it does not get the same fidelity of concentration or deconcentration. With increasing

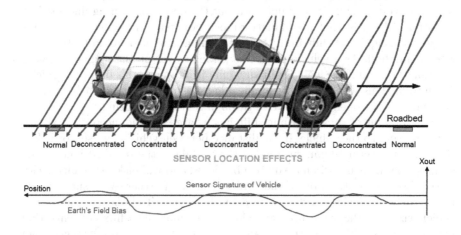

Fig. 17.38. Earth's magnetic field through a vehicle and corresponding signature of the sensor being located within the roadbed (Used with permission from Honeywell Application Note AN218)

distance from the vehicle the amount of flux density changes with vehicle presence drops off at an exponential rate.

Typical vehicle detection applications using magnetic sensors and the earth's field are:

- Railroad crossing control (for trains)
- Drive through retail (banking, fast-food, etc.)
- Automatic door/gate opening
- Traffic monitoring (speed, direction)
- Parking lot space detection
- Parking meters

Vehicle detection signature

Using the earth's magnetic field provides a magnetic background or "bias" point that substantially stays constant with a fixed sensor installation.

A typical magnetic field direction on the northern hemisphere with a truck moving southbound is shown in Fig. 17.38. The boxes represent possible sensor locations near the roadbed and the relative amounts of flux concentration they could sense. The adjoining graph shows what a signal axis sensor bridge might see when it has the sensitive axis also pointing southbound and as the truck drives past the sensor. Since the natural earth's magnetic field would bias the sensors with a slight negative voltage output increasing flux concentration would further lower the voltage and decreasing concentration would raise the voltage.

Sensors oriented sideways (horizontal, across the roadbed) and vertical would likely also shift during the vehicle passage but the bias values and signature shifts would be different. For most applications the amplitude and direction of voltage shift is not important but the detection of a significant shift in output voltage is what would matter most. For vehicle presence applications the vector magnitude shift from the earth's magnetic field would be the most reliable method. As a vehicle parks alongside or overtop the magnetic sensor location the magnitude would shift suddenly from the earth's bias (no-vehicle) magnitude. This would be most applicable for parking meters, parking space occupancy, door openers, and drive through service prompting.

The amount of sensor output shift has a large dependance on proximity of the sensors to the vehicle. If the distance between vehicle and sensor is less than about several 10 cm, such as on the middle of the roadbed lane at the surface, the signature will have quite a bit detail and will pickup the intricacies of the ferrous construction of the vehicle chassis. Further away such as one meter the vehicle signature may be a tenth in magnitude depending on the vehicle size and the signature bandwidth looks more as a flux concentration hump than a squiggle. As the distance increases the signature changes from flux concentration to deconcentration to returning to the baseline.

Vehicle direction sensing

In the previous vehicle signature example a single sensor was shown as the vehicle overpasses it on the roadbed. This is traditionally known as an x-axis system since it uses the sensitive axis in the expected vehicle direction. By employing a sensor off the side of the vehicle path the basic vehicle direction can be detected. Figure 17.39 presents a typical sensor placement for vehicle direction detection. As the vehicle approaches the sensor lines of magnetic flux begin to bend at the sensor toward the vehicle. Thus, the flux density decreases and the signature voltage from the sensor goes negative from its bias value. As the vehicle leaves the sensor the flux density chases the vehicle and a positive signature voltage results. If the vehicle backs up or returns in the opposite direction the signature plot looks like a mirror image. A second and more reliable method involves two sensors displaced by a small distance apart but with their sensitive axis' in the same direction. The intent is that a vehicle in motion will create the same signature but displaced in time.

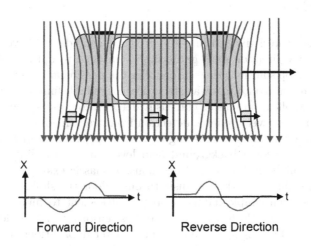

Forward Direction   Reverse Direction

Fig. 17.39. x-axis direction sensing (Used with permission from Honeywell Application Note AN218)

With the dual displaced sensors the rear sensor will hit the detection threshold a fraction of a second before the front sensor in forward traffic. And the opposite occurs in reverse direction traffic. With a known displacement distance and a reasonably precise time measurement between threshold detections a speed computation can be made as well. Figure 17.40 depicts this typical mounting.

The displacement distance of the sensors does not have to be a very large value. With today's high speed microcontrollers and precision analog circuitry speed measurement accuracy and resolution can be within single km per hour graduations.

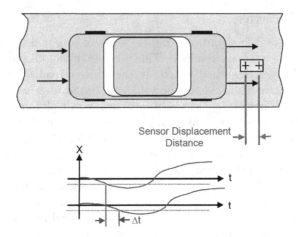

Fig. 17.40. Dual displaced sensors (Used with permission from Honeywell Application Note AN218)

Thus, magnetoresistive sensors are very good candidates for vehicle presence, speed, and direction data gathering. Compared to optical, ultrasonic, and inductive loop solutions the sensors offer the possibility of very small size installations into and onto roadway structures for reliable function. In the railroad crossing application sensors can easily detect locomotives from 20 meters away which means a very reliable detection from track roadbed positions.

Automatic door and gate keeping can be made simple and low cost using sensors by recording background field levels and controlling for vehicle waiting and vehicle parked logical decisions. A classic example is overhead door lifting for factory forklift transit through door thresholds. Vehicle detection ranges from simple inductive loop replacement to multi-lane speed and direction sensing. Both wired and wireless sensor mounts would apply in this application. Parking space and parking meter applications have similar constraints for low-cost and reliable presence/absence detection of stationary vehicles. Figure 17.41 shows a typical parking meter arrangement with three possible sensor locations.

Current Sensing

Currents in wires create magnetic fields surrounding the wires or traces on printed wiring boards. The field decreases as the reciprocal of the distance from the wire. GMR sensors can be effectively employed to sense this magnetic field. Both dc and ac currents can be detected in this manner. Bipolar ac current will be rectified by the sensors omnipolar sensitivity unless a method is used to bias the sensor away from zero. Unipolar and pulsed currents can be measured with good reproduction of fast rise time components due to the excellent high frequency response of the sensors. Since the films which the

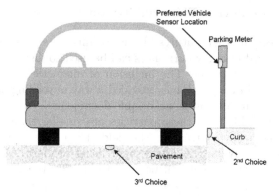

Fig. 17.41. Parking meter sensing (Used with permission from Honeywell Application Note AN218)

sensors consists of are extremely thin a response to frequencies up to 100 MHz is possible. Figure 17.42 shows the relative position of a GMR sensor and a current carrying wire to detect the current in the wire. The sensor can also be mounted immediately over a current carrying trace on a circuit board. High currents may require more separation between the sensor and the wire to keep the field within the sensor's range. Low currents may be best detected with the current being carried by a trace on the chip immediately over the GMR resistors. This application allows for current measurement without breaking or interfering with the circuit of interest.

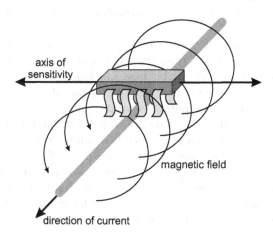

Fig. 17.42. Appropriate orientation of a GMR bridge sensor to detect the magnetic field created by a current carrying wire

Detection of Counterfeit Financial Transaction Documents

In addition to several other features like holograms the signature of magnetic information encoded in many currencies can be used to distinguish valid currency from counterfeit copies. The inclusion of magnetic particles like iron oxide as a pigment in black ink has provided a way of reading and validating not only currency but also other negotiable documents.

The magnetic fields from these particles are smaller than the terrestrial magnetic field being in contrast to the fields from stripes on credit cards which are considerably larger. Their small fields produce signatures when read by magnetic sensors that can be used to identify the denominations of currency presented to vending machines. The signature of additional magnetic information can be used to distinguish between valid currency and copies.

Unfortunately, the "magnetic" reading of currency represents a difficult task because the amount of magnetic ink is considerably small. Inductive read heads designed similarly to tape recorder heads need to be in direct contact to yield an adequate signal. But to avoid jamming in high-speed transport mechanisms it is desirable to be able to read the bill from up to 2 mm away. To achieve this goal sensitive low-field sensors such as GMR sensors are utilized.

Magnetic biasing is often important in low-field sensing procedures. In contrast to permanent magnets most magnetic materials will not have a reliable, readable signature unless a permanent magnet or the magnetic field from a current in a coil has magnetized them. This limitation is especially true for small particles of magnetic materials. The art of biasing is to be able to magnetize the magnetic particles in the ink and also any magnetic stripes on the currency to be detected so that it produces a magnetic field along the sensitive axis of the sensor while not saturating the sensor with the biasing field.

The simplest method of biasing is to pass the object to be magnetized over a permanent magnet and then transport it to the vicinity of the sensor (see Fig. 17.43). The articles to be read are being moved by a transport mechanism and passed over the magnetic sensor one at a time. A permanent magnet can be placed at some point upstream remote enough not to saturate the magnetic sensor. This set-up enables to prepare the bills or checks (see below) with their particles in a reproducible magnetic state.

The small size of GMR sensors offers the possibility of making closely spaced arrays of sensors to image a larger area rather than just obtaining a signature from a single trace along or across the material. Magnetic sensor arrays are used to achieve a magnetic image which can then be used to obtain additional information encoded in the document or object.

A second method of magnetic biasing is to make use of the fact that thin film GMR sensors are relatively unsusceptible to fields being perpendicular to their sensitive axis. A permanent magnet can be placed in close proximity to the GMR sensor with its magnetic axis perpendicular to the sensor's sensitive axis. With proper positioning the sensor will detect only a little or no field

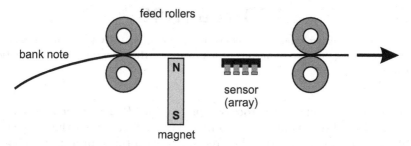

Fig. 17.43. Possible arrangement to verify a particular magnetic field signature for currency detection. In this situation the magnetic particles in the ink or toner are magnetized by a permanent magnet which is separated from the sensor or sensor array. The bank note itself is transported by feed rollers

(see Fig. 17.44(a)). When a magnetizable object approaches the end of the sensor there will be a component of the magnetic field along the sensitive axis as shown in Fig. 17.44(b). This method of biasing is often called back biasing because the magnet is usually attached to the back of the sensor.

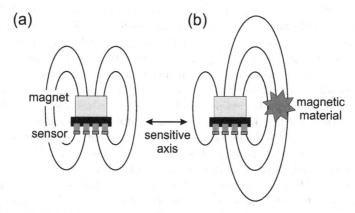

Fig. 17.44. Back biasing procedure. A magnetizable object in vicinity to the sensor induces a component of the magnetic field along the sensitive axis of the sensor (see *right part*) which is absent without this object (see *left part*)

Another financial application which small magnetic fields are detected in is the reading of magnetic ink character recognition (MICR) numbers which may be printed in lines on the bottom of a check. The stylized MICR numbers (see Fig. 17.45) produce a unique magnetic signature when documents which they appear in (e.g., checks) are sorted at high speeds. The MICR line is printed with magnetic ink or toner which results in a magnetic signal that allows to identify each unique character.

ᶦ ᴄ ᴣ ᴧ ᴌ ᴄ ᴎ ᴒ ᴗ ᴐ

Fig. 17.45. Typical magnetic ink character recognition (MICR) numbers which are printed on, e.g., checks by means of ink or toner which contain small magnetic particles

Financial documents are processed on special MICR reader and sorter machines. They first magnetize the MICR line and then read the magnetic signals. Each character, if printed correctly and with the appropriate amount of magnetic material in the ink or toner, will give a magnetic signal unique and identifiable to that character. The magnetic signal is developed from two elements. These are the character's shape, i.e. the horizontal and vertical attributes, and the magnetic content, i.e. the amount and distribution of magnetic material in the ink or toner which the character is formed from. If the shape and magnetics of the characters do not meet specified standards the machine will reject the document.

Detection of Material Defects

The sensing of eddy currents is an effective and non-destructive way of detecting cracks in conductive materials. It is also possible to detect hidden corrosion which typically produces a gradual thinning or roughening of structures. Because safety-critical systems depend on early detection of fatigue cracks to avoid major failures there is an increasing need for probes that can reliably detect very small defects. Additionally, there are increasing demands for probes which can detect deeply buried defects to avoid disassembling structures.

Eddy current testing probes combine an excitation coil that induces eddy currents in a specimen and a detection element that identifies the perturbation of the currents caused by cracks or other defects. The main components consist of either a relatively large cylindrical coil (see Fig. 17.46(a)) or a flat spiral coil with the GMR sensor located on the axis of the coil (see Fig. 17.46(b)). The use of low-field magnetic sensors represents a significant advance over more traditional inductive probes. Two key attributes open opportunities for increased use of eddy current probes: constant sensitivity over a wide range of frequencies and development of smaller sensors.

Eddy currents induced in the surface of a defect-free specimen are circular because of the circular symmetry of the field produced by the coil. The tangential component of the field created by the eddy currents is zero at the location of the sensor (see Fig. 17.47(a)). In the presence of defects the eddy currents are no longer symmetrical, and the probe provides a measure of the perturbed eddy currents caused by the underlying flaws (see Fig. 17.47(b)). The size of the coil is related to the resolution necessary to detect the defects. For large defects and for deep defects large coils surrounding the sensor are required.

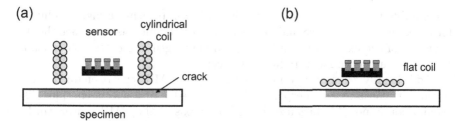

Fig. 17.46. Arrangement of coil and GMR sensor for eddy current detection of defects in conductors. (**a**) Cross section of the assembly utilizing a cylindrical coil surrounding the sensor. (**b**) Schematic diagram of a flat coil placed on the sensor package

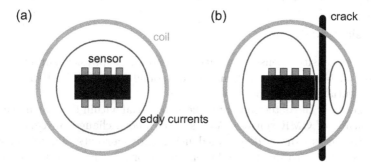

Fig. 17.47. (**a**) Magnetic fields from the excitation coils create circular eddy currents in the conducting surface below the coils. (**b**) A crack or defect alters the eddy current paths resulting in magnetic fields parallel to the conducting surface and along the sensitive axis of the sensor

Eddy currents shield the interior of the conducting material with the skin depth d:

$$d = \sqrt{\frac{\rho}{\pi \mu_0 \mu_r \nu}} \qquad (17.10)$$

with ν being the excitation frequency and ρ the resistivity of the sample. Exemplarily, we see that the frequency must be below 44 Hz for Cu in order to reach a depth of 1 cm for the eddy current. By changing the frequency we can probe different depths of the material. GMR sensors with their wide frequency response from dc into the multi-megahertz range are well suited to this application. The small size of the sensing element increases the resolution of defect location while the detector is scanned over the surface. More rapid scans can be performed using an array of detectors.

Probes that detect eddy current fields using inductive coils have less sensitivity at low frequencies. Unfortunately, this is where the device would have to operate to detect deep flaws. Small sensing coils, which are required to detect small defects, also have low sensitivity. In contrast, small GMR sensors with their high sensitivity can locally measure a magnetic field over an area comparable to the size of the sensor itself. Thus, to achieve high resolution for detecting small surface and near-surface defects one has to reduce the dimensions of the excitation coil. The minimum length of a detectable crack is roughly equal to the mean radius of the coil.

Additionally, the unidirectional sensitivity of GMR sensors enable the detection of cracks at and perpendicular to the edge of a specimen. This discrimination is possible because the sensitive axis of the GMR sensor can be rotated to be parallel to the edge. Consequently, the signal is due only to the crack. With inductive probes the edge will produce a large signal that can mask the signal produced by a crack. This capability represents a very simple solution to a difficult problem encountered in the aircraft industry namely detecting cracks that initiate at the edge of turbine disks or near the rivets.

Geophysical Surveying

Magnetically sensitive sensors can also be used for geophysical surveying. Exemplarily, airborne surveys of magnetic anomalies are utilized in order to locate potential magnetic ore bodies.

But, not only the detection of magnetic material within the soil are carried out by means of GMR sensors. As discussed above changes in conductivity of buried objects can be determined using eddy currents. Therefore, magnetic sensors allow to locate conducting pipes or even non-conducting pipes containing water.

Eddy current detection is a non-contacting method that does not require placing electrodes in order to pass currents through the ground and measuring potentials.

On the one hand the magnetic fields of interest are considerably less than the earth's magnetic field. Ground based magnetic surveys require portable equipment on the other hand. Thus, highly sensitive GMR sensors are ideal for equipment packed into remote survey areas. Additionally, arrays of sensors make simultaneous two-dimensional imaging possible.

Biosensors

Magnetic particles have been used for many years in biological assays. The magnetic component of these particles is invariably iron oxide usually in the form of maghemite (γ–Fe_2O_3). Iron oxide is the favored magnetic component because of its stability and biocompatibility. These particles range in size from a few nanometers up to a few microns and are encapsulated in plastic or ceramic spheres. Nanometer-sized particles of iron oxide are superparamagnetic and therefore only magnetic in the presence of a magnetic field; the particles immediately demagnetize when the field is removed. Thus, the beads do not attract each other and do not agglomerate.

The magnetic beads are subsequently coated with a chemical or biological species such as DNA or antibodies that selectively binds to the target analyte.

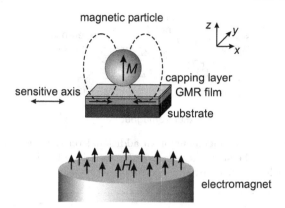

Fig. 17.48. Cross-section of a GMR sensor illustrating the method being able to detect superparamagnetic beads. A magnetizing field H magnetizes the bead which produces regions of positive and negative magnetic induction B in the plane of the underlying GMR film. Because the film is only sensitive to the x component of external magnetic fields the magnetizing field does not affect the GMR resistance. Due to a capping layer the sensor can be used in liquids

Primarily, these types of particles are applied to separate and concentrate analytes for off-line detection.

As an improvement, the distinct selectivity of sample and target can be used as a rapid sensitive detection strategy with the on-line integration of a magnetic detector. This integration is facilitated by the development of GMR sensors as the magnetic detectors.

These sensors have the unique advantage of being compatible with silicon integrated circuit fabrication technology resulting in a single detector or even multiple detectors that can be made on a single chip along with any of the required electrical circuitry.

The beads are magnetized by an electromagnet (see Fig. 17.48). If the magnetoresistive sensors lie in the xy plane and current flows through them in the x direction the sensors detect only the x component of the magnetic field. Therefore, to detect a superparamagnetic bead resting on a GMR strip a magnetic field is externally generated in the z direction causing the bead to produce a magnetic field with a detectable x component. An additional capping layer, often consisting of silicon nitride, on top of the GMR thin film device allows to operate under liquid conditions, e.g. in salt solutions.

However, since the magnetizing field is often not homogeneous, and therefore not exactly perpendicular to the GMR sensors over the entire sensor array, the sensors exhibit an offset that varies with their position in the field. To obtain an enhanced signal-to-noise ratio, a "reference" GMR sensor identical to the "main signal" sensor is additionally used nearby (see Fig. 17.49). With this detection system the presence of as few as one microbead can be detected.

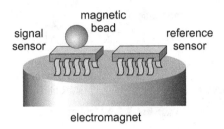

Fig. 17.49. Schematic arrangement of an additional reference GMR sensor being located in vicinity to the signal sensor in order to reduce the influence of inhomogeneities of the field produced by the electromagnet

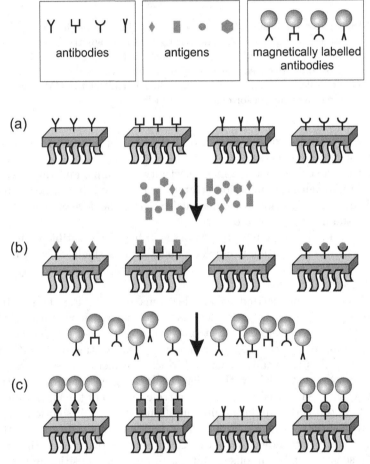

Fig. 17.50. (a) Antigens are detected by flowing them over a sensor coated with antibodies which they bind to. (b) Antigens which do not find their corresponding antibody cannot be stabilized in the solution on top of a sensor (here represented by the hexagon). (c) Subsequently, the magnetic particle-labelled antibodies bind to the antigens providing a magnetic indication of the presence of those antigens. In this example three types of antigens are detected whereas one is absent

The magnetic beads coated with a material that binds to the biological molecules to be analyzed are allowed to settle on a substrate that is selectively coated in different areas with substances that bond to specific molecules of interest. Non-binding beads can be removed by a small magnetic field. The presence of the remaining magnetic microbeads is detected by magnetic sensors within the array. Several bioassays can be simultaneously accomplished using an array of magnetic sensors each with a substance that bonds to a different biological molecule. Figure 17.50 schematically shows the bonding of the beads to the sites via the molecules to be detected.

The chemical sensitivity increases with active area, i.e. the area which binding events can occur in and be detected in. Since magnetic beads can only be detected when they bind on top of a sensor the active area is the total surface area of all sensors devoted to a particular analyte. Thus, the optimum array would consist of a large number of small sensors.

Solutions

Problems of Chapter 2

2.1 Using (2.132) we obtain for the susceptibility:

$$\chi = \frac{N\mu_0}{3VkT} \cdot g_J^2 \mu_B^2 J(J+1) \tag{18.1}$$

$$= N \frac{4\pi \times 10^{-7}}{3 \times 1 \times 1.381 \times 10^{-23} \times 300} \cdot 4$$
$$\times (9.274 \times 10^{-24})^2 \times 1 \times 2 \tag{18.2}$$

$$= N \times 6.96 \times 10^{-32} \tag{18.3}$$

The ideal gas is characterized by:

$$pV = NkT \tag{18.4}$$

which allows to determine the number of atoms or molecules to be:

$$N = \frac{pV}{kT} \tag{18.5}$$

$$= \frac{101325 \times 1}{1.381 \times 10^{-23} \times 300} \tag{18.6}$$

$$= 2.45 \times 10^{25} \tag{18.7}$$

for $T = 300$ K, $V = 1$ m^3, and $p = 1013$ mbar $= 101325$ Pa. Using (18.3) we obtain:

$$\chi = 1.7 \times 10^{-6} \tag{18.8}$$

which is positive and rather small.

2.2 From Fig. 2.5 we obtain:

$$S = \frac{1}{2} \qquad L = 3 \qquad J = \frac{5}{2}$$

Using these values we are able to calculate g_J using (2.160):

$$g_J = \frac{3}{2} + \frac{S(S+1) - L(L+1)}{2J(J+1)}$$

$$= \frac{3}{2} + \frac{1/2 \cdot 3/2 - 3 \cdot 4}{2 \cdot 5/2 \cdot 7/2}$$

$$= \frac{3}{2} + \frac{3/4 - 12}{35/2}$$

$$= 0.86$$

2.3 Term Symbols

(a) d electrons have $\ell = 2$. The number of electrons in the d shell amounts to $p = 6$. Thus, we obtain

$$2\ell + 1 = 5$$
$$|2\ell + 1 - p| = 1$$
$$|2\ell - p| = 2$$

In order to satisfy Hund's first rule we get:

$$S = 1/2 \cdot ((2\ell + 1) - |2\ell + 1 - p|) = 1/2 \cdot (5 - 1) = 2$$

Concerning Hund's second rule we obtain:

$$L = S \cdot |2\ell + 1 - p| = 2 \cdot 1 = 2$$

which results in D for the symbol. Hund's third rule reveals:

$$J = S \cdot |2\ell - p| = 2 \cdot 2 = 4$$

Therefore, the term symbol $^{2S+1}L_J$ is given by $^5D_4$.

(b) f electrons possess $\ell = 3$. The number of electrons in the f shell is $p = 7$. This allows to determine the following relations:

$$2\ell + 1 = 7$$
$$|2\ell + 1 - p| = 0$$
$$|2\ell - p| = 1$$

For Hund's first rule we obtain:

$$S = 1/2 \cdot ((2\ell + 1) - |2\ell + 1 - p|) = 1/2 \cdot (7 - 0) = 7/2$$

For Hund's second rule we get:

$$L = S \cdot |2\ell + 1 - p| = 7/2 \cdot 0 = 0$$

which results in S for the symbol. Using Hund's third rule reveals:

$$J = S \cdot |2\ell - p| = 7/2 \cdot 1 = 7/2$$

Therefore, the term symbol $^{2S+1}L_J$ is given by $^8S_{7/2}$.

Problems of Chapter 3

3.1 Ferromagnetic Alloys

(a) The Slater–Pauling curve (see Fig. 3.12) yields an electron number of 27.6 for the $Ni_x Co_{1-x}$ alloy. The number of electrons for Ni amounts to 28, that for Co to 27. Thus:

$$28x + 27 \cdot (1 - x) = 27.6 \qquad (18.9)$$

which leads to $x = 0.6$. This alloy therefore exhibits the composition $Ni_{60}Co_{40}$.

(b) For $Co_x Cr_{1-x}$ we find an electron number of 26.7. The number of electrons for Co amounts to 27, that for Cr to 24. Therefore:

$$27x + 24 \cdot (1 - x) = 26.7 \qquad (18.10)$$

which results in $x = 0.9$ and thus $Co_{90}Cr_{10}$.

(c) For $Fe_x Cr_{1-x}$ we observe an electron number of 25.1. The number of electrons for Fe amounts to 26, that for Cr to 24. Therefore:

$$26x + 24 \cdot (1 - x) = 25.1 \qquad (18.11)$$

which results in $x = 0.55$. This alloy consists of $Fe_{55}Cr_{45}$.

Problems of Chapter 5

5.1 The saturation magnetization of Ni is therefore given by $M_S = 0.6 n \mu_B$. This value is inserted into (2.96) which results in:

$$g_J = \frac{M_S}{n \mu_B J} = \frac{0.6 n \mu_B}{n \mu_B J} = \frac{0.6}{J} \qquad (18.12)$$

Now, this result is inserted into (5.16):

$$B_{\mathrm{mf}} = \frac{3kT_C}{g_J \mu_B (J + 1)} = \frac{3kJT_C}{0.6 \mu_B (J + 1)} \qquad (18.13)$$

Using the given experimental values leads to $B_{\mathrm{mf}} = 1576$ T.

5.2 Using (5.61):

$$\chi(T) = \frac{c}{T + T_N}$$

we are able to determine the susceptibility at T_N:

$$\chi(T_N) = \frac{c}{T_N + T_N} = \frac{c}{2T_N} = \chi_0 \qquad (18.14)$$

Thus, the constant c amounts to:

$$c = 2\chi_0 T_N \qquad (18.15)$$

This allows us to calculate the susceptibility to be:

$$\chi(T) = \frac{2\chi_0 T_N}{T + T_N} \tag{18.16}$$

Therefore:

$$\chi(T_3) = \chi(2T_N) = \frac{2\chi_0 T_N}{3T_N} = \frac{2}{3}\chi_0 \tag{18.17}$$

The susceptibility along the perpendicular direction at or below the Néel temperature is independent on temperature and given by (see (5.101)):

$$\chi_\perp(T) = \chi_\perp(T_N) = \chi(T_N) = \chi_0 \tag{18.18}$$

Thus:

$$\chi_\perp(T_1) = \chi_\perp(0) = \chi_0 \qquad \chi_\perp(T_2) = \chi_\perp(T_N/2) = \chi_0$$

5.3 Using the parameters:

$$a = v(\alpha c_1 + \beta c_2) \tag{18.19}$$
$$b = c_1 c_2 v^2(\alpha\beta - 1) \tag{18.20}$$
$$x = c_1 + c_2 \tag{18.21}$$
$$y = c_1 c_2 v(2 + \alpha + \beta) \tag{18.22}$$

we can rewrite (5.121) to:

$$\mu_0 \frac{1}{\chi}$$

$$= \frac{T^2 - aT + b}{xT - y} \tag{18.23}$$

$$= \frac{x^2 T^2 - ax^2 T + bx^2}{x^2(xT - y)} \tag{18.24}$$

$$= \frac{x^2 T^2 - xyT + xyT - ax^2 T - y^2 + axy + bx^2 - axy + y^2}{x^2(xT - y)} \tag{18.25}$$

$$= \frac{xT(xT - y) + (y - ax)(xT - y) + (bx^2 - axy + y^2)}{x^2(xT - y)} \tag{18.26}$$

$$= \frac{T}{x} + \frac{y - ax}{x^2} - \frac{axy - bx^2 - y^2}{x^3\left(T - \dfrac{y}{x}\right)} \tag{18.27}$$

$$= \frac{T}{x} + \frac{1}{\dfrac{x^2}{y - ax}} - \frac{\dfrac{axy - bx^2 - y^2}{x^3}}{T - \dfrac{y}{x}} \tag{18.28}$$

We see that this equation can be written as:

$$\mu_0 \frac{1}{\chi} = \frac{T}{c_1 + c_2} + \frac{1}{\chi_0} - \frac{\sigma}{T - \theta} \tag{18.29}$$

with the parameters θ, χ_0, and σ given by:

$$\theta = \frac{y}{x} \tag{18.30}$$

$$\chi_0 = \frac{x^2}{y - ax} \tag{18.31}$$

$$\sigma = \frac{axy - bx^2 - y^2}{x^3} \tag{18.32}$$

Inserting the parameters a, b, x, and y being defined above we finally obtain:

$$\theta = \frac{c_1 c_2 v (2 + \alpha + \beta)}{c_1 + c_2} \tag{18.33}$$

$$\chi_0 = -\frac{(c_1 + c_2)^2}{v(\alpha c_1^2 + \beta c_2^2 - 2c_1 c_2)} \tag{18.34}$$

$$\sigma = \frac{c_1 c_2 v^2 (c_1(\alpha + 1) - c_2(\beta + 1))^2}{(c_1 + c_2)^3} \tag{18.35}$$

5.4 Helimagnetism

(a) The prerequisites for helimagnetism (see (5.141) and (5.134)) are: $J_2 < 0$ and $|J_1| \leq 4|J_2|$. The first condition is fulfilled. The second condition can be written as:

$$\left| \frac{J_1}{J_2} \right| \leq 4 \tag{18.36}$$

Inserting results in:

$$\left| \frac{J_1}{J_2} \right| = \frac{6}{\sqrt{3}} = 3.46 \leq 4 \tag{18.37}$$

Thus, this material exhibits helical arrangement.

(b) The angle θ between adjacent layers can be calculated (see (5.133)) by:

$$\cos \theta = -\frac{J_1}{4 J_2} = \frac{-6 J_1}{-4\sqrt{3} J_1} = 0.866 \tag{18.38}$$

which leads to $\theta = 30°$.

Problems of Chapter 7

7.1 Replacing the direction cosine α_i by the angles θ and ϕ we obtain:

$$E_{\text{crys}}^{\text{tetra}} = K_0 + K_1 \alpha_3^2 + K_2 \alpha_3^4 + K_3 \left(\alpha_1^4 + \alpha_2^4 \right) \tag{18.39}$$

$$= K_0 + K_1 \cos^2 \theta + K_2 \cos^4 \theta + K_3 \sin^4 \theta (\sin^4 \phi + \cos^4 \phi) \tag{18.40}$$

Due to:

$$\sin^2 \theta + \cos^2 \theta = 1 \tag{18.41}$$

we obtain:

$$\cos^2 \theta = 1 - \sin^2 \theta \tag{18.42}$$

which results in

$$\cos^4 \theta = 1 + \sin^4 \theta - 2 \sin^2 \theta \tag{18.43}$$

Using (18.41) we get:

$$1 = (\sin^2 \phi + \cos^2 \phi)^2 \tag{18.44}$$
$$= \sin^4 \phi + \cos^4 \phi + 2 \sin^2 \phi \cos^2 \phi \tag{18.45}$$

Thus:

$$\sin^4 \phi + \cos^4 \phi = 1 - 2 \sin^2 \phi \cos^2 \phi \tag{18.46}$$

Using:

$$2 \sin \phi \cos \phi = \sin 2\phi \tag{18.47}$$

Equation (18.46) results in:

$$\sin^4 \phi + \cos^4 \phi = 1 - \frac{1}{2} \sin^2 2\phi \tag{18.48}$$

Due to:

$$\cos 2x = \cos^2 x - \sin^2 x \tag{18.49}$$

we obtain:

$$\sin^2 x = \cos^2 x - \cos 2x \tag{18.50}$$
$$= 1 - \sin^2 x - \cos 2x \tag{18.51}$$

Thus:

$$\sin^2 x = \frac{1}{2} - \frac{1}{2} \cos 2x \tag{18.52}$$

With $x = 2\phi$ we obtain:

$$\sin^2 2\phi = \frac{1}{2} - \frac{1}{2} \cos 4\phi \tag{18.53}$$

which leads to:

$$\sin^4 \phi + \cos^4 \phi = \frac{3}{4} + \frac{1}{4} \cos 4\phi \tag{18.54}$$

using (18.46). Inserting of (18.42), (18.43), and (18.54) into (18.40) results in:

$$E_{\text{crys}}^{\text{tetra}} = K_0 + K_1(1 - \sin^2 \theta) + K_2(1 + \sin^4 \theta - 2 \sin^2 \theta)$$
$$+ K_3 \sin^4 \theta \left(\frac{3}{4} + \frac{1}{4} \cos 4\phi \right) \tag{18.55}$$
$$= K_0 + K_1 + K_2 - K_1 \sin^2 \theta - 2K_2 \sin^2 \theta + K_2 \sin^4 \theta$$
$$+ \frac{3}{4} K_3 \sin^4 \theta + \frac{1}{4} K_3 \sin^4 \theta \cos 4\phi \tag{18.56}$$
$$= K_0' + K_1' \sin^2 \theta + K_2' \sin^4 \theta + K_3' \sin^4 \theta \cos 4\phi \tag{18.57}$$

with

$$K_0' = K_0 + K_1 + K_2 \tag{18.58}$$

$$K_1' = -K_1 - 2K_2 \tag{18.59}$$

$$K_2' = K_2 + \frac{3}{4}K_3 \tag{18.60}$$

$$K_3' = \frac{1}{4}K_3 \tag{18.61}$$

7.2 Due to:

$$\sin^2 x + \cos^2 x = 1 \tag{18.62}$$

we obtain:

$$\sin^2 x = 1 - \cos^2 x \tag{18.63}$$

which results in

$$\sin^4 x = 1 + \cos^4 x - 2\cos^2 x \tag{18.64}$$

Both equations lead to:

$$\sin^6 x = 1 - 3\cos^2 x + 3\cos^4 x - \cos^6 x \tag{18.65}$$

Now we use:

$$0 = \cos^6 x + \cos^4 x - \cos^6 x - \cos^2 x - \cos^6 x + 2\cos^4 x$$
$$-1 + 3\cos^2 x - 3\cos^4 x + \cos^6 x + 1 - 2\cos^2 x \tag{18.66}$$
$$= \cos^6 x + (1 - \cos^2 x)\cos^4 x - (1 + \cos^4 x - 2\cos^2 x)\cos^2 x$$
$$-(1 - 3\cos^2 x + 3\cos^4 x - \cos^6 x) + (1 - \cos^2 x) - \cos^2 x \tag{18.67}$$

Replacing the expressions in brackets by (18.63)–(18.65) results in:

$$0 = \cos^6 x + \sin^2 x \cos^4 x - \sin^4 x \cos^2 x - \sin^6 x + \sin^2 x - \cos^2 x \tag{18.68}$$

Next, we will transform the following expression Y:

$$Y = \cos^6 x - 15\sin^2 x \cos^4 x + 15\sin^4 x \cos^2 x - \sin^6 x \tag{18.69}$$

By adding three times (18.68) we obtain:

$$Y = 4\cos^6 x - 12\sin^2 x \cos^4 x + 12\sin^4 x \cos^2 x$$
$$-4\sin^6 x + 3\sin^2 x - 3\cos^2 x \tag{18.70}$$
$$= 4\sin^4 x \cos^2 x + 4\cos^6 x - 8\sin^2 x \cos^4 x - 3\cos^2 x$$
$$-4\sin^6 x - 4\sin^2 x \cos^4 x + 8\sin^4 x \cos^2 x + 3\sin^2 x \tag{18.71}$$
$$= (\cos^2 x - \sin^2 x)(4(\sin^4 x + \cos^4 x - 2\sin^2 x \cos^2 x) - 3) \tag{18.72}$$

Next, we make use of the following addition theorem:

$$2\cos nx \cos mx = \cos(n - m)x + \cos(n + m)x \tag{18.73}$$

Setting $n = 1$ and $m = 1$ in (18.73) we get:

$$2 \cos^2 x = 1 + \cos 2x \qquad (18.74)$$

Using (18.62) we obtain:

$$\cos 2x = \cos^2 x - \sin^2 x \qquad (18.75)$$

and therefore:

$$\cos^2 2x = \cos^4 x + \sin^4 x - 2 \sin^2 x \cos^2 x \qquad (18.76)$$

Setting $n = 2$ and $m = 2$ in (18.73) we get:

$$2 \cos^2 2x = 1 + \cos 4x \qquad (18.77)$$

which results in:

$$\cos 4x = 2 \cos^2 2x - 1 \qquad (18.78)$$

Setting $n = 4$ and $m = 2$ in (18.73) we get:

$$2 \cos 4x \cos 2x = \cos 2x + \cos 6x \qquad (18.79)$$

which results in:

$$\cos 6x = 2 \cos 2x \cos 4x - \cos 2x \qquad (18.80)$$

Now, we use (18.75) and (18.76) to transform (18.72) into:

$$Y = \cos 2x \, (4 \cos^2 2x - 3) \qquad (18.81)$$
$$= 4 \cos^3 x - 3 \cos 2x \qquad (18.82)$$
$$= 2 \cos 2x \, (2 \cos^2 2x - 1) - \cos 2x \qquad (18.83)$$

Using (18.78) we obtain:

$$Y = 2 \cos 2x \cos 4x - \cos 2x \qquad (18.84)$$

which finally leads to:

$$Y = \cos 6x \qquad (18.85)$$

using (18.80). Comparison with (18.69) shows us that we have proven:

$$\cos^6 x - 15 \sin^2 x \cos^4 x + 15 \sin^4 x \cos^2 x - \sin^6 x = \cos 6x \qquad (18.86)$$

Now, we are able to calculate the anisotropy energy density using (7.1)–(7.3):

$$\begin{aligned}
E_{\text{crys}}^{\text{hex}} &= K_0 + K_1(\alpha_1^2 + \alpha_2^2) + K_2(\alpha_1^2 + \alpha_2^2)^2 \\
&\quad + K_3(\alpha_1^2 + \alpha_2^2)^3 + K_4(\alpha_1^2 - \alpha_2^2)(\alpha_1^4 - 14\alpha_1^2\alpha_2^2 + \alpha_2^4) \qquad (18.87) \\
&= K_0 + K_1(\sin^2 \theta \cos^2 \phi + \sin^2 \theta \sin^2 \phi) \\
&\quad + K_2(\sin^2 \theta \cos^2 \phi + \sin^2 \theta \sin^2 \phi)^2 \\
&\quad + K_3(\sin^2 \theta \cos^2 \phi + \sin^2 \theta \sin^2 \phi)^3 \\
&\quad + K_4(\sin^2 \theta \cos^2 \phi - \sin^2 \theta \sin^2 \phi) \\
&\quad \times (\sin^4 \theta \cos^4 \phi - 14 \sin^4 \theta \sin^2 \phi \cos^2 \phi + \sin^4 \theta \sin^4 \phi) \qquad (18.88)
\end{aligned}$$

By means of (18.62) we obtain:

$$
\begin{aligned}
E_{\text{crys}}^{\text{hex}} &= K_0 + K_1 \sin^2 \theta + K_2 \sin^4 \theta + K_3 \sin^6 \theta \\
&\quad + K_4 \sin^6 \theta (\cos^2 \phi - \sin^2 \phi)(\sin^4 \phi + \cos^4 \phi - 14 \sin^2 \phi \cos^2 \phi) \\
&= K_0 + K_1 \sin^2 \theta + K_2 \sin^4 \theta + K_3 \sin^6 \theta \\
&\quad + K_4 \sin^6 \theta (\cos^6 \phi - 15 \sin^2 \phi \cos^4 \phi + 15 \sin^4 \phi \cos^2 \phi - \sin^6 \phi) \\
&= K_0 + K_1 \sin^2 \theta + K_2 \sin^4 \theta + K_3 \sin^6 \theta + K_4 \sin^6 \theta \cos 6\phi \quad (18.89)
\end{aligned}
$$

using (18.86).

7.3 Spin Flop Transition

(a) At the critical field $B_{\text{spin-flop}}$ the energy of the antiferromagnetic phase is equal to that of the spin-flop phase which leads to:

$$
-AM^2 - \Delta = -AM^2 - \frac{M^2 B_{\text{spin-flop}}^2}{2AM^2 - \Delta} \tag{18.90}
$$

Thus, we obtain:

$$
B_{\text{spin-flop}}^2 = \frac{2AM^2\Delta - \Delta^2}{M^2} \tag{18.91}
$$

which gives us:

$$
B_{\text{spin-flop}} = \sqrt{2A\Delta - \left(\frac{\Delta}{M}\right)^2} \tag{18.92}
$$

(b) Plotting the energy of the antiferromagnetic and spin-flop phase in dependence of the magnetic field B (shown below) allows to determine $B_{\text{spin-flop}}$ graphically which corresponds to the crossing point.

$$
E_{\text{ind}} \cdot 1/aV = -G(\alpha_1 \alpha_2 \beta_1 \beta_2 + \alpha_2 \alpha_3 \beta_2 \beta_3 + \alpha_1 \alpha_3 \beta_1 \beta_3) \tag{18.93}
$$

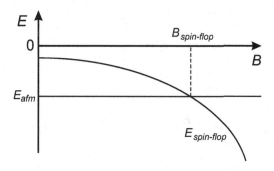

7.4 Due to the bcc structure the constant F amounts to $F = 0$ (see (7.70)). The induced anisotropy energy can thus be given by (see (7.67)):

with α_i being the direction cosine of the magnetization during measurement and β_i that during annealing. The $(1\bar{1}0)$ plane can be characterized by the angle θ, being that between the z-axis and the plane, and $\phi = 45°$. Due to $\sin\phi = \cos\phi = 1/\sqrt{2}$ we obtain for the direction cosine α_i (see (7.1)–(7.3)):

$$\alpha_1 = \frac{1}{\sqrt{2}}\sin\theta \tag{18.94}$$

$$\alpha_2 = \frac{1}{\sqrt{2}}\sin\theta \tag{18.95}$$

$$\alpha_3 = \cos\theta \tag{18.96}$$

Setting $E_{ind}^* = -E_{ind}/aVG$ we get:

$$E_{ind}^* = \frac{1}{2}\beta_1\beta_2 \sin^2\theta + \frac{1}{\sqrt{2}}\beta_2\beta_3 \sin\theta\cos\theta + \frac{1}{\sqrt{2}}\beta_1\beta_3 \sin\theta\cos\theta \tag{18.97}$$

(a) Annealing along the [001] direction
Due to $\theta_0 = 0$ and $\phi_0 = 0$ we obtain for the direction cosine β_i (see (7.1)–(7.3)):

$$\beta_1 = 0 \tag{18.98}$$
$$\beta_2 = 0 \tag{18.99}$$
$$\beta_3 = 1 \tag{18.100}$$

Inserting into (18.97) results in:

$$E_{[001]}^* = 0 \tag{18.101}$$

(b) Annealing along the [110] direction
Due to $\theta_0 = 90°$ and $\phi_0 = 45°$ we get:

$$\beta_1 = \frac{1}{\sqrt{2}} \tag{18.102}$$

$$\beta_2 = \frac{1}{\sqrt{2}} \tag{18.103}$$

$$\beta_3 = 0 \tag{18.104}$$

Therefore:

$$E_{[110]}^* = \frac{1}{4}\sin^2\theta \tag{18.105}$$

Using the relation

$$\sin(x - y) = \sin x \cos y - \cos x \sin y \tag{18.106}$$

we get:

$$\sin(\theta - \theta_0) = \sin\theta \cos\theta_0 - \cos\theta \sin\theta_0 \tag{18.107}$$

which simplifies to (due to $\theta_0 = 90°$):

$$\sin(\theta - \theta_0) = -\cos\theta \tag{18.108}$$

Thus:

$$\sin^2(\theta - \theta_0) = \cos^2\theta = 1 - \sin^2\theta \tag{18.109}$$

which gives:

$$\sin^2\theta = 1 - \sin^2(\theta - \theta_0) \tag{18.110}$$

Inserting into (18.105) results in:

$$E^*_{[110]} = \frac{1}{4} - \frac{1}{4}\sin^2(\theta - \theta_0) \tag{18.111}$$

(c) Annealing along the [111] direction
Due to $\theta_0 = 54.7°$ and $\phi_0 = 45°$ we get:

$$\beta_1 = \sqrt{\frac{2}{3}} \times \sqrt{\frac{1}{2}} = \frac{1}{\sqrt{3}} \tag{18.112}$$

$$\beta_2 = \sqrt{\frac{2}{3}} \times \sqrt{\frac{1}{2}} = \frac{1}{\sqrt{3}} \tag{18.113}$$

$$\beta_3 = \frac{1}{\sqrt{3}} \tag{18.114}$$

Additionally, the following relations are valid:

$$\sin 2\theta_0 = \frac{2\sqrt{2}}{3} \tag{18.115}$$

$$\cos 2\theta_0 = -\frac{1}{3} \tag{18.116}$$

Inserting into (18.97) results in:

$$E^*_{[111]} = \frac{1}{6}\sin^2\theta + \frac{\sqrt{2}}{3}\sin\theta\cos\theta \tag{18.117}$$

Using

$$\cos(x \pm y) = \cos x \cos y \mp \sin x \sin y \tag{18.118}$$

we obtain for $x = y$:

$$\sin^2 x = \cos^2 x - \cos 2x = 1 - \sin^2 x - \cos 2x \tag{18.119}$$

which results in:

$$\sin^2 x = \frac{1}{2}(1 - \cos 2x) \tag{18.120}$$

Inserting the relation:

$$\sin 2x = 2\sin x \cos x \tag{18.121}$$

and (18.120) into (18.117) leads to

$$E^*_{[111]} = \frac{1}{12} - \frac{1}{12}\cos 2\theta + \frac{\sqrt{2}}{6}\sin 2\theta \tag{18.122}$$

$$= \frac{1}{12} + \frac{1}{4}\left(-\frac{1}{3}\cos 2\theta + \frac{2\sqrt{2}}{3}\sin 2\theta\right) \tag{18.123}$$

$$= \frac{1}{12} + \frac{1}{4}\left(\cos 2\theta \cos 2\theta_0 + \sin 2\theta \sin 2\theta_0\right) \tag{18.124}$$

The last relation was obtained using (18.115) and (18.116). Making use of (18.118) we get referring to (18.124):

$$E^*_{[111]} = \frac{1}{12} + \frac{1}{4}\cos 2(\theta - \theta_0) \tag{18.125}$$

Rewriting (18.120) to:

$$\cos 2x = 1 - 2\sin^2 x \tag{18.126}$$

results in:

$$E^*_{[111]} = \frac{1}{3} - \frac{1}{2}\sin^2(\theta - \theta_0) \tag{18.127}$$

Finally, we can compare the different induced magnetic anisotropy constants neglecting constant parts and obtain for the ratio:

$$E^*_{[001]} : E^*_{[110]} : E^*_{[111]} = 0 : -\frac{1}{4}\sin^2(\theta - \theta_0) : -\frac{1}{2}\sin^2(\theta - \theta_0) \tag{18.128}$$

$$= 0 : 1 : 2 \tag{18.129}$$

Problems of Chapter 9

9.1 Using $\boldsymbol{M} = (M_x, M_y, M_z)$ and $\boldsymbol{H} = (H_x, H_y, H_z)$ we obtain:

$$-\boldsymbol{M} \times (\boldsymbol{M} \times \boldsymbol{H})$$

$$= -\begin{pmatrix} M_x \\ M_y \\ M_z \end{pmatrix} \times \begin{pmatrix} M_y H_z - M_z H_y \\ M_z H_x - M_x H_z \\ M_x H_y - M_y H_x \end{pmatrix}$$

$$= -\begin{pmatrix} M_y(M_x H_y - M_y H_x) - M_z(M_z H_x - M_x H_z) \\ M_z(M_y H_z - M_z H_y) - M_x(M_x H_y - M_y H_x) \\ M_x(M_z H_x - M_x H_z) - M_y(M_y H_z - M_z H_y) \end{pmatrix}$$

$$= -\begin{pmatrix} M_x M_y H_y - M_y^2 H_x - M_z^2 H_x + M_x M_z H_z \\ M_y M_z H_z - M_z^2 H_y - M_x^2 H_y + M_x M_y H_x \\ M_x M_z H_x - M_x^2 H_z - M_y^2 H_z + M_y M_z H_y \end{pmatrix}$$

$$= \begin{pmatrix} (M_y^2 + M_z^2 + M_x^2 - M_x^2)H_x - M_xM_yH_y - M_xM_zH_z \\ (M_x^2 + M_z^2 + M_y^2 - M_y^2)H_y - M_xM_yH_x - M_yM_zH_z \\ (M_x^2 + M_y^2 + M_z^2 - M_z^2)H_z - M_xM_zH_x - M_yM_zH_y \end{pmatrix}$$

$$= \begin{pmatrix} M_x^2H_x + M_y^2H_x + M_z^2H_x - M_x^2H_x - M_xM_yH_y - M_xM_zH_z \\ M_x^2H_y + M_y^2H_y + M_z^2H_y - M_xM_yH_x - M_y^2H_y - M_yM_zH_z \\ M_x^2H_z + M_y^2H_z + M_z^2H_z - M_xM_zH_x - M_yM_zH_y - M_z^2H_z \end{pmatrix}$$

$$= (M_x^2 + M_y^2 + M_z^2) \begin{pmatrix} H_x \\ H_y \\ H_z \end{pmatrix} - (M_xH_x + M_yH_y + M_zH_z) \begin{pmatrix} M_x \\ M_y \\ M_z \end{pmatrix}$$

$$= M^2 \cdot H - ((M \cdot H)M)$$

9.2 First, we calculate the parameter α (see (9.64)):

$$\alpha = \frac{4\pi\lambda}{\gamma M} \tag{18.130}$$

$$= \frac{4\pi \times 2 \times 10^8}{1.761 \times 10^{11} \times 2} \tag{18.131}$$

$$= 0.007 \tag{18.132}$$

Thus:

$$\alpha^2 = 5.1 \times 10^{-5} \ll 1 \tag{18.133}$$

The resonance frequency is given by (see (9.78)):

$$\omega_0 = \gamma\mu_0 H \tag{18.134}$$

$$= 1.761 \times 10^{11} \times 4\pi \times 10^{-7} \times (-200)\,\mathrm{s}^{-1} \tag{18.135}$$

$$= -4.4 \times 10^7\,\mathrm{s}^{-1} \tag{18.136}$$

For the corresponding relaxation time τ_0 we therefore obtain (see (9.79)):

$$\tau_0 = \frac{1}{\alpha\omega_0} \tag{18.137}$$

$$= \frac{1}{0.007 \times (-4.4 \times 10^7)}\,\mathrm{s} \tag{18.138}$$

$$= -3.2 \times 10^{-6}\,\mathrm{s} \tag{18.139}$$

Using the approximation for α^2 we are able to calculate the angular frequency ω (see (9.76)) to be:

$$\omega = \frac{\omega_0}{1 + \alpha^2} \approx \omega_0 = -4.4 \times 10^7\,\mathrm{Hz} \tag{18.140}$$

and the related time constant τ (see (9.77)) to be:

$$\tau = \tau_0 \cdot (1 + \alpha^2) \approx \tau_0 = -3.2 \times 10^{-6}\,\mathrm{s} \tag{18.141}$$

Using (9.75):

$$\tan \frac{\theta}{2} = \tan \frac{\theta_0}{2} \cdot e^{-t/\tau} \tag{18.142}$$

we get:

$$-\frac{t}{\tau} = \ln \tan \frac{\theta}{2} - \ln \tan \frac{\theta_0}{2} \tag{18.143}$$

which enables to calculate the time which is required to rotate the magnetization from $\theta_0 = 30°$ to $\theta = 150°$:

$$t = -\tau \left(\ln \tan \frac{\theta}{2} - \ln \tan \frac{\theta_0}{2} \right) \tag{18.144}$$

$$= 3.2 \times 10^{-6} \, \text{s} \times (\ln \tan 75° - \ln \tan 15°) \tag{18.145}$$

$$= 8.4 \times 10^{-6} \, \text{s} \tag{18.146}$$

Using (18.140) the precession frequency ν results in:

$$\nu = \frac{\omega}{2\pi} = 7.0 \times 10^6 \, \text{Hz} \tag{18.147}$$

Thus, we can calculate the number of precession rotations n that occur in this
time interval t using the result of (18.146):

$$n = \nu t = 7.0 \times 10^6 \, \text{Hz} \times 8.4 \times 10^{-6} \, \text{s} = 59 \tag{18.148}$$

Problems of Chapter 12

12.1 Let us consider that tangent line that passes through the tip point of
the reduced applied magnetic field h with coordinates (h_x, h_y). The point of
contact with the astroid is assumed to possess the coordinates (h_{x1}, h_{y1}). The
angle between the tangent line and the easy magnetization direction along the
x-axis is given by β.

The slope s of this tangent line can be expressed as:

$$s = \frac{h_y - h_{y1}}{h_x - h_{x1}} \tag{18.149}$$

which is equivalent to:

$$h_y - h_{y1} = s(h_x - h_{x1}) \tag{18.150}$$

Because the point of contact is a point of the astroid we know (see (12.16)):

$$h_{x1}^{2/3} + h_{y1}^{2/3} = 1 \tag{18.151}$$

Using the implicit differentiation with respect to h_{x1} we obtain:

$$\frac{2}{3} h_{x1}^{-1/3} + \frac{2}{3} h_{y1}^{-1/3} \frac{dh_{y1}}{dh_{x1}} = 0 \tag{18.152}$$

which leads to:

$$\frac{dh_{y1}}{dh_{x1}} = -\left(\frac{h_{y1}}{h_{x1}}\right)^{1/3} \tag{18.153}$$

This derivative describes the slope of the astroid curve at the point with the coordinates (h_{x1}, h_{y1}). Thus, we have:

$$\frac{dh_{y1}}{dh_{x1}} = s \tag{18.154}$$

The slope can be expressed using the angle β as:

$$s = \tan\beta \tag{18.155}$$

Thus, we obtain:

$$\frac{h_{y1}}{h_{x1}} = -\tan^3\beta \tag{18.156}$$

Using this expression and (18.151) we derive:

$$h_{x1} = -\cos^3\beta \tag{18.157}$$
$$h_{y1} = \sin^3\beta \tag{18.158}$$

Substituting (18.155), (18.157), and (18.158) into (18.150) we get:

$$h_y - \sin^3\beta = \frac{\sin\beta}{\cos\beta}(h_x + \cos^3\beta) \tag{18.159}$$

which results in:

$$h_x \sin\beta - h_y \cos\beta + \sin\beta \cos^3\beta + \cos\beta \sin^3\beta = 0 \tag{18.160}$$
$$h_x \sin\beta - h_y \cos\beta + \sin\beta \cos\beta(\cos^2\beta + \sin^2\beta) = 0 \tag{18.161}$$
$$h_x \sin\beta - h_y \cos\beta + \sin\beta \cos\beta = 0 \tag{18.162}$$

This proves the condition to be satisfied:

$$h_x \sin\theta - h_y \cos\theta + \sin\theta \cos\theta = 0 \tag{18.163}$$

(see (12.5)) by setting $\beta = \theta$.

References

1. V.L. Moruzzi: *Calculated electronic properties of metals* (Pergamon Press, New York 1978)
2. R.M. Bozorth: *Ferromagnetism* (Wiley, Hoboken 2003)
3. M.F. Collins, D.A. Wheeler: *Magnetic moments and the degree of order in cobalt-nickel alloys*, Proc. Phys. Soc. **82**, 633–635 (1963)
4. M.F. Collins, J.B. Forsyth: *The magnetic moment distribution in some transition metal alloys*, Phil. Mag. **8**, 401–410 (1963)
5. V. Cannella, J.A. Mydosh: *Magnetic ordering in gold-iron alloys*, Phys. Rev. B **6**, 4220–4237 (1972)
6. E. Kneller: *Ferromagnetismus* (Springer-Verlag, Berlin 1962)
7. J.W. Shih: *Magnetic properties of iron-cobalt single crystals*, Phys. Rev. **46**, 139–142 (1934)
8. F.J.A. den Broeder, W. Hoving, P.J.H. Bloemen: *Magnetic anisotropy of multilayers*, J. Magn. Magn. Mater. **93**, 562–570 (2002)
9. C. Chappert, P. Bruno: *Magnetic anisotropy in metallic ultrathin films and related experiments on cobalt films*, J. Appl. Phys. **64**, 5736–5741 (1988)
10. A. Hubert, R. Schäfer: *Magnetic domains* (Springer-Verlag, Berlin 2000)
11. Y. Tomono: *Magnetic after effect of cold rolled iron (I)*, J. Phys. Soc. Jpn. **7**, 174–179 (1952)
12. F.G. Brockman.P.H. Dowling, W.G. Steneck: *Dimensional effects resulting from a high dielectric constant found in a ferromagnetic ferrite*, Phys. Rev. **77**, 85–93 (1950)
13. V.S. Stepanyuk, W. Hergert, K. Wildberger et al: *Magnetism of 3d, 4d, and 5d transition-metal impurities on Pd(001) and Pt(001) surfaces*, Phys. Rev. B **53**, 2121–2125 (1996)
14. M. Garnier, K. Breuer, D. Purdie et al: *Applicability of the single impurity model to photoemission spectroscopy of heavy Fermion Ce compounds*, Phys. Rev. Lett. **78**, 4127–4130 (1997)
15. H.C. Manoharan, C.P. Lutz, D.M. Eigler: *Quantum mirages formed by coherent projection of electronic structure*, Nature **403**, 512–515 (2000)
16. www.almaden.ibm.com/almaden/media/image_mirage.html
17. A.J. Heinrich, J.A. Gupta, C.P. Lutz et al: *Single-atom spin-flip spectroscopy*, Science **306**, 466–469 (2004)

18. J.T. Lau, A. Föhlisch, R. Nietubyč et al: *Size-dependent magnetism of deposited small iron clusters studied by X-Ray magnetic circular dichroism*, Phys. Rev. Lett. **89**, 057201 (2002)

19. I. Cabria, B. Nonas, R. Zeller et al: *Orbital magnetism of transition-metal adatoms and clusters on the Ag and Au(001) surfaces*, Phys. Rev. B **65**, 054414 (2002)

20. B. Lazarovits, L. Szunyogh, P. Weinberger: *Fully relativistic calculation of magnetic properties of Fe, Co, and Ni adclusters on Ag(100)*, Phys. Rev. B **65**, 104441 (2002)

21. V.S. Stepanyuk, W. Hergert, P. Rennert et al: *Magnetic nanostructures on the fcc Fe/Cu(100) surface*, Phys. Rev. B **61**, 2356–2361 (2000)

22. P. Gambardella, S. Rusponi, M. Veronese et al: *Giant magnetic anisotropy of single cobalt atoms and nanoparticles*, Science **300**, 1130–1133 (2003)

23. V. Bellini, N. Papanikolaou, R. Zeller et al: *Magnetic 4d monoatomic rows on Ag vicinal surfaces*, Phys. Rev. B **64**, 094403 (2001)

24. K. Wildberger, V.S. Stepanyuk, P. Lang et al: *Magnetic nanostructures: 4d clusters on Ag(001)*, Phys. Rev. Lett. **75**, 509–512 (1995)

25. R. Schäfer: Magnetische Mikrostrukturen. In: *30. Ferienkurs des FZ Jülich: Magnetische Schichtsysteme*, ed by P.H. Dederichs, P. Grünberg (FZ Jülich 1999)

26. R.P. Cowburn, D.K. Koltsov, A.O. Adeyeye et al: *Single-domain circular nanomagnets*, Phys. Rev. Lett. **83**, 1042–1045 (1999)

27. M. Bode: *Spin-polarized scanning tunnelling microscopy*, Rep. Prog. Phys. **66**, 523–582 (2003)

28. M. Kläui, C.A.F. Vaz, J.A.C. Bland et al: *Direct observation of spin configurations and classification of switching processes in mesoscopic ferromagnetic rings*, Phys. Rev. B **68**, 134426 (2003)

29. J. Bekaert, D. Buntinx, C. van Haesendonck et al: *Noninvasive magnetic imaging and magnetization measurement of isolated mesoscopic Co rings*, Appl. Phys. Lett. **81**, 3413–3415 (2002)

30. T. Shinjo, T. Okuno, R. Hassdorf et al: *Magnetic vortex core observation in circular dots of permalloy*, Science **289**, 930–932 (2000)

31. A. Wachowiak, J. Wiebe, M. Bode et al: *Direct observation of internal spin structure of magnetic vortex cores*, Science **298**, 577–580 (2002)

32. M. Bode, O. Pietzsch, A. Kubetzka et al: *Shape-dependent thermal switching behavior of superparamagnetic nanoislands*, Phys. Rev. Lett. **92**, 067201 (2004)

33. I.M.L. Billas, J.A. Becker, A. Châtelain et al: *Magnetic moments of iron clusters with 25 to 700 atoms and their dependence on temperature*, Phys. Rev. Lett. **71**, 4067–4070 (1993)

34. A. Hirt, D. Gerion, I.M.L. Billas et al: *Thermal properties of ferromagnetic clusters*, Z. Phys. D **40**, 160–163 (1997)

35. M. Jamet, W. Wernsdorfer, Ch. Thirion et al: *Magnetic anisotropy in single clusters*, Phys. Rev. B **69**, 024401 (2004)

36. S.H. Baker, C. Binns, K.W. Edmonds et al: *Enhancements in magnetic moments of exposed and Co-coated Fe nanoclusters as a function of cluster size*, J. Magn. Magn. Mater. **247**, 19–25 (2002)

37. J. Bansmann, S.H. Baker, C. Binns et al: *Magnetic and structural properties of isolated and assembled clusters*, Surf. Sci. Rep. **56**, 189–275 (2005)

38. S. Sun, C.B. Murray: *Synthesis of monodisperse cobalt nanocrystals and their assembly into magnetic superlattices*, J. Appl. Phys. **85**, 4325–4330 (1999)

39. S. Sun, C.B. Murray, D. Weller et al: *Monodisperse FePt nanoparticles and ferromagnetic FePt nanocrystal superlattices*, Science **287**, 1989–1992 (2000)

40. T. Schmitte, K. Theis-Bröhl, V. Leiner et al: *Magneto-optical study of the magnetization reversal process of Fe nanowires*, J. Phys.: Condens. Matter **14**, 7525–7538 (2002)

41. T. Schmitte, K. Westerholt, H. Zabel: *Magneto-optical Kerr effect in the diffracted light of Fe gratings*, J. Appl. Phys. **92**, 4524–4530 (2002)

42. B. Hausmanns: *Magnetoresistance and magnetization reversal processes in single magnetic nanowires.* PhD Thesis, University of Duisburg-Essen (2003)

43. C.L. Dennis, R.P. Borges, L.D. Buda et al: *The defining length scales of mesomagnetism: a review*, J. Phys.: Condens. Matter **14**, R1175–R1262 (2002)

44. P. Gambardella: *Magnetism in monatomic metal wires*, J. Phys.: Condens. Matter **15**, S2533–S2546 (2003)

45. P. Gambardella, A. Dallmeyer, K. Maiti et al: *Ferromagnetism in one-dimensional monatomicmetal chains*, Nature **416**, 301–304 (2002)

46. P. Gambardella, A. Dallmeyer, K. Maiti et al: *Oscillatory magnetic anisotropy in one-dimensional atomic wires*, Phys. Rev. Lett. **93**, 077203 (2004)

47. R. Bergholz, U. Gradmann: *Structure and magnetism of oligatomic Ni(111)-films on Re(0001)*, J. Magn. Magn. Mater. **45**, 389–398 (1984)

48. U. Gradmann: Magnetism in ultrathin transition metal films. In: *Handbook of magnetic materials*, Vol. 7, ed by K.H.J. Buschow (Elsevier, Amsterdam 1993)

49. M.P. Seah, W.A. Dench: *Quantitative electron spectroscopy of surfaces*, Surf. Interf. Anal. **1**, 2–11 (1979)

50. C. Carbone: Photoemission from magnetic films. In: *30. Ferienkurs des FZ Jülich: Magnetische Schichtsysteme*, ed by P.H. Dederichs, P. Grünberg (FZ Jülich 1999)

51. E. Bergter, U. Gradmann, R. Bergholz: *Magnetometry of the Ni(111)/Cu(111) interface*, Sol. State Comm. **53**, 565–567 (1985)

52. T.S. Sherwood, S.R. Mishra, A.P. Popov et al: *Magnetization of ultrathin Fe films deposited on Gd(0001)*, J. Vac. Sci. Technol. A **16**, 1364–1367 (1998)

53. W.H. Meiklejohn, C.P. Bean: *New magnetic anisotropy*, Phys. Rev. **102**, 1413–1414 (1956)

54. R. Radu, M. Etzkorn, R. Siebrecht et al: *Interfacial domain formation during magnetization reversal in exchange biased CoO/Co bilayers*, Phys. Rev. B **67**, 134409 (2003)

55. Y. Zhao, T. Shishidou, A.J. Freeman: *Ruderman–Kittel–Kasuya–Yosida-like ferromagnetism in $Mn_x Ge_{1-x}$*, Phys. Rev. Lett. **90**, 047204 (2003)

56. P. Bruno, Ch. Chappert: *Oscillatory coupling between ferromagnetic layers separated by a non-magnetic metal spacer*, Phys. Rev. Lett. **67**, 1602–1605 (1991)

57. R. Coehoorn: *Period of oscillatory exchange interactions in Co/Cu and Fe/Cu multilayer systems*, Phys. Rev. B **44**, 9331–9337 (1991)

58. P.J.H. Bloemen, H.W. van Kesteren, H.J.M. Swagten et al: *Oscillatory interlayer exchange coupling in Co/Ru multilayers and bilayers*, Phys. Rev. B **50**, 13505–13514 (1994)

59. J. Unguris, R.J. Celotta, D.T. Pierce: *Determination of the exchange coupling strengths for Fe/Au/Fe*, Phys. Rev. Lett. **79**, 2734–2737 (1997)

60. J.A. Wolf, Q. Leng, R. Schreiber et al: *Interlayer coupling of Fe films across Cr interlayers and its relation to growth and structure*, J. Magn. Magn. Mater. **121**, 253–258 (1993)

61. J. Unguris, R.J. Celotta, D.T. Pierce: *Observation of two different oscilla-tion periods in the exchange coupling of Fe/Cr/Fe(100)*, Phys. Rev. Lett. **67**, 140–143 (1991)

62. J. Unguris, R.J. Celotta, D.T. Pierce: *Magnetism in Cr thin films on Fe(100)*, Phys. Rev. Lett. **69**, 1125–1128 (1992)

63. P. Bödeker, A. Schreyer, H. Zabel: *Spin-density waves and reorientation effects in thin epitaxial Cr films covered with ferromagnetic and paramagnetic layers*, Phys. Rev. B **59**, 9408–9431 (1999)

64. P. Bödeker, A. Hucht, A. Schreyer et al: *Reorientation of spin density waves in Cr(001) films induced by Fe(001) cap layers*, Phys. Rev. Lett. **81**, 914–917 (1998)

65. R. Schäfer: *Magneto-optical domain studies in coupled magnetic multilayers*, J. Magn. Magn. Mater. **148**, 226–231 (1995)

66. M. Rührig, R Schäfer, A. Hubert et al: *Domain observations on Fe/Cr/Fe layered structures. Evidence for a biquadratic coupling effect*, phys. stat. sol. (a) **125**, 635–656 (1991)

67. B.R. Coles: *Spin disorder effects in the electrical resistivities of metals and alloys*, Adv. Phys. **7**, 40–71 (1958)

68. G. Binasch, P. Grünberg, F. Saurenbach et al: *Enhanced magnetoresistance in layered magnetic structures with antiferromagnetic interlayer exchange*, Phys. Rev. B **39**, 4828–4830 (1989)

69. M.N. Baibich, J.M. Broto, A. Fert et al: *Giant magnetoresistance of (001)Fe/(001)Cr magnetic superlattices*, Phys. Rev. Lett. **61**, 2472–2475 (1988)

70. J. Barnas, A. Fuss, R.E. Camley et al: *Novel magnetoresistance effect in layered magnetic structures: theory and experiment*, Phys. Rev. B **42**, 8110–8120 (1990)

71. R.M. Kusters, J. Singleton, D.A. Keen et al: *Magnetoresistance measure-ments on the magnetic semiconductor $N_{0.5}, Pb_{0.5} MnO_3$*, Physica B **155**, 362–365 (1989)

72. W. Buckel: *Superconductivity*, (VCH, Weinheim 1996)

73. P.M. Tedrow, R. Meservey: *Spin polarization of electrons tunneling from films of Fe, Co, Ni, and Gd*, Phys. Rev. B **7**, 318–326 (1973)

74. R. Meservey and P.M. Tedrow: *Spin polarized electron tunneling*, Phys. Rep. **238**, 173–243 (1994)

75. R. Meservey, P.M. Tedrow, P. Fulde: *Magnetic field splitting of the quasiparti-cle states in superconducting aluminum films*, Phys. Rev. Lett. **25**, 1270–1272 (1970)

76. P.M. Tedrow, R. Meservey: *Spin-dependent tunneling into ferromagnetic nickel*, Phys. Rev. Lett. **26**, 192–195 (1971)

77. M. Julliere: *Tunnelling between ferromagnetic films*, Phys. Lett. A **54**, 225–226 (1975)

78. J.S. Moodera, J. Nowak, R.J.M. van de Veerdonk: *Interface magnetism and spin wave scattering in ferromagnet-insulator-ferromagnet tunnel junctions*, Phys. Rev. Lett. **80**, 2941–2944 (1998)

79. J.S. Moodera, E.F. Gallagher, K. Robinson et al: *Optimum tunnel barrier in ferromagnetic – insulator – ferromagnetic tunneling structures*, Appl. Phys. Lett. **70**, 3050–3052 (1997)

80. R. Jansen, J.S. Moodera: *Influence of barrier impurities on the magnetoresis-tance in ferromagnetic tunnel junctions*, J. Appl. Phys. **83**, 6682–6684 (1998)

81. R. Wiesendanger, H.-J. Güntherodt, G. Güntherodt et al: *Observation of vacuum tunneling of spin polarized electrons with the scanning tunneling microscope*, Phys. Rev. Lett. **65**, 247–250 (1990)

82. R. Ravlić, M. Bode, A. Kubetzka et al: *Correlation of dislocation and domain structure of Cr(001) investigated by spin polarized scanning tunneling microscopy*, Phys. Rev. B **67**, 174411 (2003)

83. A. Cho: *Microchips that never forget*, Science **296**, 246–249 (2002)

84. W.J. Gallagher, S.S.P. Parkin: *Development of the magnetic tunnel junction MRAM at IBM: from first junctions to a 16-Mb MRAM demonstrator chip*, IBM J. Res. & Dev. **50**, 5–23 (2006)

85. T.M. Maffitt, J.K. DeBrosse, J.A. Gabric et al: *Design considerations for MRAM*, IBM J. Res. & Dev. **50**, 25–39 (2006)

86. J.M. Slaughter, M. DeHerrera, H. Dürr: Magnetoresistive RAM. In: *Nanoelectronics and information technology*, Vol. 2, ed by R. Waser (Wiley-VCH, Weinheim 2005)

87. A. Dietzel: Hard disk drives. In: *Nanoelectronics and information technology*, Vol. 2, ed by R. Waser (Wiley-VCH, Weinheim 2005)

88. R. Coehoorn: Giant magnetoresistance and magnetic interactions in exchange-biased spin valves. In: *Handbook of magnetic materials*, Vol. 15, ed by K.H.J. Buschow (Elsevier, Amsterdam 2003)

89. R. Hsiao: *Plasma-etching processes for ULSI semiconductor circuits*, IBM J. Res. & Dev. **43**, 89–102 (1999)

90. D.A. Thompson, J.S. Best: *The future of magnetic data storage technology*, IBM J. Res. & Dev. **44**, 311–322 (2000)

91. K.-M.H. Lenssen, G.H.J. Somers, J.B.A.D. van Zon: *Magnetoresistive sensors for string instruments*, J. Appl. Phys. **91**, 7777–7779 (2002)

Abbreviations

1D	one-dimensional
1T1MTJ	one transistor one magnetic tunnelling junction
2D	two-dimensional
3D	three-dimensional
ABS	anti-lock braking system
AF	antiferromagnet
AFM	atomic force microscopy, atomic force microscope
AMR	anisotropic magnetoresistance
bcc	body centered cubic
BCS	Bardeen, Cooper, Schrieffer
CIP	current in plane
CMR	colossal magnetoresistance
CPP	current perpendicular to plane
DL	double layer
DOS	density of states
DRAM	dynamic random access memory
EEPROM	electrically erasable programmable read-only memory
fc	flux changes
fcc	face centered cubic
FeRAM	ferroelectric random access memory
FET	field effect transistor
FM	ferromagnet, ferromagnetic metal
FWHM	full-width at half-maximum
GMR	giant magnetoresistance
hcp	hexagonally closed packed
HOPG	highly oriented pyrolytic graphite
HRSEM	high resolution scanning electron microscopy, high resolution scanning electron microscope
I	insulator
IEC	interlayer exchange coupling
IETS	inelastic electron tunnelling spectroscopy

JMR	junction magnetoresistance
LDOS	local density of states
MOKE	magneto-optical Kerr effect
MR	magnetoresistivity, magnetoresistance
MRAM	magnetic random access memory
N	normal conductor
PEEM	photoemission electron microscopy,
	photoemission electron microscope
RKKY	Ruderman, Kittel, Kasuya, Yosida
SAF	synthetic antiferromagnet
SC	superconductor
SEMPA	secondary electron microscopy with polarization analysis
SPSTM	spin polarized scanning tunnelling microscopy,
	spin polarized scanning tunnelling microscope
STM	scanning tunnelling microscopy,
	scanning tunnelling microscope
TEM	transmission electron microscopy,
	transmission electron microscope
TMJ	tunnelling magnetic junction
TMR	tunnelling magnetoresistance
XMCD	x-ray magnetic circular dichroism
XPT	cross-point

Symbols

a	length of the semimajor axis
A	area
A	exchange stiffness constant
$\boldsymbol{A}$	magnetic vector potential
$\boldsymbol{A}$	vector perpendicular to an area A
b	length of the semiminor axis
$\boldsymbol{B}$	magnetic induction, magnetic flux density
B_i	magneto elastic constants
B_J	Brillouin function
$\boldsymbol{B}_{\mathrm{mf}}$	magnetic flux density of a molecular field
$\boldsymbol{B}_{\text{spin-flop}}$	critical field between the antiferromagnetic and spin-flop phase
c_{Curie}	Curie constant
c_H	specific heat
C_i	i-fold rotational symmetry
c_{ij}	elastic constants
d	thickness
d	(inner) diameter
d	dimensionality of the crystal lattice
D	outer diameter
D	dimensionality of the spin lattice
d_c	critical thickness
e	magnitude of the electron charge
e	eccentricity
E	energy
$\mathcal{E}$	deformation tensor
E^{BW}	Bloch wall energy per area
E_{crys}	magneto crystalline anisotropy energy per volume, magneto crystalline anisotropy energy density

$E_{\text{crys}}^{\text{BW}}$	magneto crystalline anisotropy energy of a Bloch wall per area
$E_{\text{crys}}^{\text{cubic}}$	E_{crys} of a cubic system
$E_{\text{crys}}^{\text{hex}}$	E_{crys} of a hexagonal system
$E_{\text{crys}}^{\text{tetra}}$	E_{crys} of a tetragonal system
E_{el}	elastic energy density
$E_{\text{el}}^{\text{cubic}}$	elastic energy density of a cubic system
$E_{\text{el}}^{\text{hex}}$	elastic energy density of a hexagonal system
E_{exch}	exchange energy
$E_{\text{exch}}^{\text{BW}}$	exchange energy of a Bloch wall per area
E_F	Fermi energy
E_{IEC}	energy density due to interlayer exchange coupling
E_{ind}	induced magnetic anisotropy energy
E_{kin}	kinetic energy
E_{pot}	potential energy
$E_{\text{spin-flop}}$	energy density of the spin-flop phase
E_S	energy of a singlet state
E_{str}	stray field energy
E_T	energy of a triplet state
E_{Zeeman}	Zeeman energy
F	Helmholtz free energy
$f(E)$	Fermi function
g	g-factor
G	(differential) conductance
$G^{\uparrow\uparrow}$	conductance for a parallel magnetization
$G^{\uparrow\downarrow}$	conductance for an antiparallel magnetization
G_{ap}	conductance for an antiparallel magnetization
g_J	Landé $-$ g-factor
g_J	g-factor of the total angular momentum
g_L	g-factor of the orbital angular momentum
G_{p}	conductance for a parallel magnetization
g_S	g-factor of the spin angular momentum
$g(E)$	density of states in energy space
$g(k)$	density of states in wave vector space
$g(U)$	differential conductance without Zeeman splitting
h	(step) height
h	reduced magnetic field
h	Planck's constant
$\hbar$	Planck's constant $/2\pi$
$\mathcal{H}$	Hamiltonian
H_c	coercive field, coercivity
$\boldsymbol{H}$	magnetic field
$\boldsymbol{H}_{\text{demag}}$	demagnetizing field
H_{EB}	exchange bias field
$\mathcal{H}_i$	part of the Hamiltonian

H_r	field at resonance
H_s	saturation field
$\mathcal{H}_{\text{spin}}$	spin dependent term of the Hamiltonian
$\mathcal{H}^{\text{dia}}$	diamagnetic term of the Hamiltonian
$\mathcal{H}^{\text{para}}$	paramagnetic term of the Hamiltonian
I	current
$I^{\uparrow\uparrow}$	(tunnelling) current for a parallel magnetization
$I^{\uparrow\downarrow}$	(tunnelling) current for an antiparallel magnetization
J	total angular momentum quantum number
J	exchange constant
J	exchange integral
$\mathcal{J}$	exchange constant
$\boldsymbol{J}$	total angular momentum
J_{BL}	bilinear coupling constant
J_{BQ}	biquadratic coupling constant
J_{ij}	exchange constant between spin i and spin j
J_{RKKY}	exchange constant of the RKKY interaction
k	Boltzmann's constant
k	wave vector
k_F	Fermi wave vector
K_i	anisotropy constants
K_i'	anisotropy constants
K_i^{cubic}	anisotropy constants of cubic systems
K_i^{tetra}	anisotropy constants of tetragonal systems
K^{eff}	effective anisotropy constant
K^{S}	surface contribution of the anisotropy constant
K^{V}	volume contribution of the anisotropy constant
$K_{\text{shape}}^{\text{V}}$	volume contribution of the shape anisotropy constant
ℓ	length
ℓ	orbital angular momentum quantum number
L	orbital angular momentum quantum number
$\boldsymbol{L}$	torque
$\boldsymbol{L}$	orbital angular momentum
$L(y)$	Langevin function
m	electron mass
$\boldsymbol{m}$	$= \boldsymbol{M}/|\boldsymbol{M}|$ magnetization direction
M	tunnelling matrix element
$\boldsymbol{M}$	magnetization
$\boldsymbol{M}_A$	magnetization of sublattice A
$\boldsymbol{M}_B$	magnetization of sublattice B
m_e	effective electron mass
M_i	initial jump of the magnetization
$\boldsymbol{M}_i$	magnetization of sublattice i
m_J	magnetic quantum number
m_L	orbital magnetic quantum number

$M_n(t)$	time dependent subsequent change of the magnetization
M_r	remanence
m_S	spin magnetic quantum number
$\boldsymbol{M}_S$	saturation magnetization
m_T	magnetic dipole term
n	number (of electrons) per volume
n	density of states
$\mathcal{N}$	demagnetizing tensor
$n^\uparrow$	number of spin up (majority) electrons per volume
$n^\uparrow(E)$	density of states of spin up (majority) electrons
$n^\downarrow$	number of spin down (minority) electrons per volume
$n^\downarrow(E)$	density of states of spin down (minority) electrons
N_a	elements of the demagnetization tensor
n_c	number of nearest neighbors, coordination number
n_d	number of d electrons per volume
n_h	number of d holes per volume
n_h	number of d holes per volume
n_{SC}	density of states of quasiparticles in a superconductor
N_X	number of atoms of type X
N_{XXi}	number of XX bonds in direction of bond i
$\mathcal{O}(x^\alpha)$	term of αth order in x
p	number of electrons in a specific electron shell
$\boldsymbol{p}$	momentum
P	spin polarization
P_t	tunnelling spin polarization
q	wave vector
$\boldsymbol{Q}$	(critical) spanning vector
$\boldsymbol{r}$	position vector
R	electrical resistance
$R^\uparrow$	electrical resistance for spin up (majority) electrons
$R^\downarrow$	electrical resistance for spin down (minority) electrons
$R_\parallel$	electrical resistance for a parallel orientation between current and magnetization
$R_\perp$	electrical resistance for a perpendicular orientation between current and magnetization
R_{p}	electrical resistance for a parallel alignment of the magnetization in neighbored layers
R_{ap}	electrical resistance for an antiparallel alignment of the magnetization in neighbored layers
S	spin angular momentum quantum number
S	entropy
S	sensitivity
$\boldsymbol{S}$	spin angular momentum
t	time
t	thickness

T	temperature
T^*	critical temperature
T_B	blocking temperature
T_C	Curie temperature
T_C^+	temperature T near T_C but $T \geq T_C$
T_C^-	temperature T near T_C but $T \leq T_C$
T_{comp}	compensation temperature
T_K	Kondo temperature
T_N	Neel temperature
T_s	substrate temperature
T_{SR}	spin reorientation temperature
$\langle T_Z \rangle$	magnetic dipole term
U	voltage
U	Coulomb energy
V	volume
w	width
w	probability
W_L	Lilley's domain wall width
W_m	domain wall width based on the magnetization profile
Z	partition function
α	damping parameter
α	critical exponent characterizing the specific heat c_H
α_i	direction cosine
β	critical exponent characterizing the order parameter M
β_i	direction cosine
γ	critical exponent characterizing the susceptibility χ
γ	gyromagnetic ratio
δ	critical exponent characterizing the influence of an external magnetic field H on the magnetization for $T = T_C$
δ	domain wall width
δ	phase difference
Δ	difference
Δ	exchange splitting
Δ	Zeeman energy
Δ	half the gap width of a superconductor
ϵ	deformation parameter
ϵ_{ij}	elements of the deformation tensor
θ	angular variable
θ	Weiss temperature, paramagnetic Curie temperature
θ_{ann}	angle of magnetization during annealing
θ_K^{rem}	angle of the Kerr rotation measured at remanence
θ_K^{sat}	angle of the Kerr rotation measured at saturation
λ	molecular field constant
λ	strain, relative change in length
λ	(oscillation) period

λ	relaxation frequency
λ	wave length
λ	inelastic mean free path
λ_+	inelastic mean free path of spin up (majority) electrons
λ_-	inelastic mean free path of spin down (minority) electrons
λ_{eff}	effective period
λ_{ijk}	magnetostriction constants of cubic systems
λ_X	magnetostriction constants of hexagonal systems
μ	chemical potential
μ'	real part of the permeability
μ''	imaginary part of the permeability
$\boldsymbol{\mu}$	magnetic moment
μ_0	magnetic permeability of free space
μ_B	Bohr magneton
μ_{eff}	effective magnetic moment
μ_r	relative magnetic permeability
ν	frequency
ν	critical exponent characterizing the correlation length ξ
ξ	correlation length
ξ	ratio of χ_n to χ_i
ρ	resistivity
ρ^+	resistivity for a parallel alignment of the spins
ρ^-	resistivity for an antiparallel alignment of the spins
ρ_{ap}	resistivity for an antiparallel alignment of the magnetization in neighbored layers
ρ_{p}	resistivity for a parallel alignment of the magnetization in neighbored layers
σ	conductivity, specific conductance
σ_{ap}	conductivity for an antiparallel alignment of the magnetization in neighbored layers
σ_{p}	conductivity for a parallel alignment of the magnetization in neighbored layers
τ	relaxation time
ϕ	angular variable
Φ	angular variable
χ	rotation angle
χ	spin part of the wave function
χ	(magnetic) susceptibility
χ'	real part of the susceptibility
χ''	imaginary part of the susceptibility
$\chi_{\parallel}$	component of the susceptibility being parallel to $\boldsymbol{H}$
$\chi_{\perp}$	component of the susceptibility being perpendicular to $\boldsymbol{H}$
χ_i	initial change of the susceptibility
χ_n	susceptibility after changing the magnetization
χ_{poly}	magnetic susceptibility of polycrystalline material

χ_S	spin part of a wave function describing a singlet state
χ_T	spin part of a wave function describing a triplet state
χ^{coll}	susceptibility of systems exhibiting collective magnetism
χ^{dia}	diamagnetic susceptibility
χ^{Langevin}	Langevin paramagnetic susceptibility
χ^{para}	paramagnetic susceptibility
χ^{Pauli}	Pauli paramagnetic susceptibility
ψ	(spatial part of the) wave function
ψ_S	spatial part of a wave function describing a singlet state
ψ_T	spatial part of a wave function describing a triplet state
ω	angular frequency
∇	nabla operator

Constants

c	speed of light	$2.998 \times 10^8 \ \mathrm{ms}^{-1}$
e	electron charge	$1.602 \times 10^{-19} \ \mathrm{C}$
h	Planck's constant	$6.626 \times 10^{-34} \ \mathrm{Js}$
$\hbar$	Planck's constant $/2\pi$	$1.055 \times 10^{-34} \ \mathrm{Js}$
k	Boltzmann's constant	$1.381 \times 10^{-23} \ \mathrm{JK}^{-1}$
m	electron mass	$9.109 \times 10^{-31} \ \mathrm{kg}$
γ	electron gyromagnetic ratio	$1.761 \times 10^{11} \ \mathrm{As \ kg}^{-1}$
ϵ_0	dielectric constant	$8.854 \times 10^{-12} \ \mathrm{AsV}^{-1}\mathrm{m}^{-1}$
μ_0	magnetic permeability of free space	$4\pi \times 10^{-7} \ \mathrm{VsA}^{-1}\mathrm{m}^{-1}$
μ_B	Bohr magneton	$9.274 \times 10^{-24} \ \mathrm{Am}^2$

Units in the SI and cgs System

Table 18.4. Units in the SI system und the cgs system (g: gramme, G: Gauss, Oe: Oerstedt, emu: electromagnetic unit)

Quantity	Symbol	SI unit		cgs unit	
length	x	10^{-2}	m	$= 1$	cm
mass	m	10^{-3}	kg	$= 1$	g
force	F	10^{-5}	N	$= 1$	dyne
energy	E	10^{-7}	J	$= 1$	erg
magnetic induction	$\boldsymbol{B}$	10^{-4}	T	$= 1$	G
magnetic field	$\boldsymbol{H}$	$10^3/4\pi$	A/m	$= 1$	Oe
magnetic moment	$\boldsymbol{\mu}$	10^{-3}	J/T	$= 1$	erg/G
magnetization	$\boldsymbol{M}$	10^3	A/m	$= 1$	Oe
magnetic susceptibility	χ	4π		$= 1$	emu/cm^3

Index

1T1MTJ architecture, 295

ABS, 306
active target, 319
adatom, 168
agglomeration, 207
alloy, 36, 97, 105
AMR, 262, 306
AMR angle sensor, 311
AMR sensor, 306, 307
angle
    loss, 138
angle sensor, 306
    AMR, 311
    GMR, 313
angular momentum
    orbital, 8, 21
    spin, 8, 21
    total, 14, 20, 21
angular momentum quantum number
    orbital, 21
    spin, 21
    total, 14, 21
anisotropic magnetoresistance, 262, 306
anisotropic magnetoresistive effect, 262, 306
anisotropy
    exchange, 233, 235
    induced magnetic, 105
    magnetic, 89, 242
    magnetic interface, 112, 203
    magnetic surface, 112, 203, 205
    magneto crystalline, 90, 263

non-uniaxial magnetic, 126
orbital, 262
roll-magnetic, 107
shape, 102, 211, 263
stress, 108
uniaxial magnetic, 124, 176, 213, 219, 234, 242, 314
anisotropy constant, 203
    shape, 200
    surface, 200
anisotropy energy, 197
    induced magnetic, 106
    magnetic, 219
    shape, 254
anti-lock braking system, 306
antibody, 334
antiferromagnet, 4
    artificial, 294
    synthetic, 294, 295
    topological, 61, 247, 289
antiferromagnetic coupling, 232, 240, 291, 313
antiferromagnetic exchange coupling, 269
antiferromagnetic order, 4, 55
antiferromagnetic spacer layer, 247
antiferromagnetism, 4, 55
antigen, 334
antisymmetric wave function, 42
application, 293
    biomedical, 209, 332
    MRAM, 293
    read head, 301